student study

ART NOTEBOOK

ESSENTIALS

OF HUMAN ANATOMY AND PHYSIOLOGY

FIFTH EDITION

JOHN W. HOLE, JR.

WCB
Wm. C. Brown Publishers
Dubuque, IA Bogota Boston Buenos Aires Caracas Chicago
Guilford, CT London Madrid Mexico City Sydney Toronto

Wm. C. Brown Communications, Inc.

President and Chief Executive Officer *G. Franklin Lewis*
Corporate Senior Vice President, President of WCB Manufacturing *Roger Meyer*
Corporate Senior Vice President and Chief Financial Officer *Robert Chesterman*

The credits section for this book begins on page 139 and
is considered an extension of the copyright page.

A Times Mirror Company

ISBN 0–697–24396-6

Printed in the United States of America by Wm. C. Brown Communications, Inc.,
2460 Kerper Boulevard, Dubuque, IA 52001

10 9 8 7 6 5 4 3 2 1

TO THE STUDENT

The Student Study Art Notebook is designed to help you in your study of human anatomy and physiology. The notebook contains art taken directly from the text and overhead transparencies; thus you can take notes during lectures, or jot down comments as you are reading through the chapters.

The notebook is perforated and 3-hole punched so, if you wish, you can remove sheets and put them in a binder with other study or lecture notes. Any blank pages at the end of this notebook can be used for additional notes or drawings.

We hope this notebook, used along with your text, helps to make the study of the human body easier for you.

DIRECTORY OF NOTEBOOK FIGURES

TO ACCOMPANY HOLE,

ESSENTIALS OF HUMAN ANATOMY AND PHYSIOLOGY, 5TH ED.

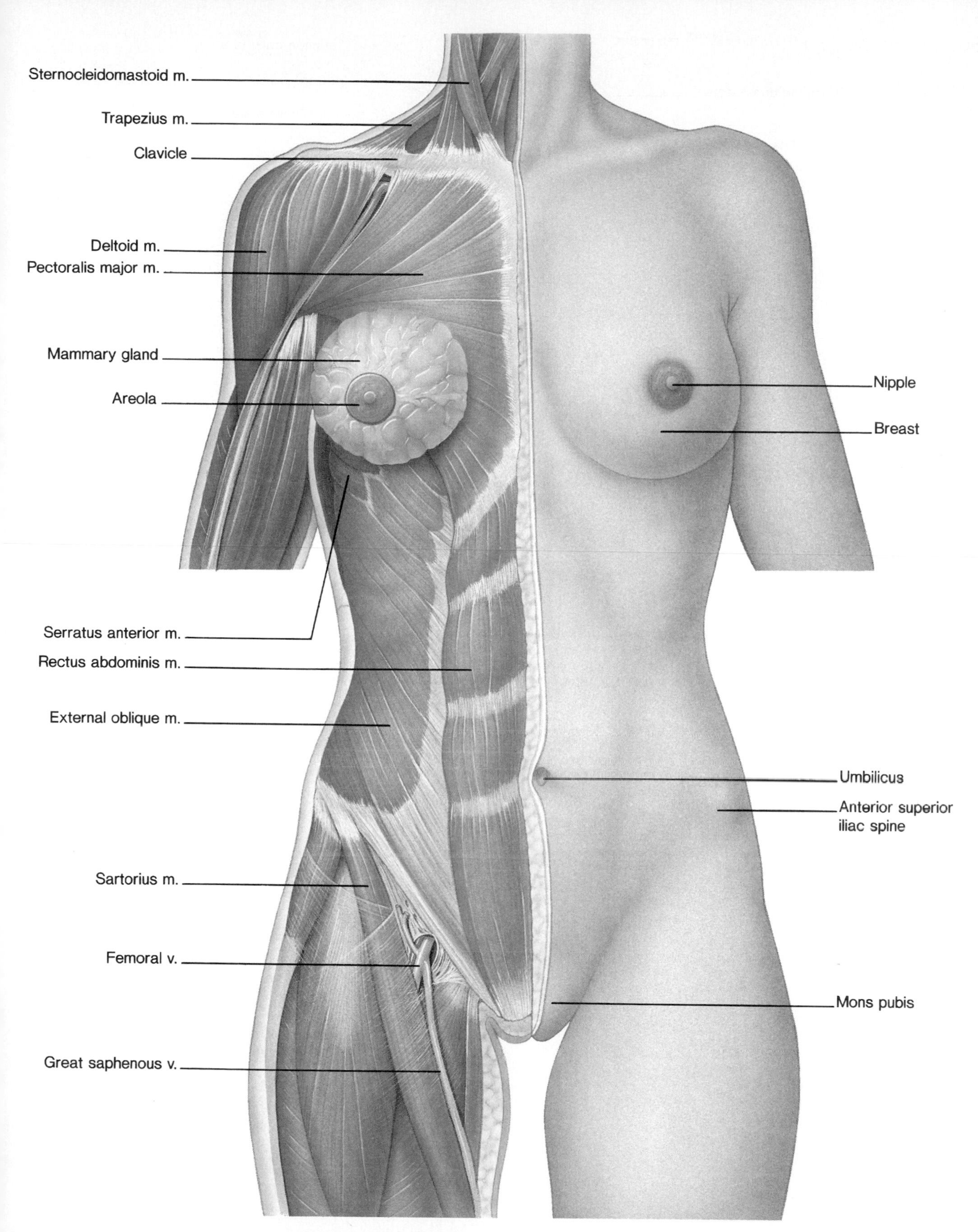
Sternocleidomastoid m.
Trapezius m.
Clavicle
Deltoid m.
Pectoralis major m.
Mammary gland
Areola
Nipple
Breast
Serratus anterior m.
Rectus abdominis m.
External oblique m.
Umbilicus
Anterior superior iliac spine
Sartorius m.
Femoral v.
Mons pubis
Great saphenous v.

Human Torso, Anterior Surface and Superficial Muscles
Reference Plate 1

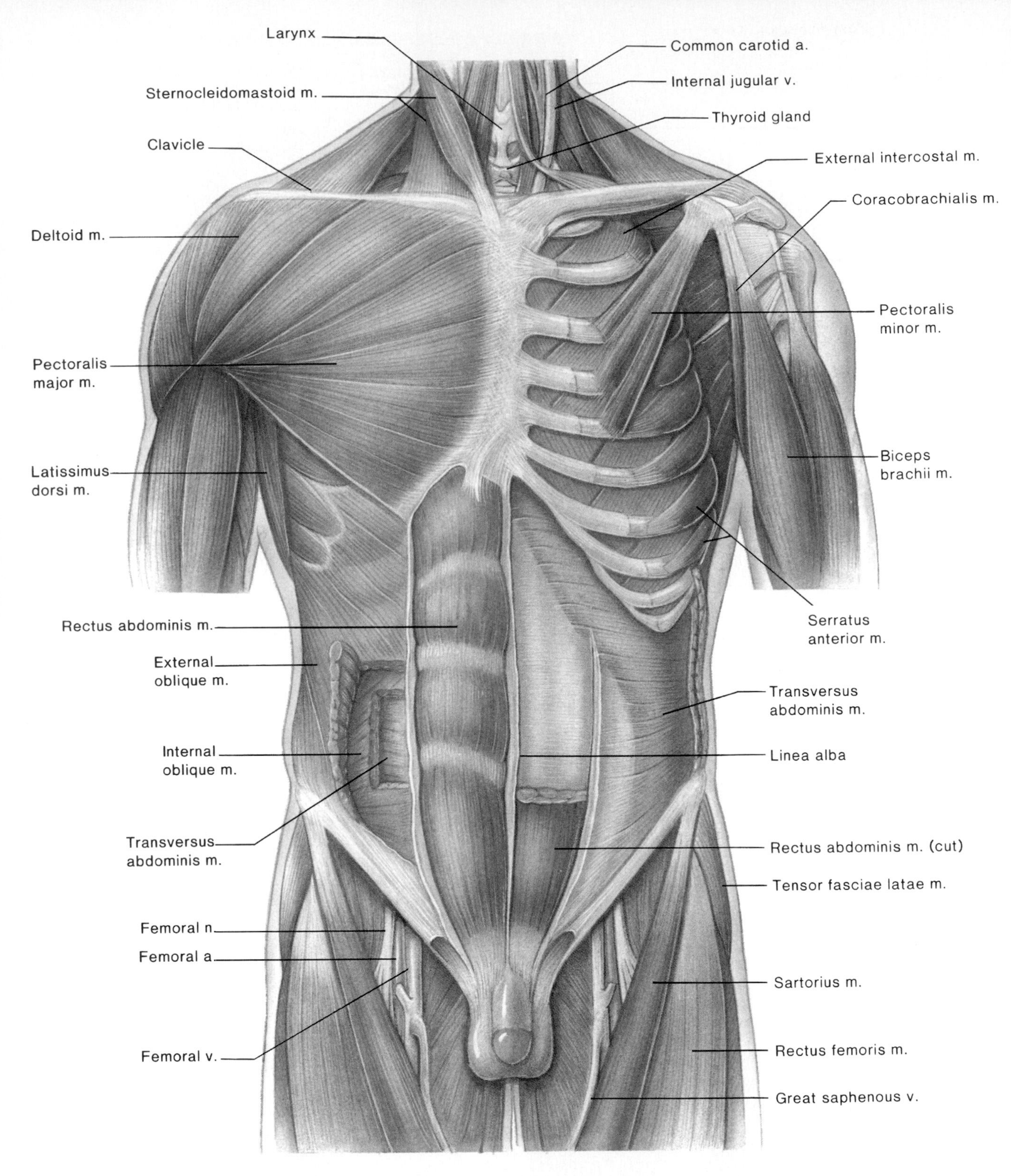
Larynx
Common carotid a.
Sternocleidomastoid m.
Internal jugular v.
Thyroid gland
Clavicle
External intercostal m.
Coracobrachialis m.
Deltoid m.
Pectoralis minor m.
Pectoralis major m.
Biceps brachii m.
Latissimus dorsi m.
Serratus anterior m.
Rectus abdominis m.
External oblique m.
Transversus abdominis m.
Internal oblique m.
Linea alba
Transversus abdominis m.
Rectus abdominis m. (cut)
Tensor fasciae latae m.
Femoral n.
Femoral a.
Sartorius m.
Rectus femoris m.
Femoral v.
Great saphenous v.

Human Torso, Deeper Muscle Layers
Reference Plate 2

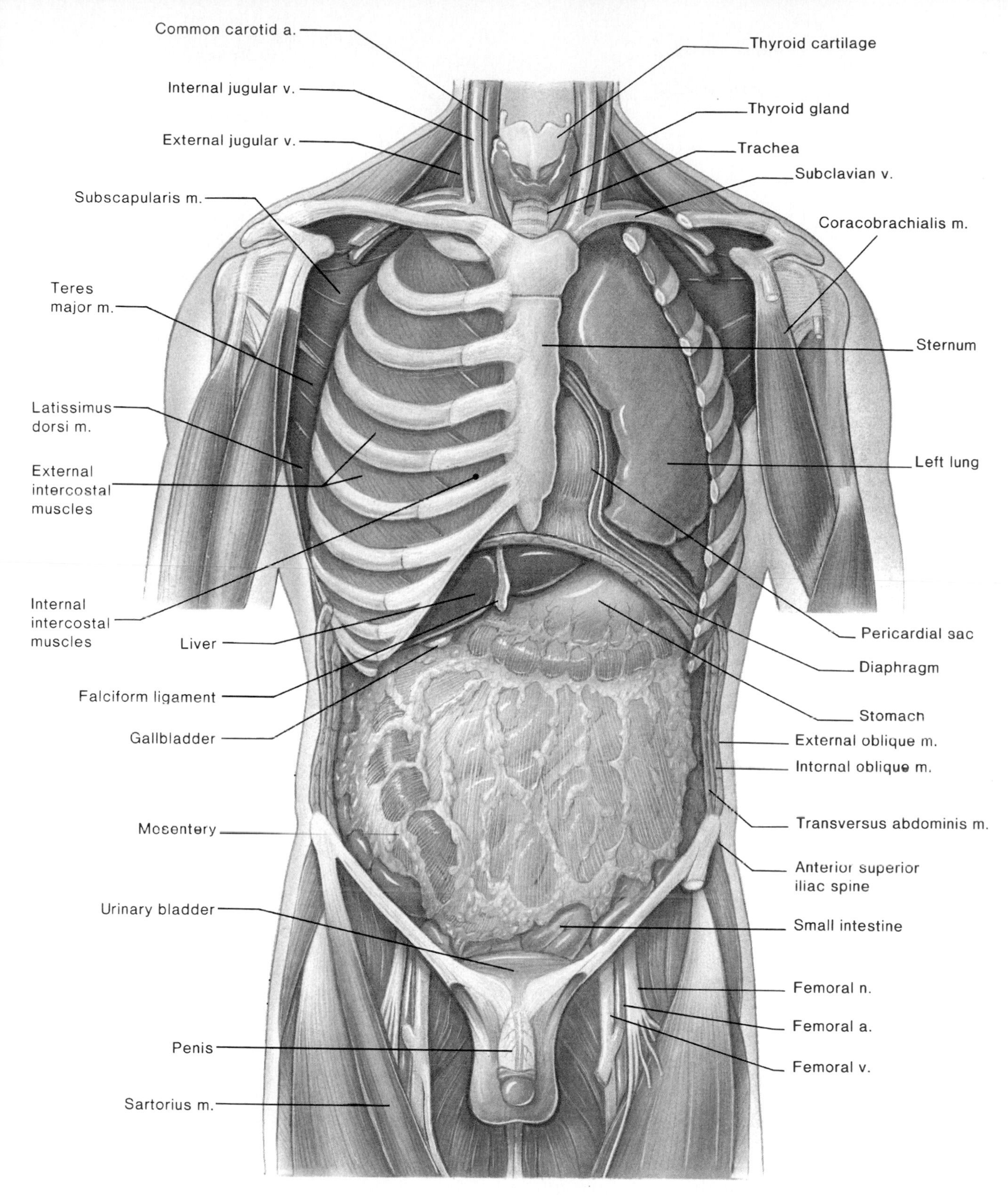
Common carotid a.
Internal jugular v.
External jugular v.
Subscapularis m.
Teres major m.
Latissimus dorsi m.
External intercostal muscles
Internal intercostal muscles
Liver
Falciform ligament
Gallbladder
Mesentery
Urinary bladder
Penis
Sartorius m.
Thyroid cartilage
Thyroid gland
Trachea
Subclavian v.
Coracobrachialis m.
Sternum
Left lung
Pericardial sac
Diaphragm
Stomach
External oblique m.
Internal oblique m.
Transversus abdominis m.
Anterior superior iliac spine
Small intestine
Femoral n.
Femoral a.
Femoral v.

Human Torso, Abdominal Viscera
Reference Plate 3

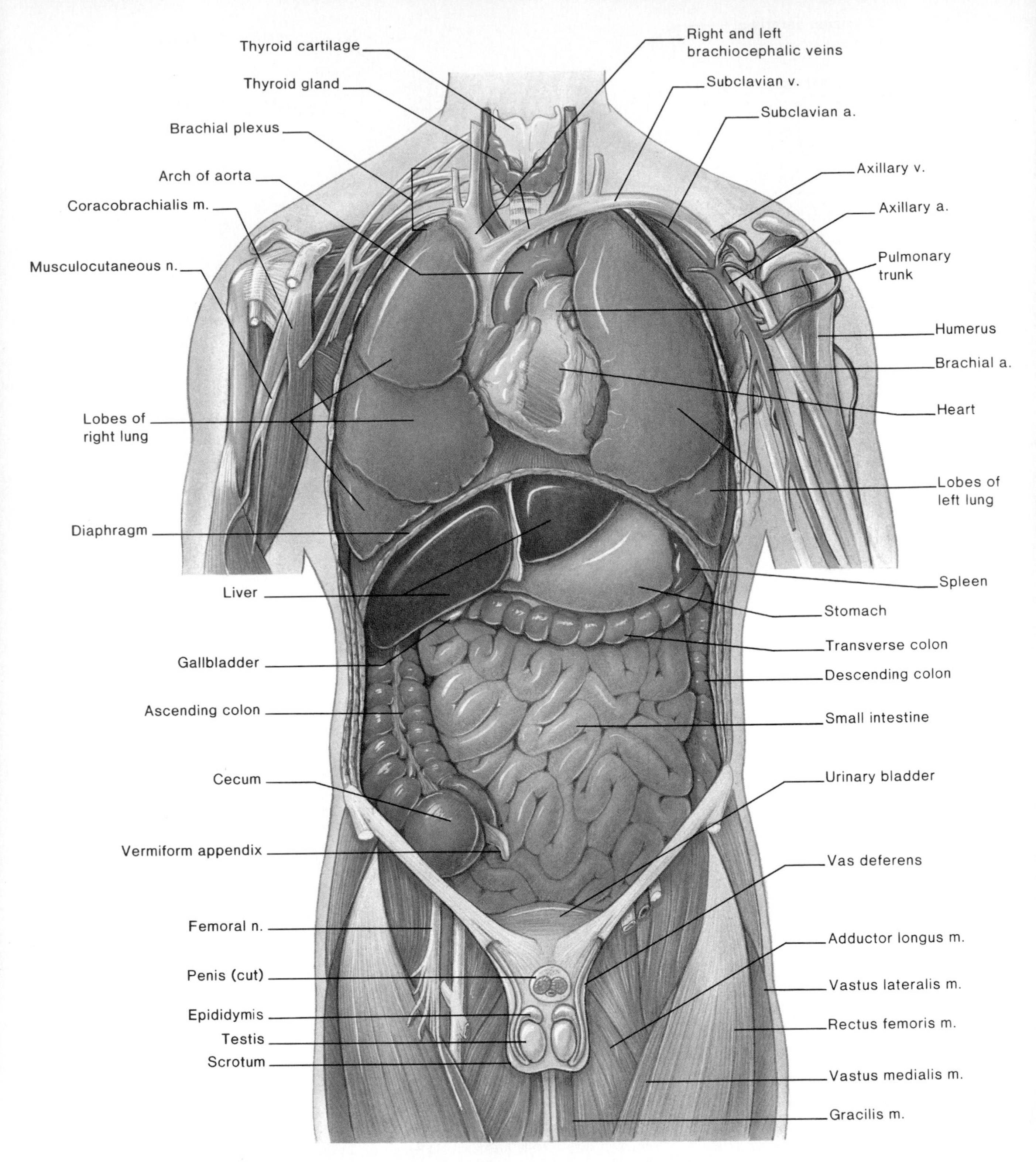
Thyroid cartilage
Thyroid gland
Brachial plexus
Arch of aorta
Coracobrachialis m.
Musculocutaneous n.
Lobes of right lung
Diaphragm
Liver
Gallbladder
Ascending colon
Cecum
Vermiform appendix
Femoral n.
Penis (cut)
Epididymis
Testis
Scrotum
Right and left brachiocephalic veins
Subclavian v.
Subclavian a.
Axillary v.
Axillary a.
Pulmonary trunk
Humerus
Brachial a.
Heart
Lobes of left lung
Spleen
Stomach
Transverse colon
Descending colon
Small intestine
Urinary bladder
Vas deferens
Adductor longus m.
Vastus lateralis m.
Rectus femoris m.
Vastus medialis m.
Gracilis m.

Human Torso, Thoracic Viscera
Reference Plate 4

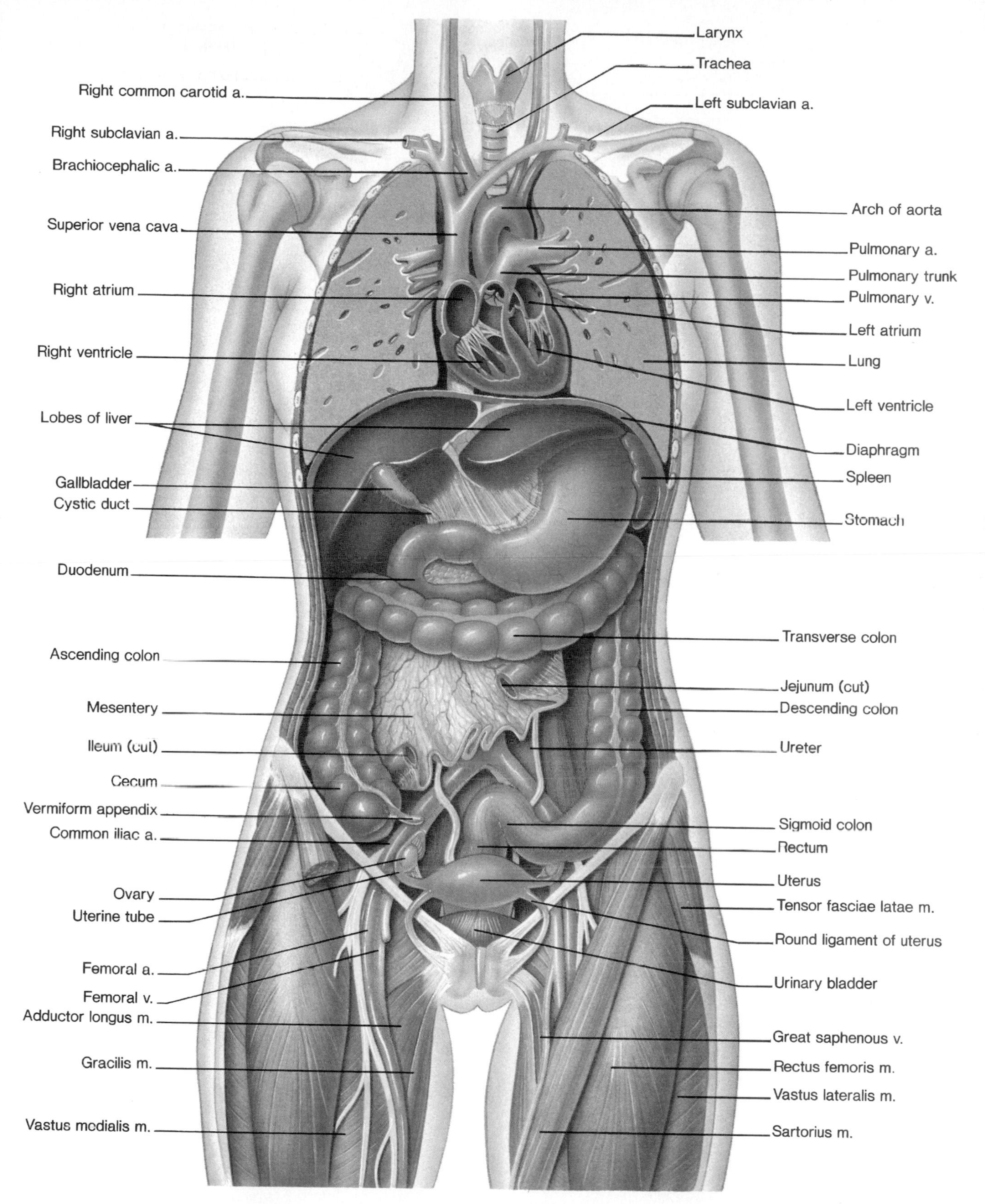
Larynx
Trachea
Left subclavian a.
Right common carotid a.
Right subclavian a.
Brachiocephalic a.
Arch of aorta
Superior vena cava
Pulmonary a.
Pulmonary trunk
Pulmonary v.
Right atrium
Left atrium
Right ventricle
Lung
Left ventricle
Lobes of liver
Diaphragm
Gallbladder
Spleen
Cystic duct
Stomach
Duodenum
Transverse colon
Ascending colon
Jejunum (cut)
Mesentery
Descending colon
Ileum (cut)
Ureter
Cecum
Vermiform appendix
Sigmoid colon
Common iliac a.
Rectum
Uterus
Ovary
Tensor fasciae latae m.
Uterine tube
Round ligament of uterus
Femoral a.
Urinary bladder
Femoral v.
Adductor longus m.
Great saphenous v.
Gracilis m.
Rectus femoris m.
Vastus lateralis m.
Vastus medialis m.
Sartorius m.

Human Torso, Lungs, Heart, and Small Intestine Sectioned
Reference Plate 5

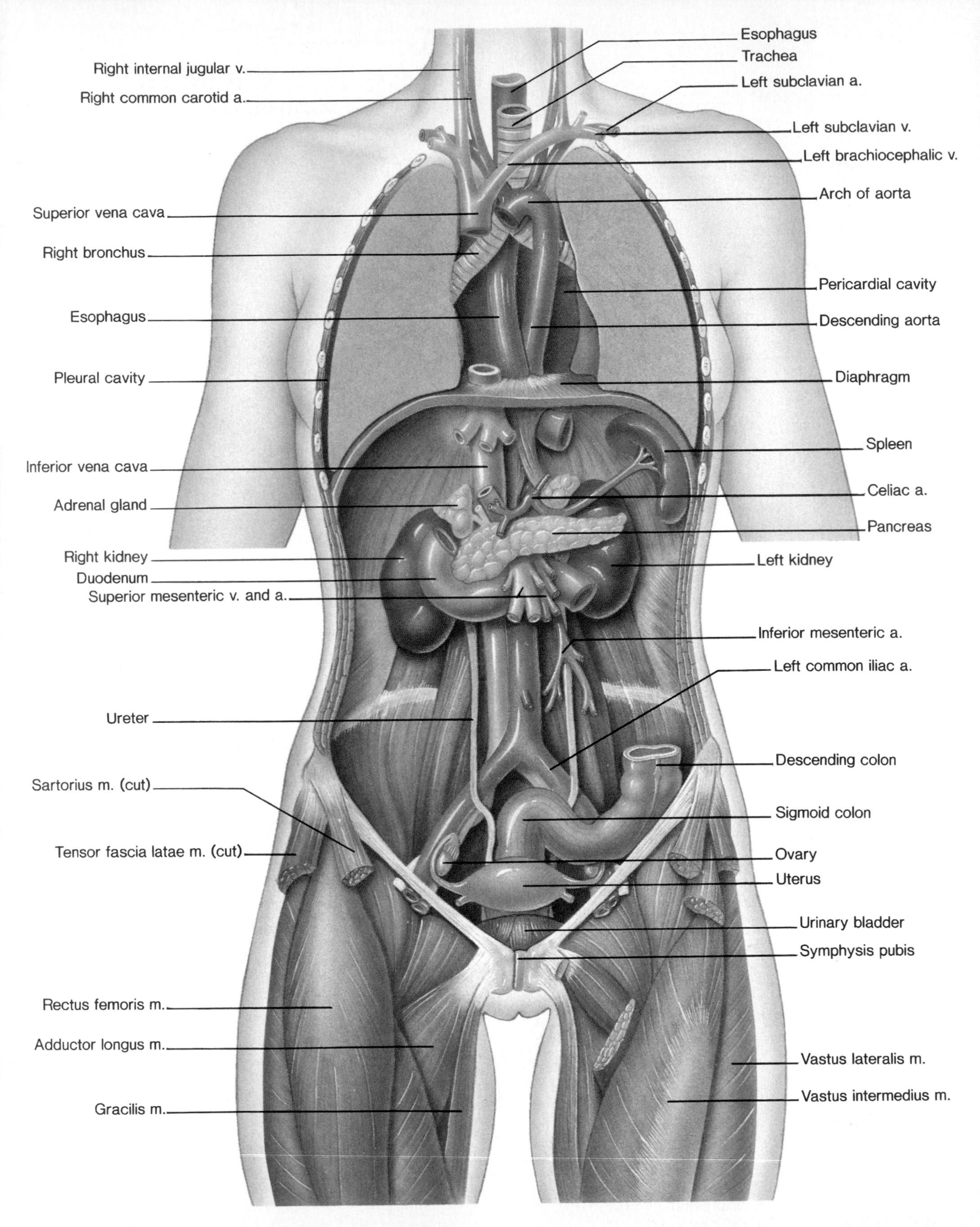

Right internal jugular v.
Right common carotid a.
Superior vena cava
Right bronchus
Esophagus
Pleural cavity
Inferior vena cava
Adrenal gland
Right kidney
Duodenum
Superior mesenteric v. and a.
Ureter
Sartorius m. (cut)
Tensor fascia latae m. (cut)
Rectus femoris m.
Adductor longus m.
Gracilis m.
Esophagus
Trachea
Left subclavian a.
Left subclavian v.
Left brachiocephalic v.
Arch of aorta
Pericardial cavity
Descending aorta
Diaphragm
Spleen
Celiac a.
Pancreas
Left kidney
Inferior mesenteric a.
Left common iliac a.
Descending colon
Sigmoid colon
Ovary
Uterus
Urinary bladder
Symphysis pubis
Vastus lateralis m.
Vastus intermedius m.

Human Torso, Heart, Stomach and Part of Intestine and Lungs Removed
Reference Plate 6

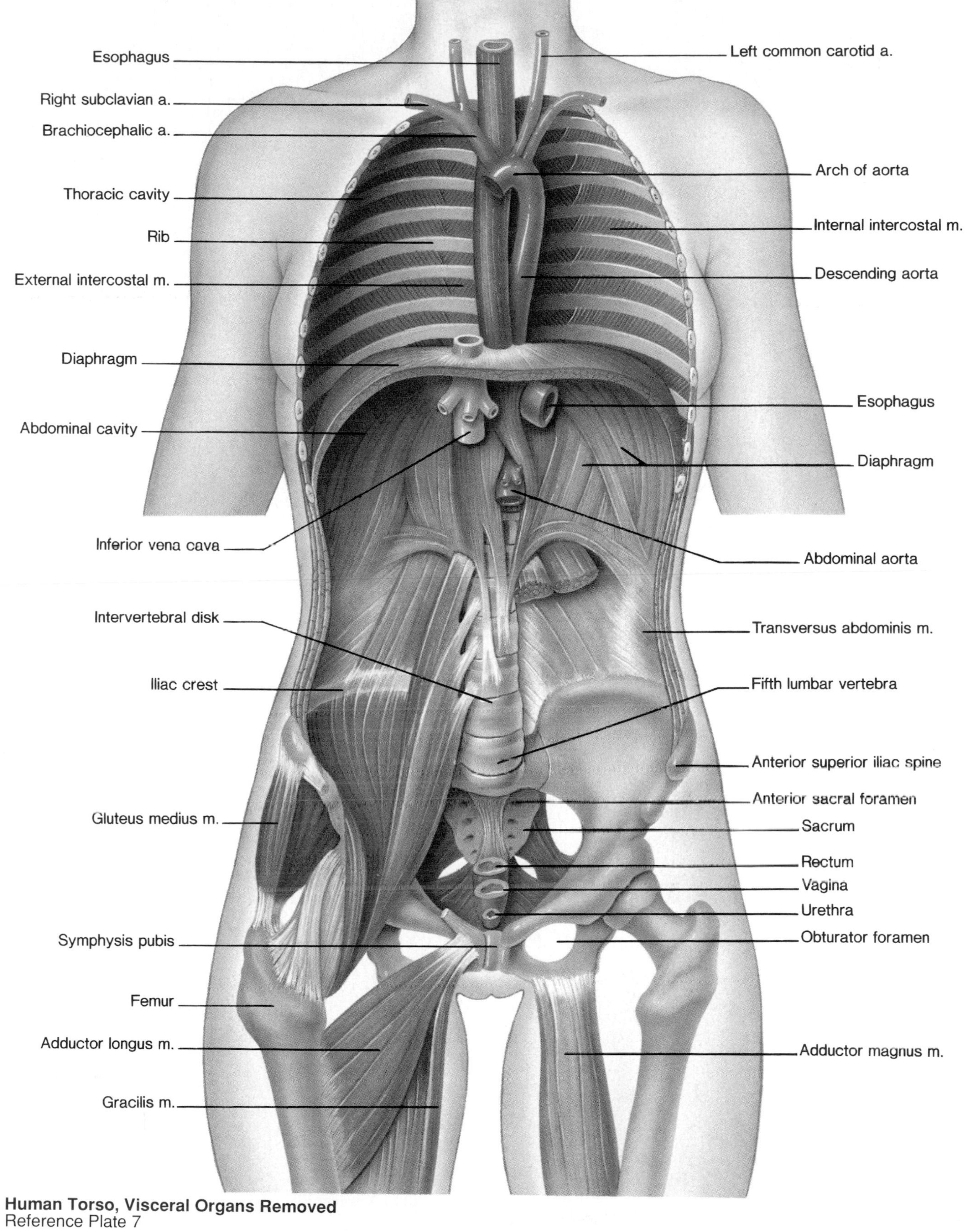

Esophagus
Right subclavian a.
Brachiocephalic a.
Thoracic cavity
Rib
External intercostal m.
Diaphragm
Abdominal cavity
Inferior vena cava
Intervertebral disk
Iliac crest
Gluteus medius m.
Symphysis pubis
Femur
Adductor longus m.
Gracilis m.
Left common carotid a.
Arch of aorta
Internal intercostal m.
Descending aorta
Esophagus
Diaphragm
Abdominal aorta
Transversus abdominis m.
Fifth lumbar vertebra
Anterior superior iliac spine
Anterior sacral foramen
Sacrum
Rectum
Vagina
Urethra
Obturator foramen
Adductor magnus m.

Human Torso, Visceral Organs Removed
Reference Plate 7

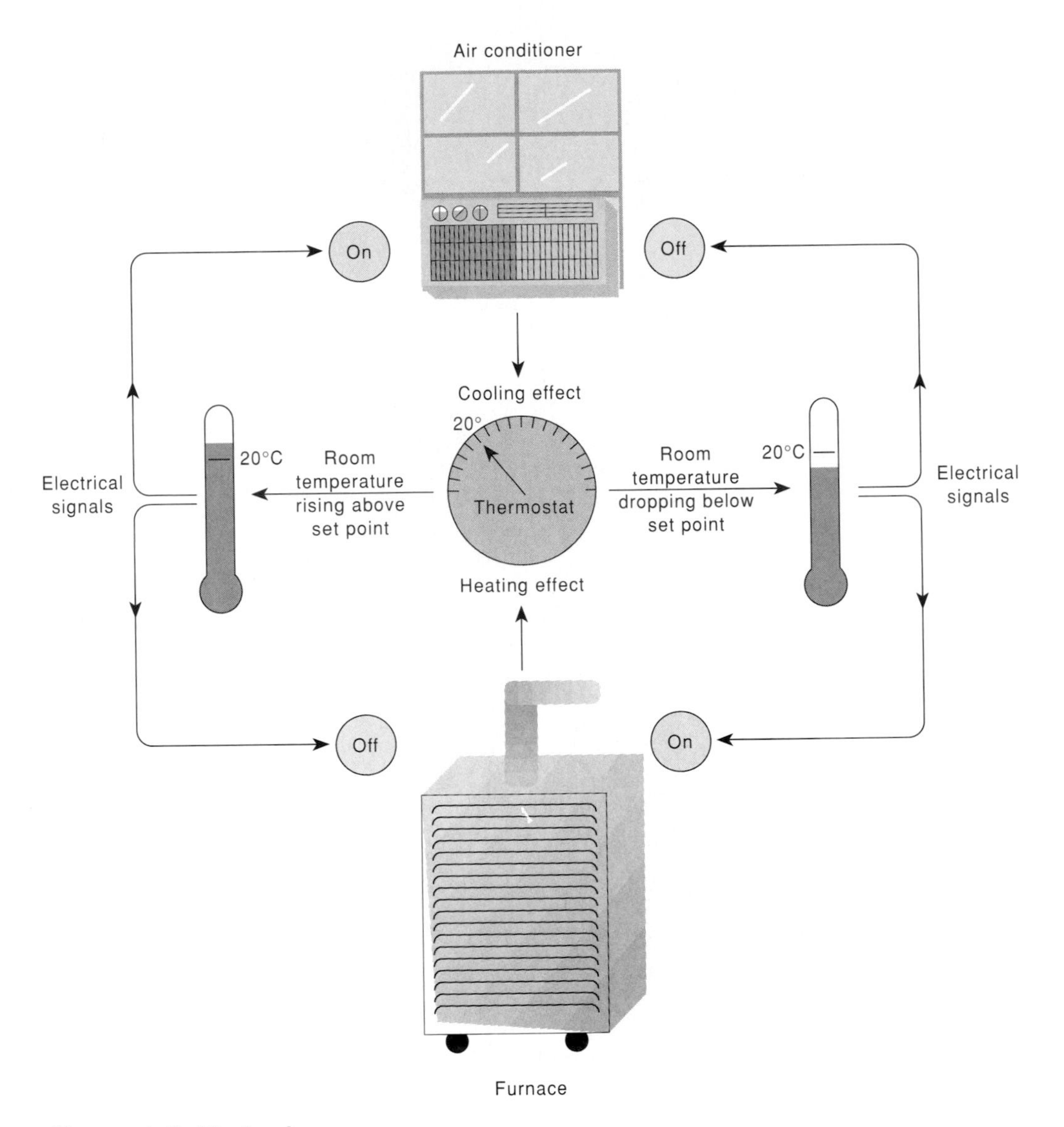
Air conditioner
On
Off
Cooling effect
20°
Electrical
signals
20°C
Room
temperature
rising above
set point
Thermostat
Room
temperature
dropping below
set point
20°C
Electrical
signals
Heating effect
Off
On
Furnace

Homeostatic Mechanism
Figure 1.3

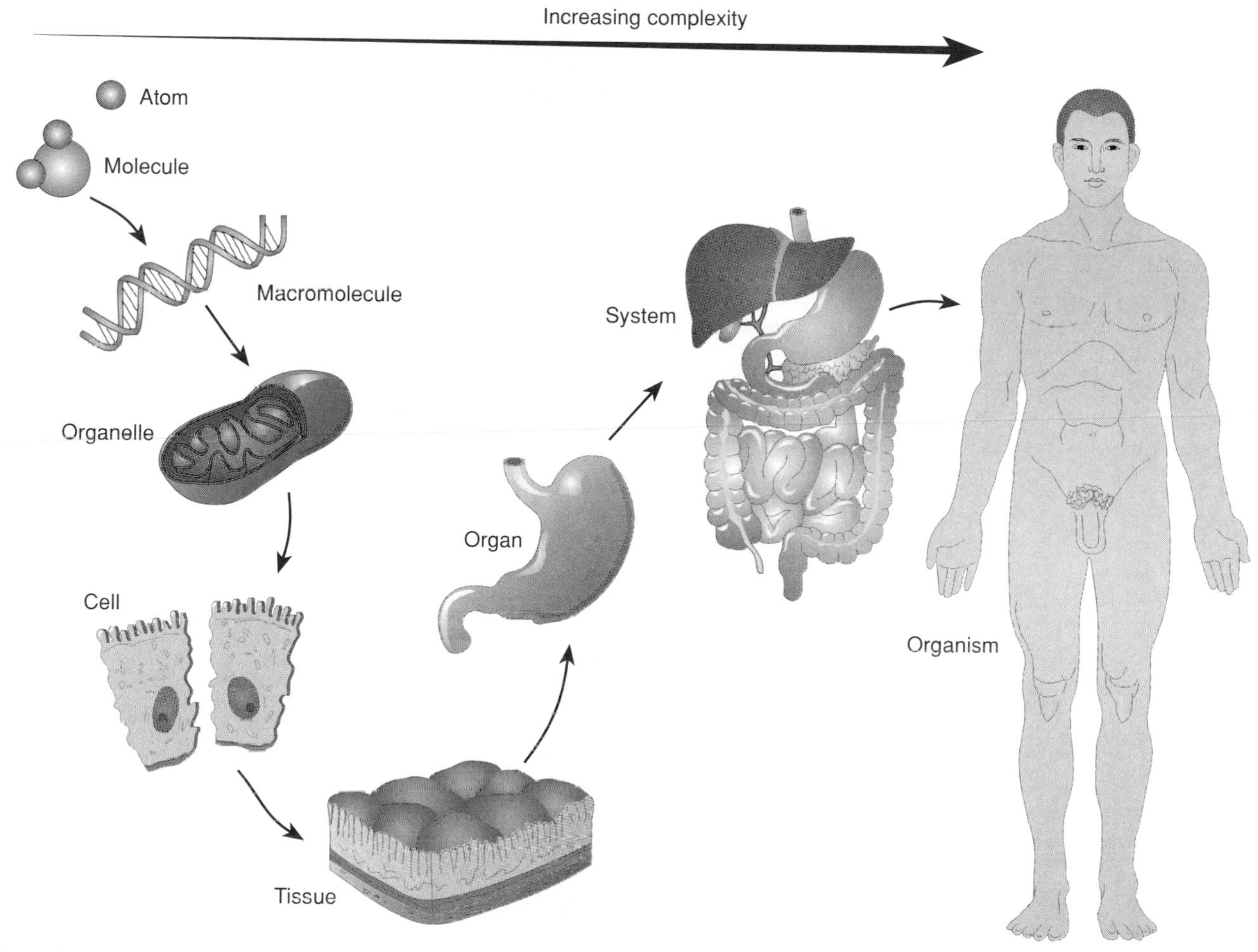
Increasing complexity
Atom
Molecule
Macromolecule
Organelle
Cell
Tissue
Organ
System
Organism

Levels of Organization
Figure 1.5

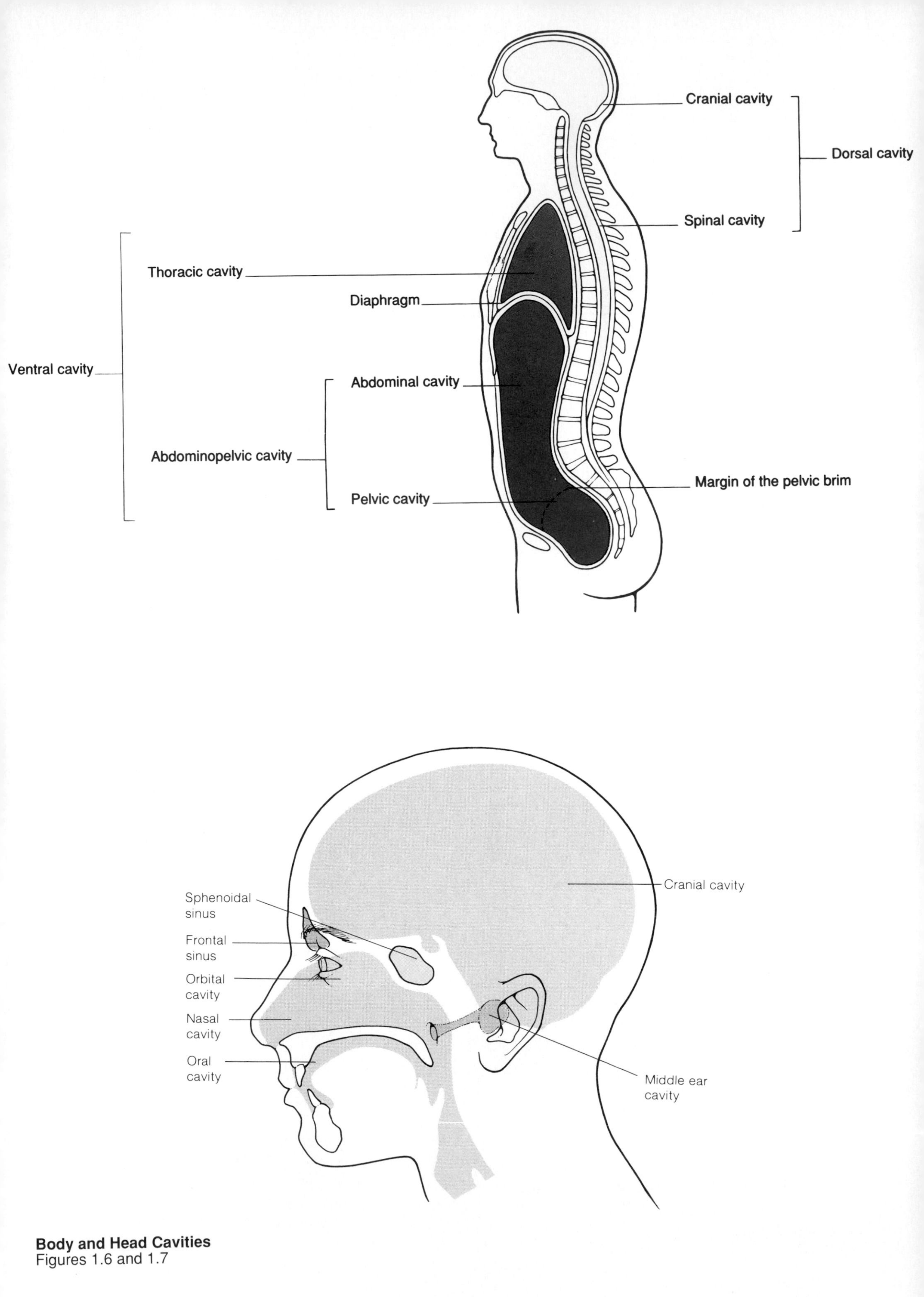
Cranial cavity
Dorsal cavity
Spinal cavity
Thoracic cavity
Diaphragm
Ventral cavity
Abdominal cavity
Abdominopelvic cavity
Margin of the pelvic brim
Pelvic cavity
Cranial cavity
Sphenoidal sinus
Frontal sinus
Orbital cavity
Nasal cavity
Oral cavity
Middle ear cavity

Body and Head Cavities
Figures 1.6 and 1.7

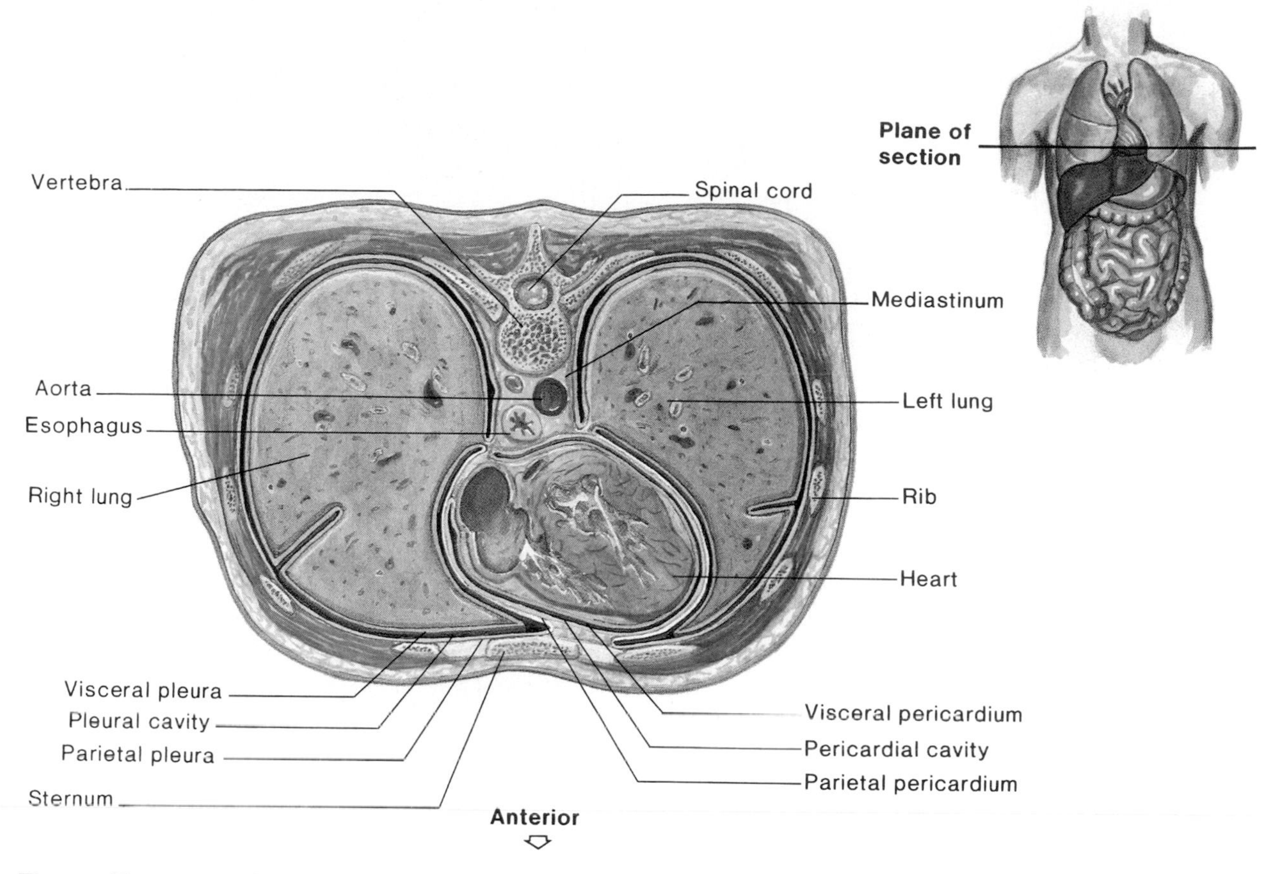
Plane of section
Vertebra
Spinal cord
Mediastinum
Aorta
Esophagus
Left lung
Right lung
Rib
Heart
Visceral pleura
Pleural cavity
Parietal pleura
Sternum
Visceral pericardium
Pericardial cavity
Parietal pericardium
Anterior

Thorax, Transverse Section
Figure 1.8

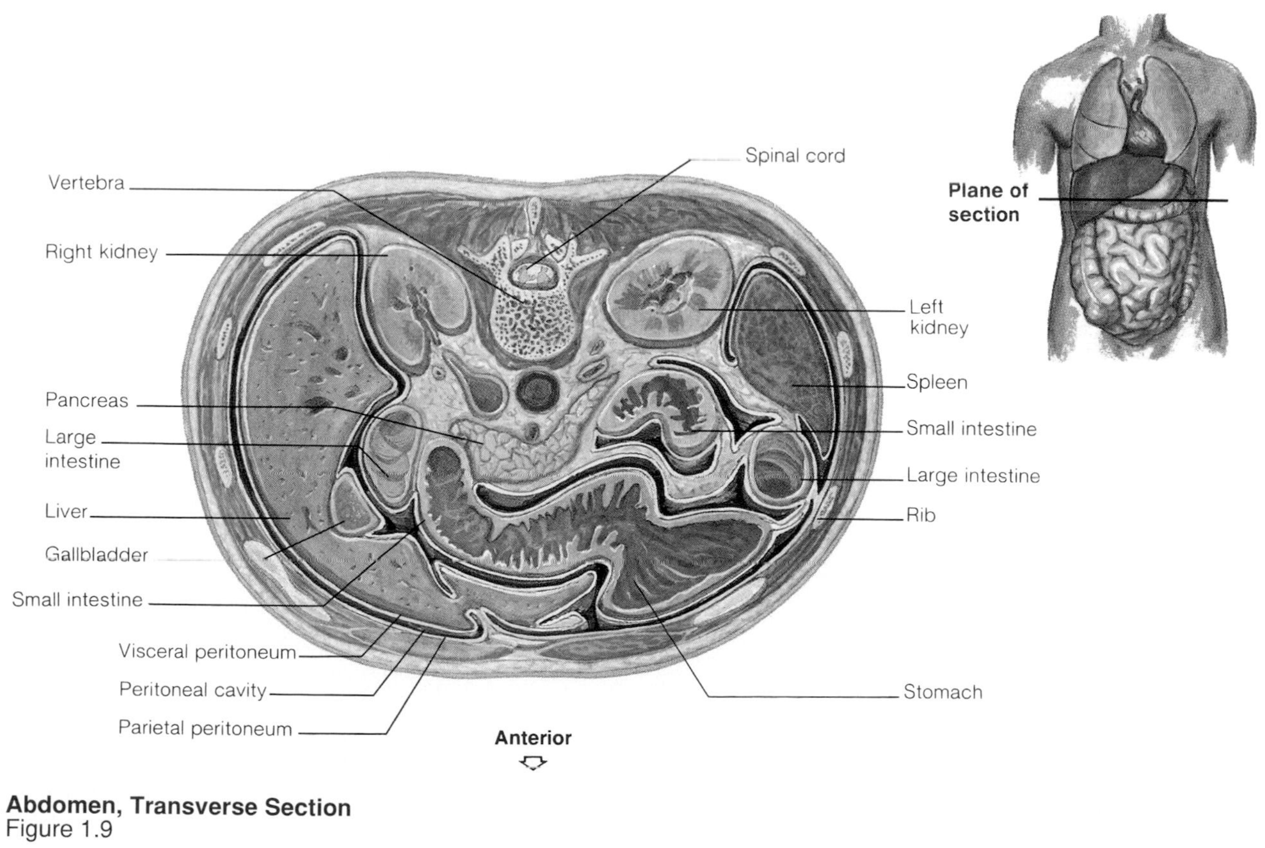
Spinal cord
Vertebra
Plane of section
Right kidney
Left kidney
Spleen
Pancreas
Small intestine
Large intestine
Large intestine
Liver
Rib
Gallbladder
Small intestine
Visceral peritoneum
Peritoneal cavity
Stomach
Parietal peritoneum
Anterior

Abdomen, Transverse Section
Figure 1.9

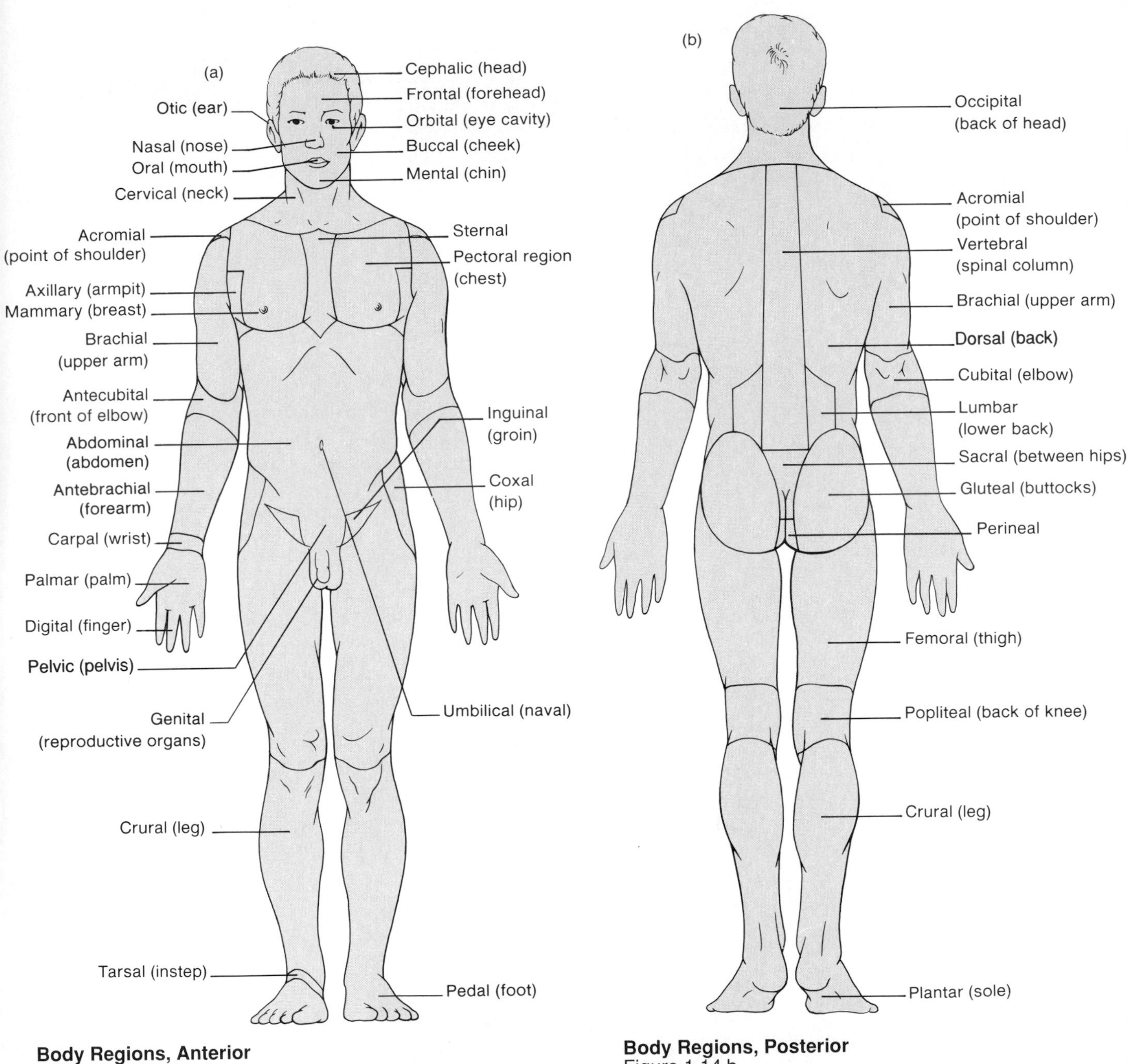

(a)
Cephalic (head)
Frontal (forehead)
Otic (ear)
Orbital (eye cavity)
Nasal (nose)
Buccal (cheek)
Oral (mouth)
Mental (chin)
Cervical (neck)
Acromial
(point of shoulder)
Sternal
Pectoral region
(chest)
Axillary (armpit)
Mammary (breast)
Brachial
(upper arm)
Antecubital
(front of elbow)
Inguinal
(groin)
Abdominal
(abdomen)
Coxal
(hip)
Antebrachial
(forearm)
Carpal (wrist)
Palmar (palm)
Digital (finger)
Pelvic (pelvis)
Umbilical (naval)
Genital
(reproductive organs)
Crural (leg)
Tarsal (instep)
Pedal (foot)
(b)
Occipital
(back of head)
Acromial
(point of shoulder)
Vertebral
(spinal column)
Brachial (upper arm)
Dorsal (back)
Cubital (elbow)
Lumbar
(lower back)
Sacral (between hips)
Gluteal (buttocks)
Perineal
Femoral (thigh)
Popliteal (back of knee)
Crural (leg)
Plantar (sole)

Body Regions, Anterior
Figure 1.14 a

Body Regions, Posterior
Figure 1.14 b

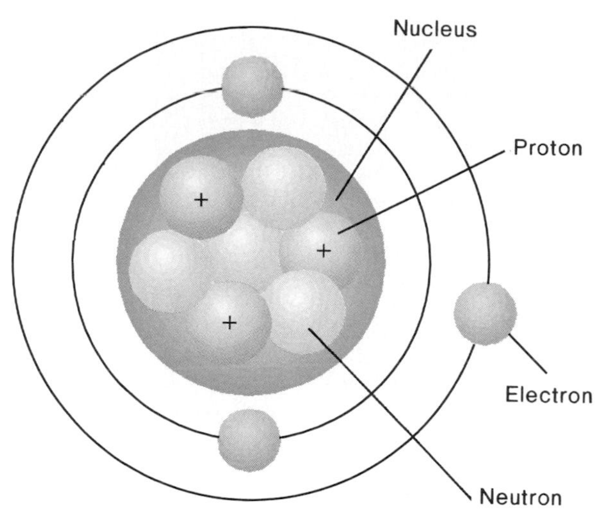
Nucleus
Proton
Electron
Neutron
Lithium atom

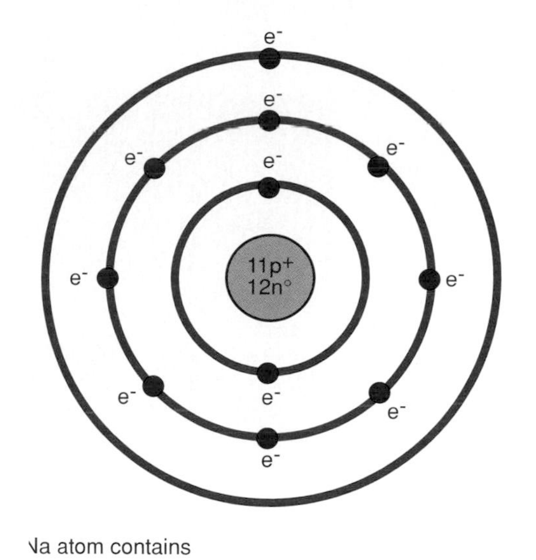
11p+
12n°
\a atom contains
11 electrons (e⁻)
11 protons (p+)
12 neutrons (n°)
Atomic number = 11

Atomic Structure
Figures 2.1 and 2.5

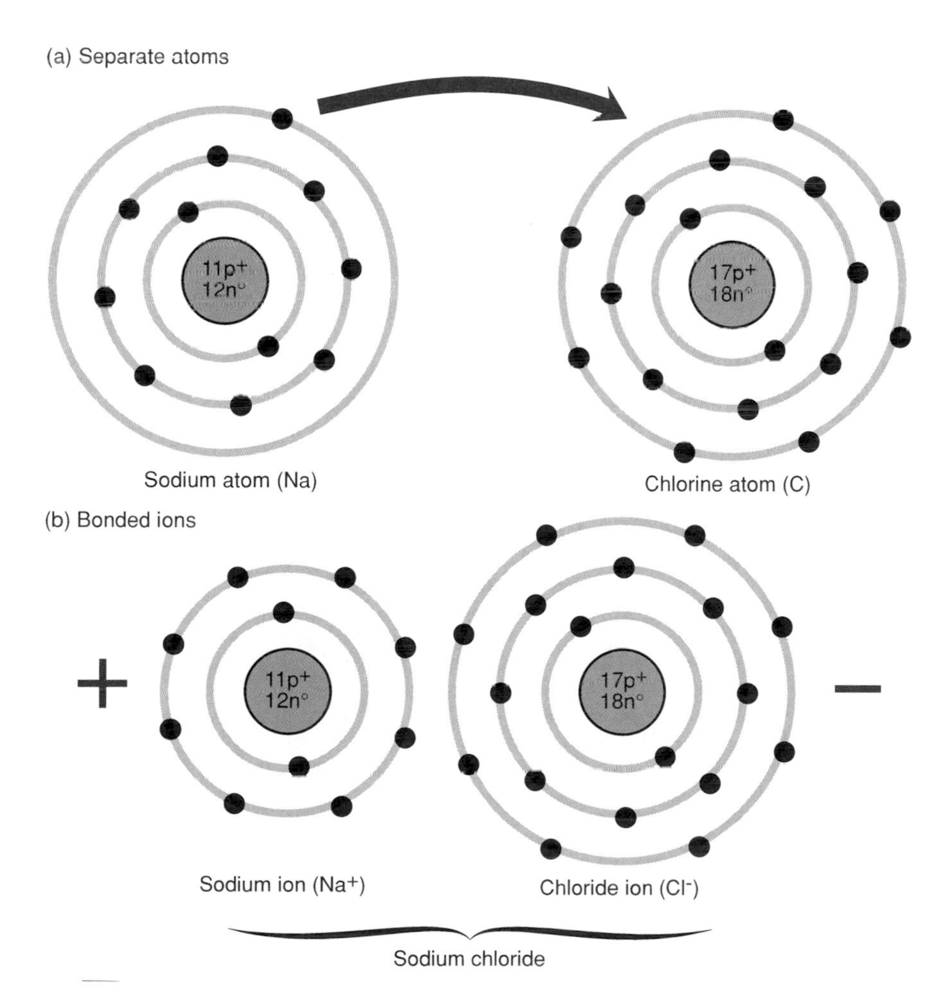
(a) Separate atoms
11p+
12n°
17p+
18n°
Sodium atom (Na)
Chlorine atom (C)
(b) Bonded ions
+
−
Sodium ion (Na+)
Chloride ion (Cl-)
Sodium chloride

Electrovalent Bond
Figure 2.6

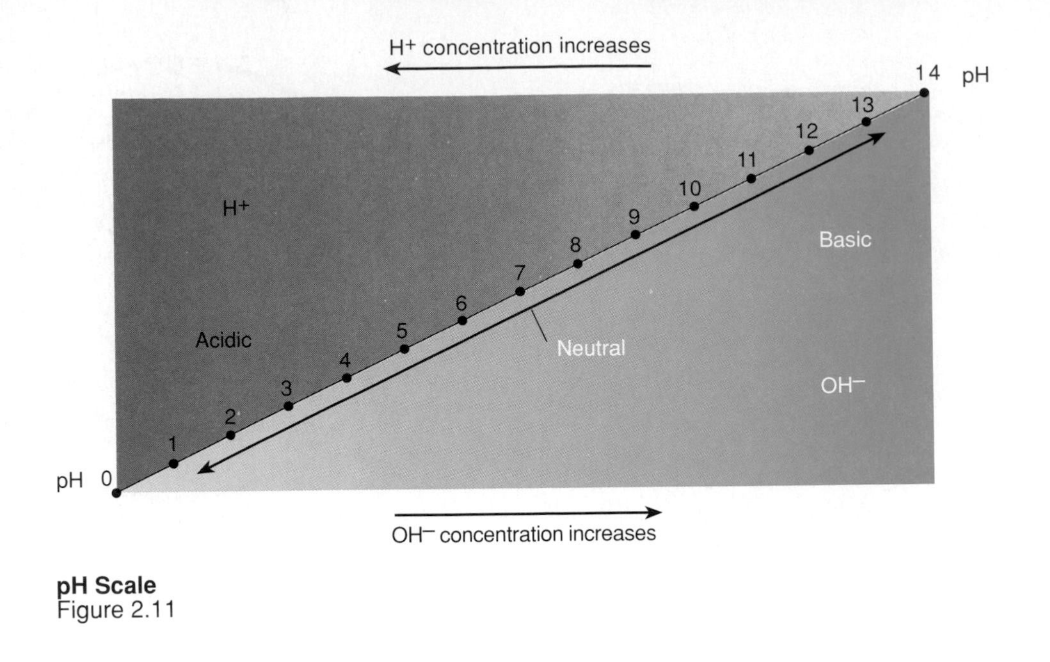
H+ concentration increases
14 pH
13
12
11
10
9
8
7
6
5
4
3
2
1
pH 0
H+
Acidic
Neutral
Basic
OH−
OH− concentration increases

pH Scale
Figure 2.11

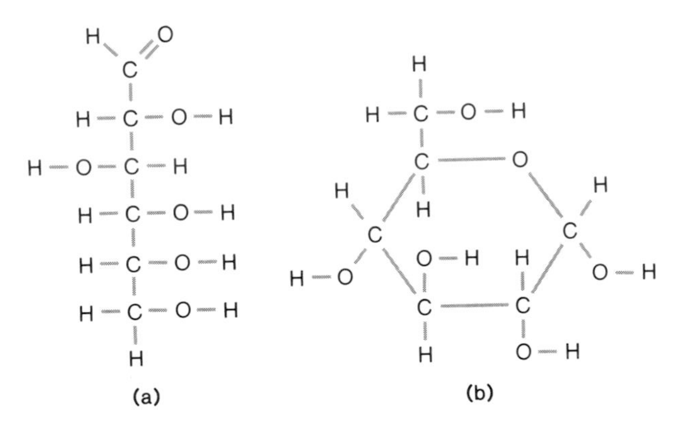
(a)
(b)

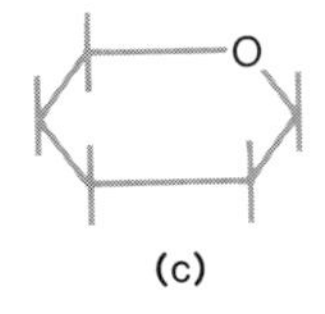
(c)

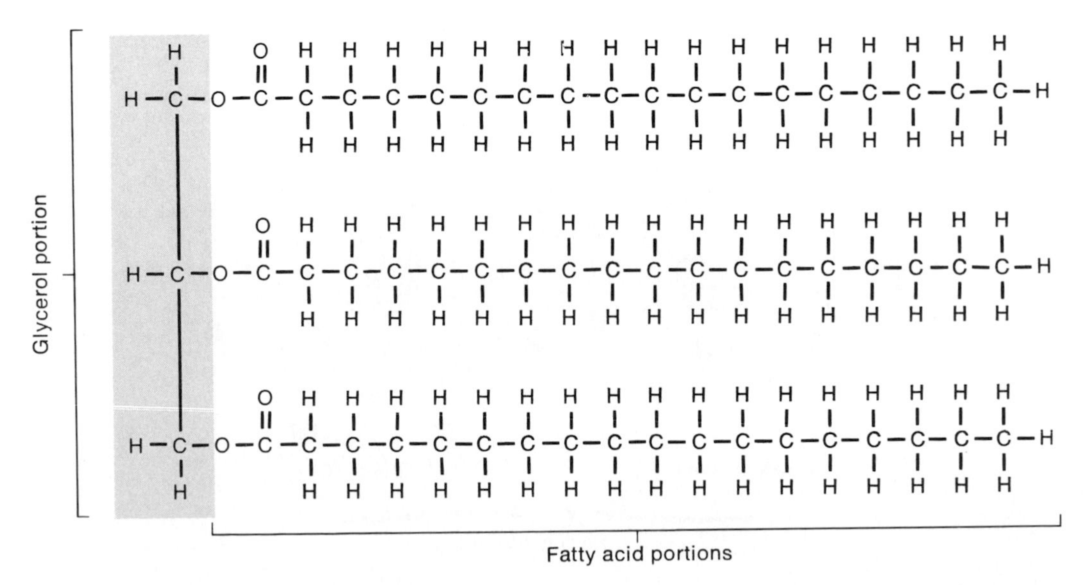
Glycerol portion
Fatty acid portions

Glucose and Triglyceride Molecules
Figures 2.13 and 2.15

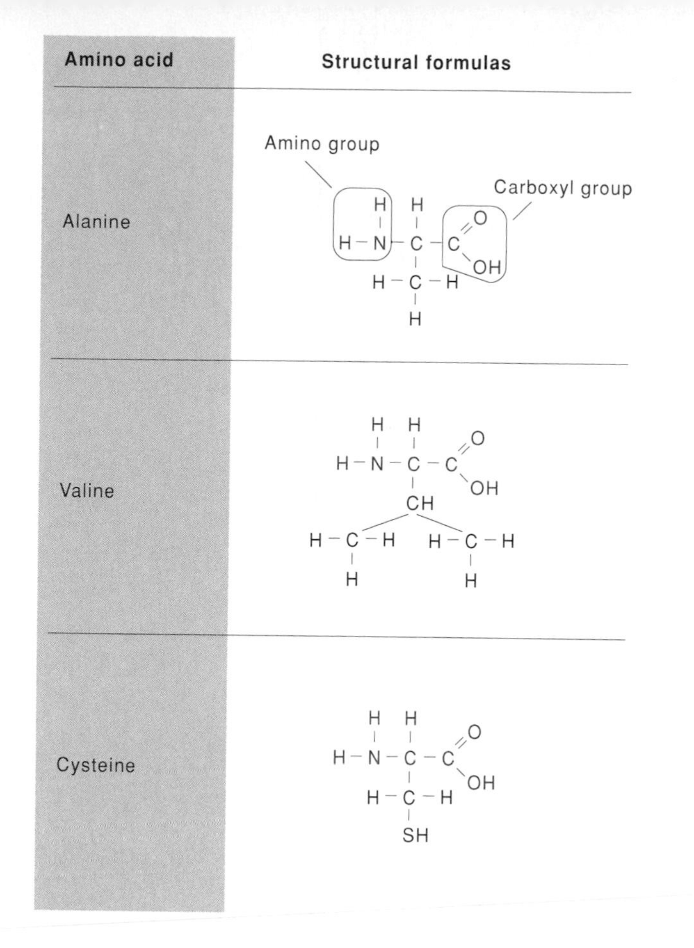
Amino acid
Structural formulas
Alanine
Amino group
Carboxyl group
Valine
Cysteine

Amino Acids
Figure 2.16

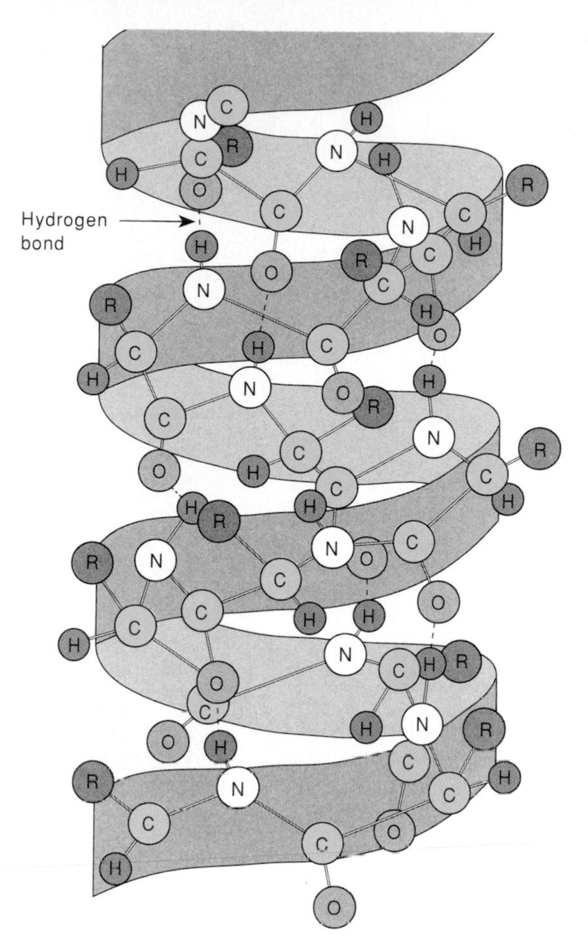
Hydrogen bond

Portion of a Protein Molecule
Figure 2.18

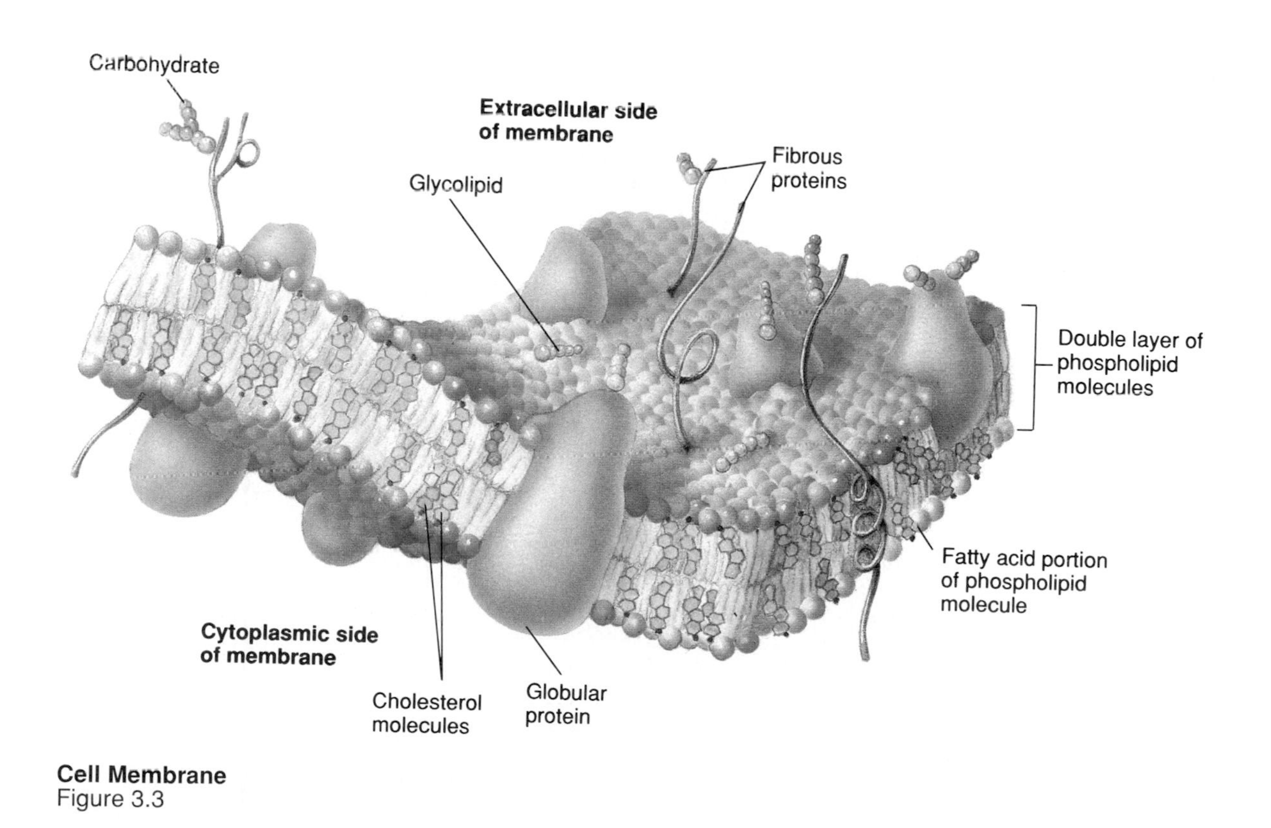
Carbohydrate
Extracellular side of membrane
Glycolipid
Fibrous proteins
Double layer of phospholipid molecules
Fatty acid portion of phospholipid molecule
Cytoplasmic side of membrane
Cholesterol molecules
Globular protein

Cell Membrane
Figure 3.3

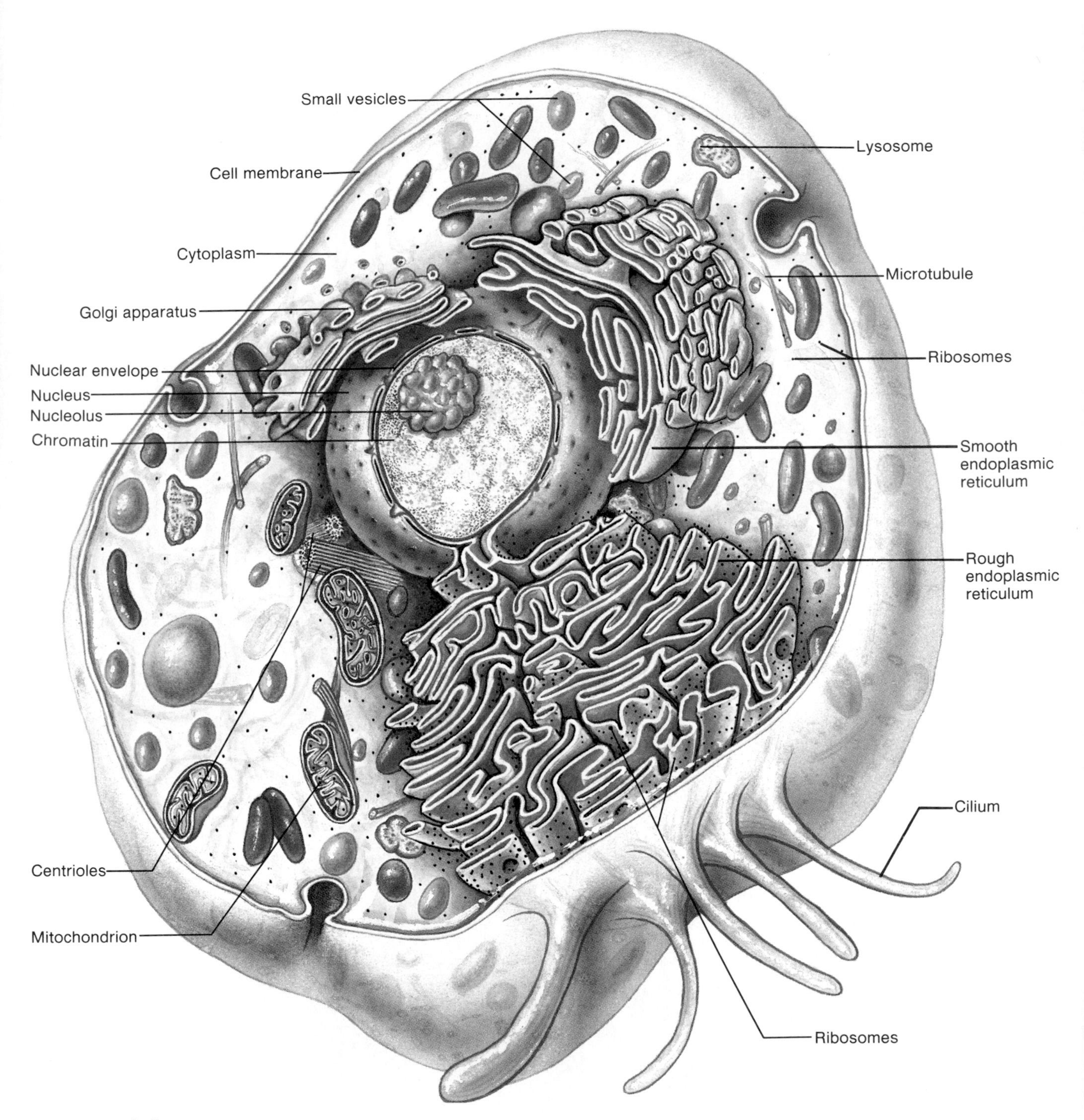
Small vesicles
Lysosome
Cell membrane
Cytoplasm
Microtubule
Golgi apparatus
Ribosomes
Nuclear envelope
Nucleus
Nucleolus
Chromatin
Smooth endoplasmic reticulum
Rough endoplasmic reticulum
Cilium
Centrioles
Mitochondrion
Ribosomes

Composite Cell
Figure 3.2

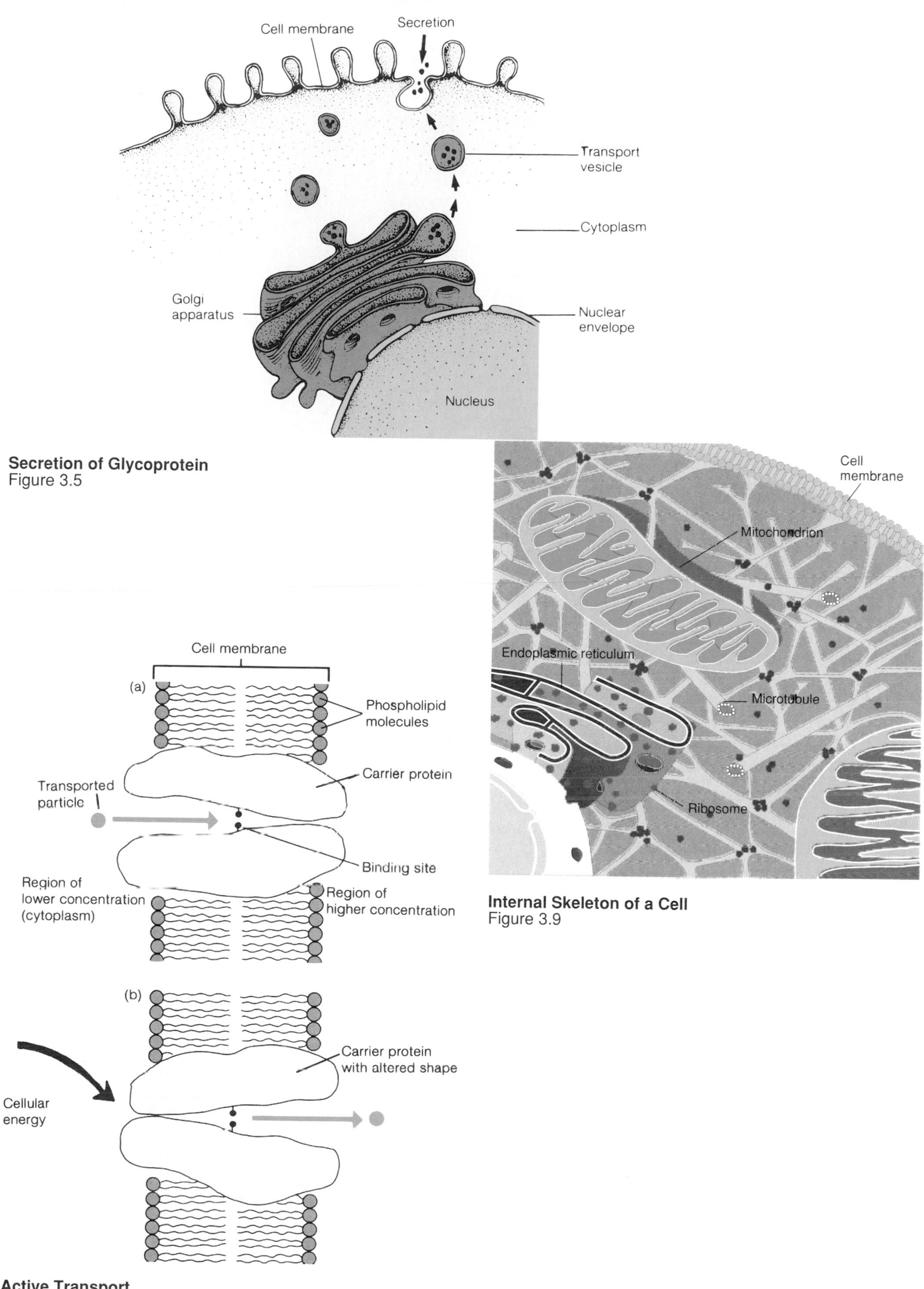
Cell membrane
Secretion
Transport vesicle
Cytoplasm
Golgi apparatus
Nuclear envelope
Nucleus
Cell membrane
Mitochondrion
Endoplasmic reticulum
Microtubule
Ribosome
Cell membrane
(a)
Phospholipid molecules
Carrier protein
Transported particle
Binding site
Region of lower concentration (cytoplasm)
Region of higher concentration
(b)
Carrier protein with altered shape
Cellular energy

Secretion of Glycoprotein
Figure 3.5

Internal Skeleton of a Cell
Figure 3.9

Active Transport
Figure 3.16

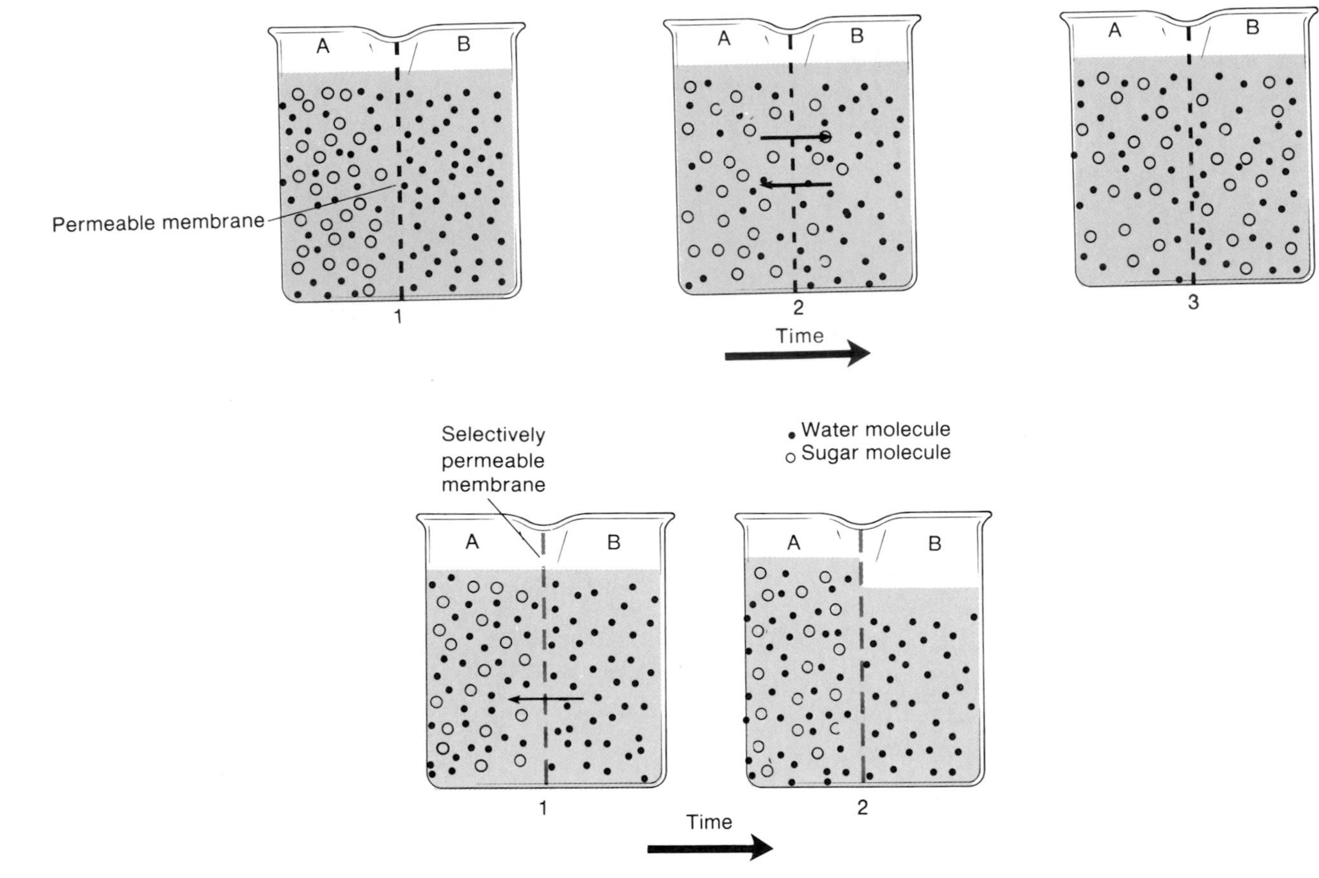
Water molecule
Sugar molecule
A
B
Permeable membrane
1
2
3
Time
Selectively permeable membrane
Water molecule
Sugar molecule
A
B
1
2
Time

Diffusion and Osmosis
Figures 3.12 (1–3) and 3.14 (1–2)

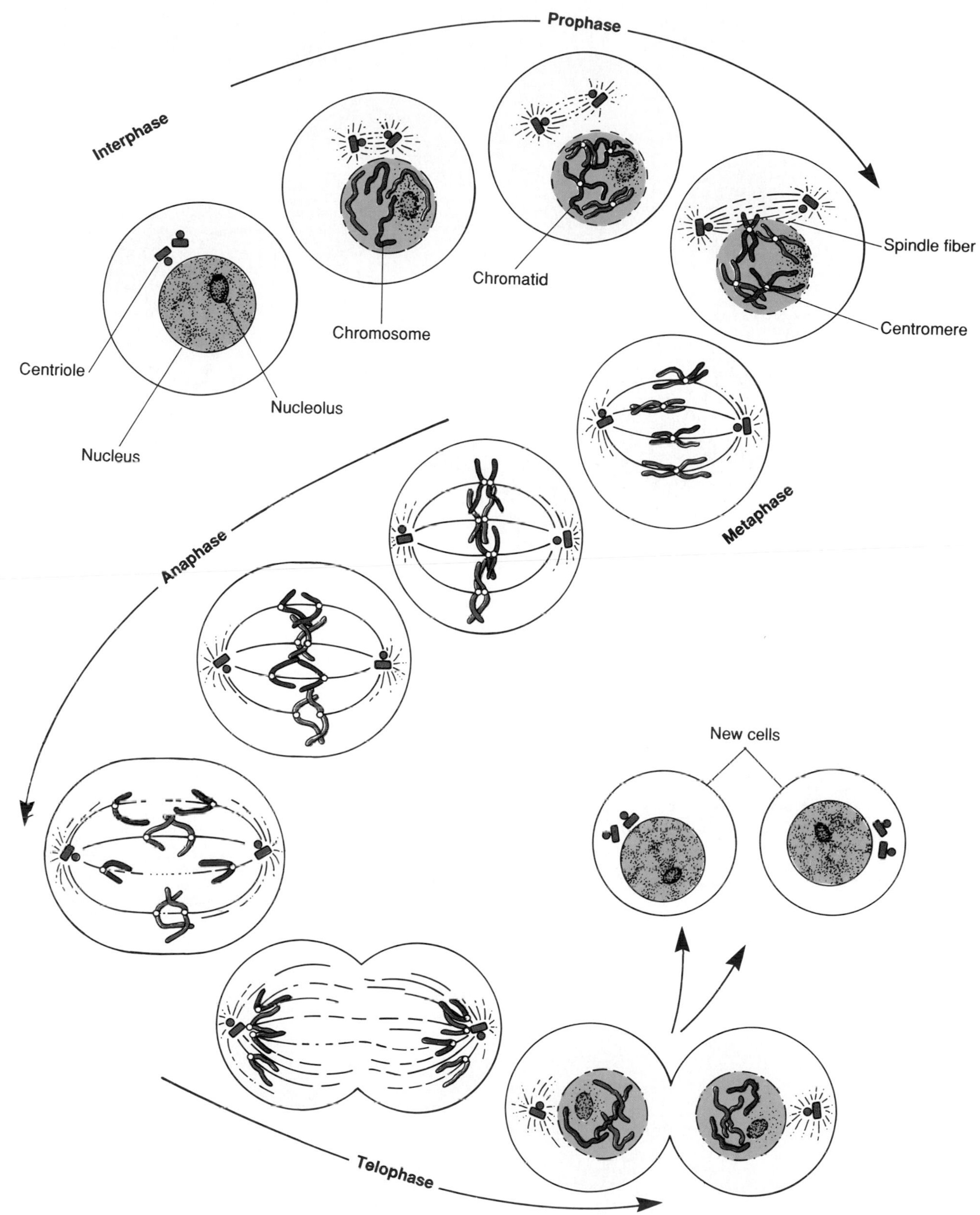
Prophase
Interphase
Spindle fiber
Chromatid
Centromere
Chromosome
Centriole
Nucleolus
Nucleus
Metaphase
Anaphase
New cells
Telophase

Mitosis
Figure 3.18

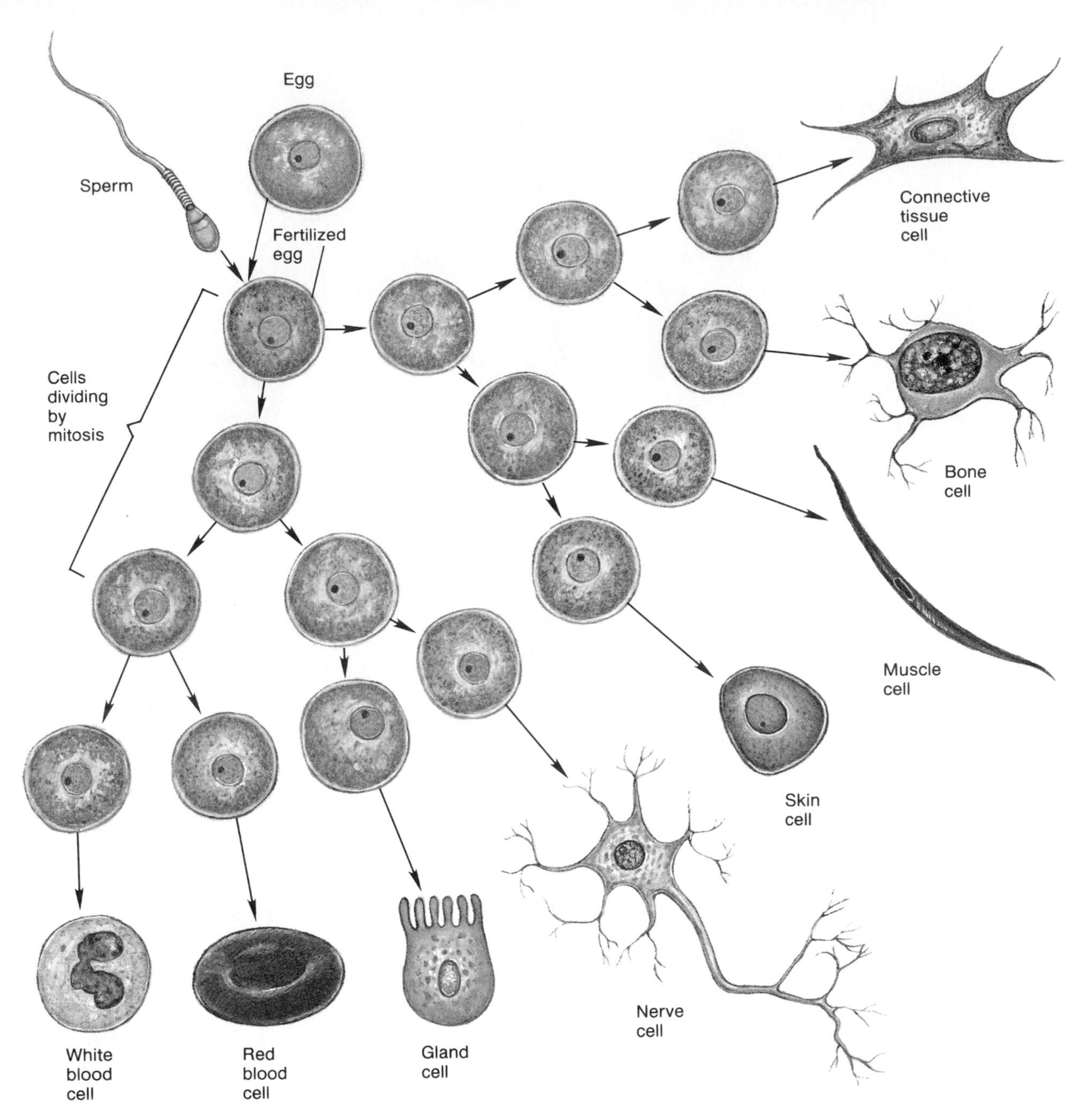
Egg
Sperm
Fertilized
egg
Connective
tissue
cell
Cells
dividing
by
mitosis
Bone
cell
Muscle
cell
Skin
cell
Nerve
cell
White
blood
cell
Red
blood
cell
Gland
cell

Cellular Differentiation
Figure 3.20

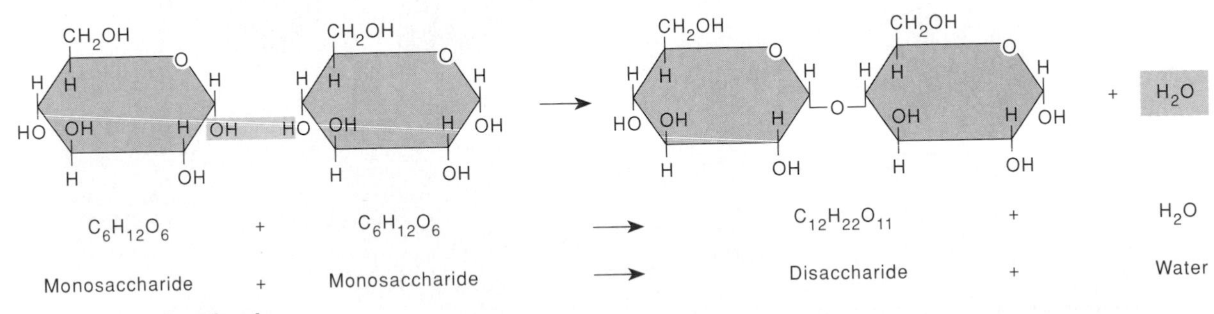
CH2OH
O
H
OH
HO
+
H2O
C6H12O6
C12H22O11
Monosaccharide
Disaccharide
Water

Dehydration Synthesis
Figures 4.1

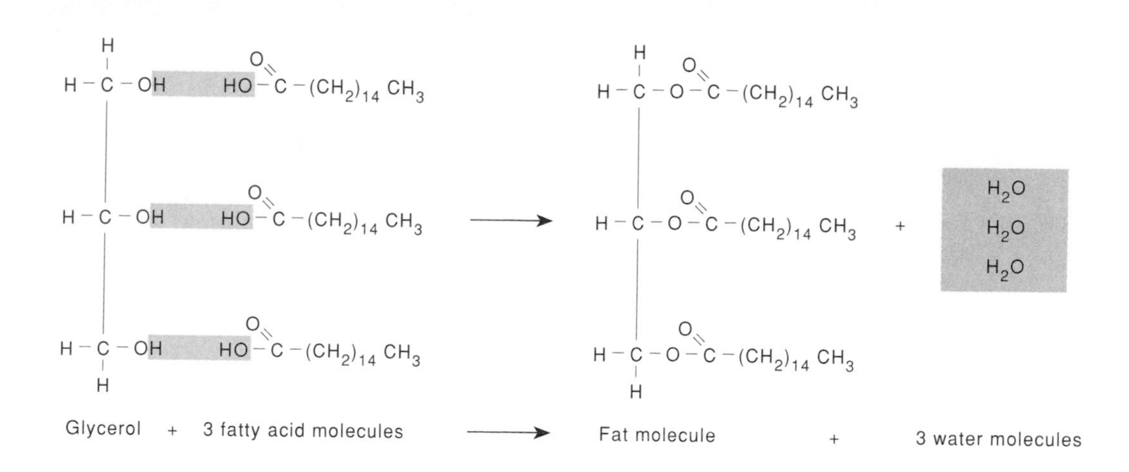
H
H − C − OH HO − C − $(CH_2)_{14}$ CH_3
H − C − OH HO − C − $(CH_2)_{14}$ CH_3
H − C − OH HO − C − $(CH_2)_{14}$ CH_3
H − C − O − C − $(CH_2)_{14}$ CH_3
H_2O
H_2O
H_2O
Glycerol + 3 fatty acid molecules
Fat molecule + 3 water molecules

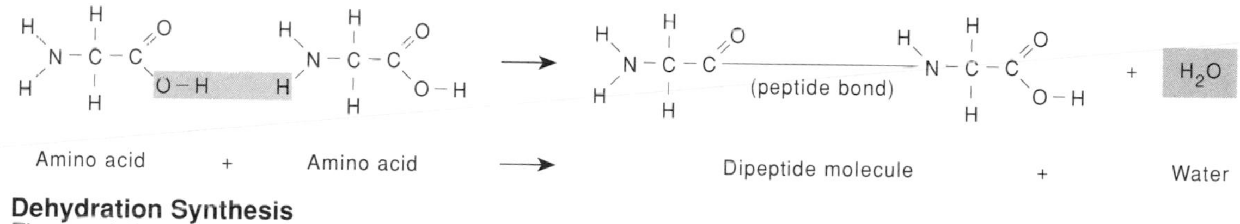
(peptide bond)
H_2O
Amino acid + Amino acid
Dipeptide molecule + Water

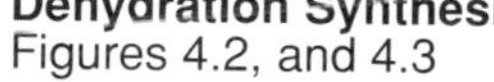
Dehydration Synthesis
Figures 4.2, and 4.3

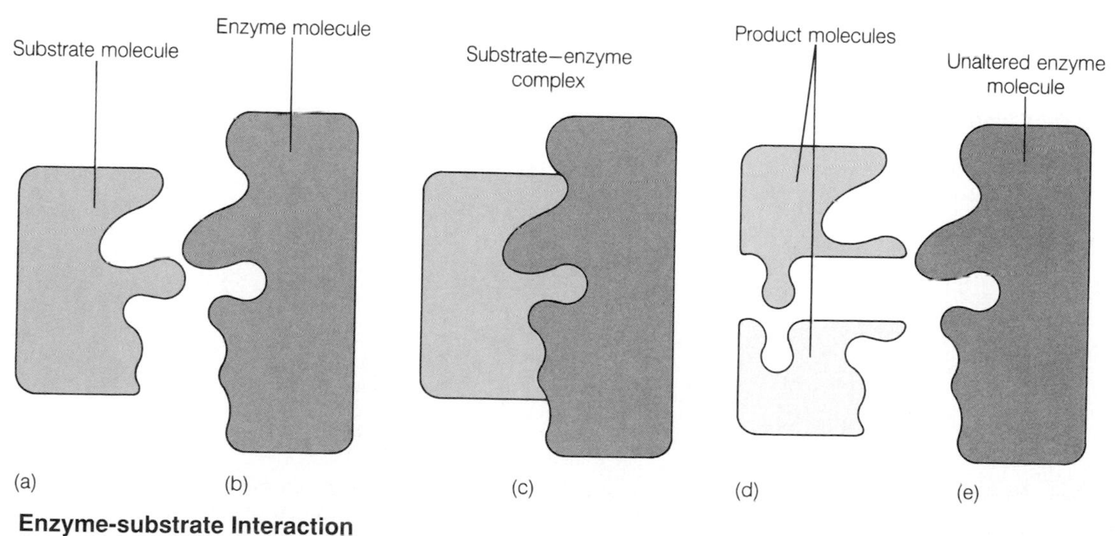
Substrate molecule
Enzyme molecule
Substrate−enzyme complex
Product molecules
Unaltered enzyme molecule
(a)
(b)
(c)
(d)
(e)

Enzyme-substrate Interaction
Figure 4.4

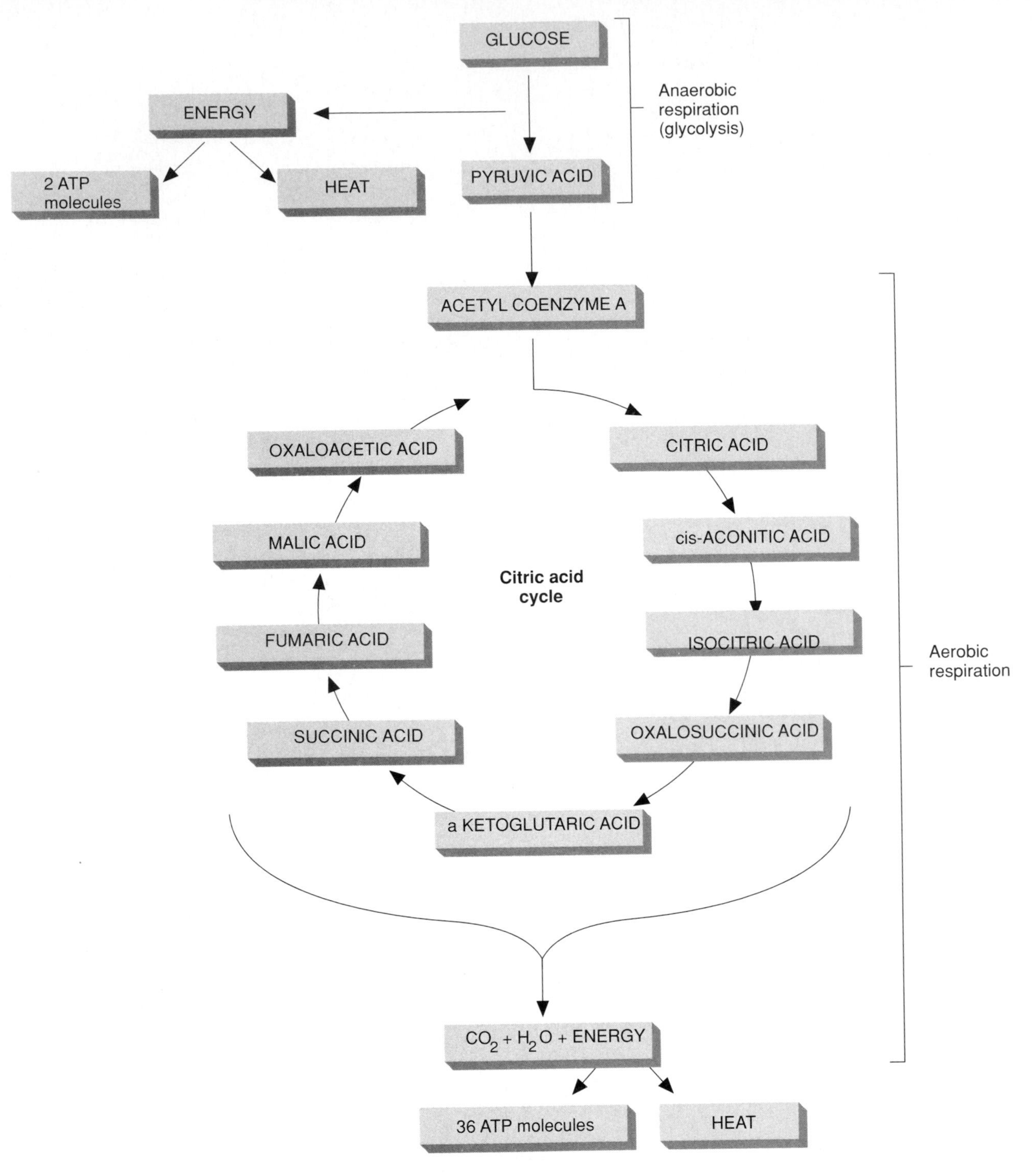
GLUCOSE
Anaerobic respiration (glycolysis)
ENERGY
2 ATP molecules
HEAT
PYRUVIC ACID
ACETYL COENZYME A
OXALOACETIC ACID
CITRIC ACID
MALIC ACID
cis-ACONITIC ACID
Citric acid cycle
FUMARIC ACID
ISOCITRIC ACID
Aerobic respiration
SUCCINIC ACID
OXALOSUCCINIC ACID
a KETOGLUTARIC ACID
CO_2 + H_2O + ENERGY
36 ATP molecules
HEAT

Citric Acid Cycle
Figure 4.8

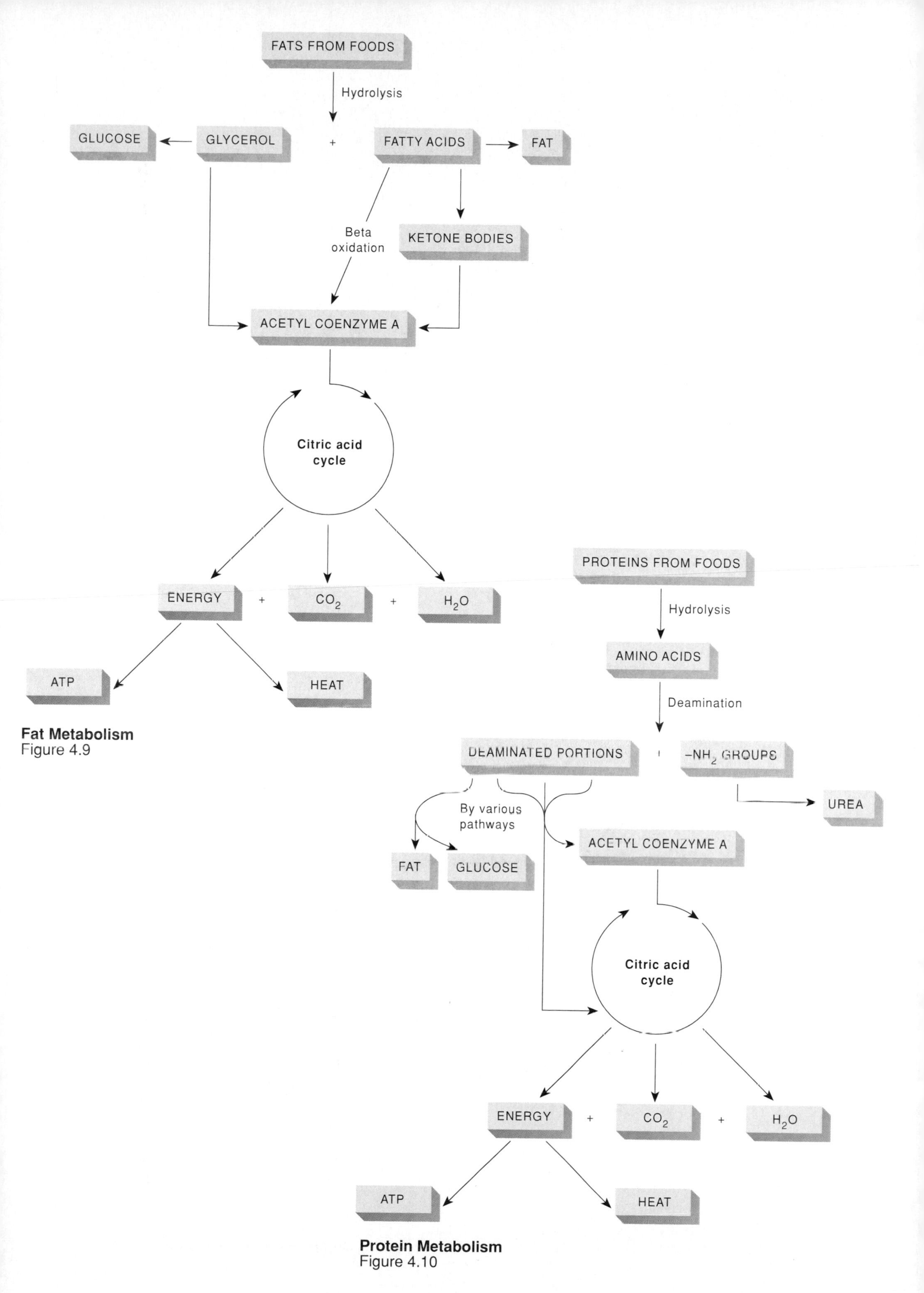
FATS FROM FOODS
Hydrolysis
GLUCOSE
GLYCEROL
+
FATTY ACIDS
FAT
Beta
oxidation
KETONE BODIES
ACETYL COENZYME A
Citric acid
cycle
ENERGY
+
CO_2
+
H_2O
ATP
HEAT
PROTEINS FROM FOODS
Hydrolysis
AMINO ACIDS
Deamination
DEAMINATED PORTIONS
$-NH_2$ GROUPS
UREA
By various
pathways
FAT
GLUCOSE
ACETYL COENZYME A
Citric acid
cycle
ENERGY
+
CO_2
+
H_2O
ATP
HEAT

Fat Metabolism
Figure 4.9

Protein Metabolism
Figure 4.10

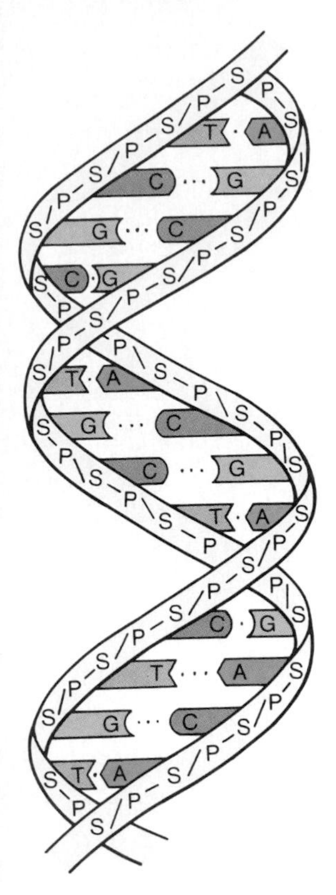

Portion of a DNA Double Helix
Figure 4.12

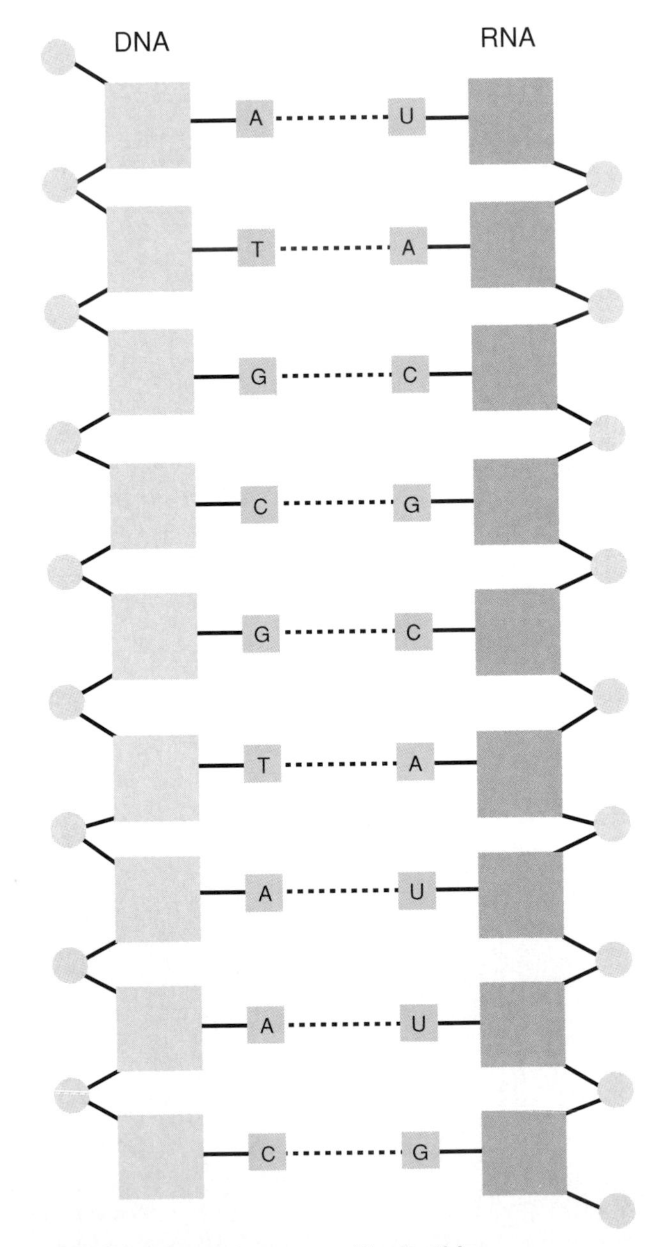
DNA
RNA
A
U
T
A
G
C
C
G
G
C
T
A
A
U
A
U
C
G

DNA-RNA Complementary Nucleotides
Figure 4.15

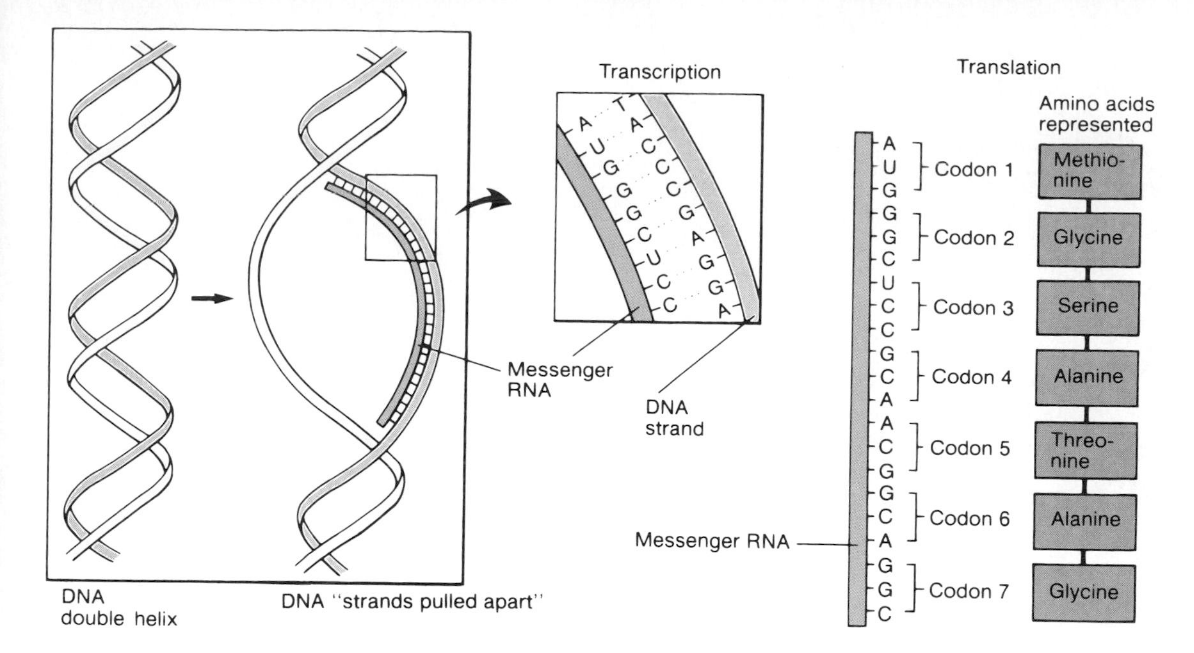
Transcription
A T
U A
G C
G C
G C
C G
U A
C G
C G
A
Messenger RNA
DNA strand
DNA double helix
DNA "strands pulled apart"
Translation
Amino acids represented
A U G Codon 1 Methio-nine
G G C Codon 2 Glycine
U C C Codon 3 Serine
G C A Codon 4 Alanine
A C G Codon 5 Threo-nine
G C A Codon 6 Alanine
G G C Codon 7 Glycine
Messenger RNA

DNA-RNA Transcription
Figure 4.17

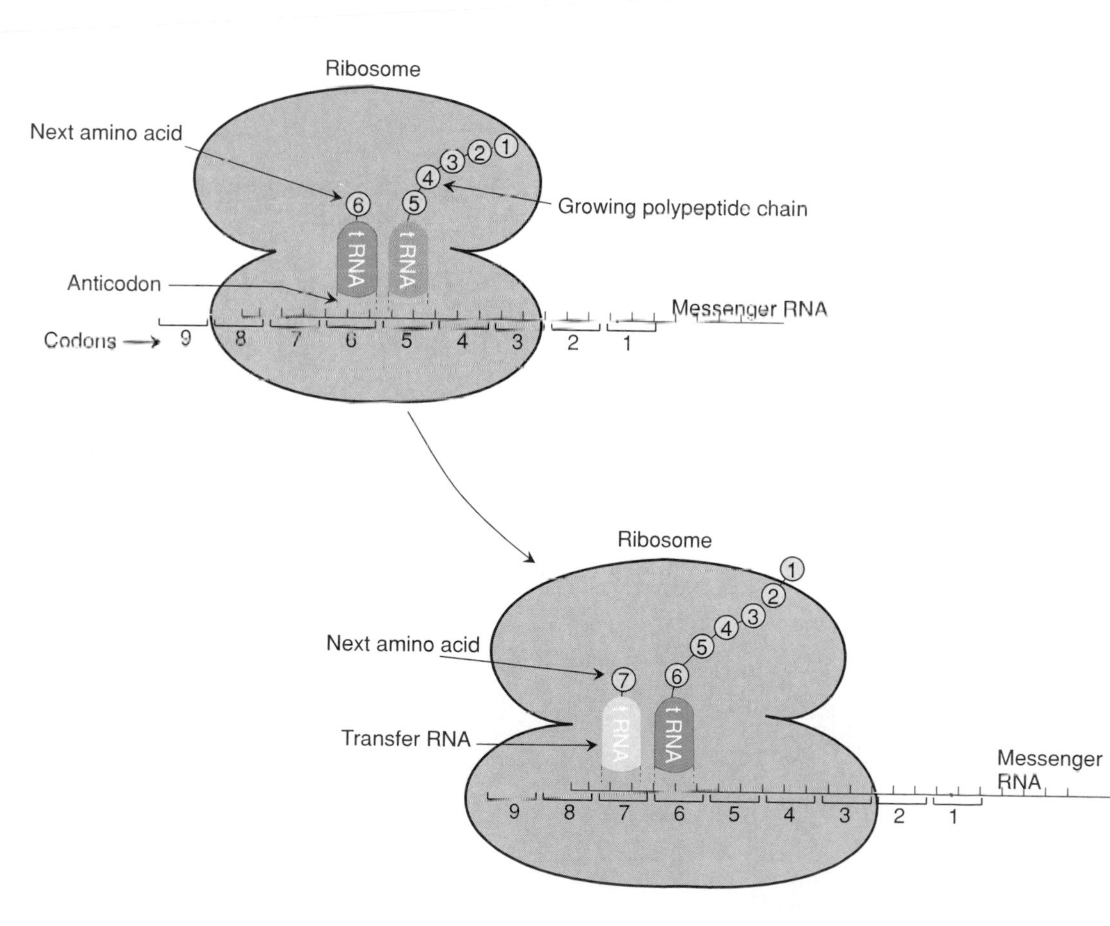
Ribosome
Next amino acid
Growing polypeptide chain
Anticodon
t RNA
Messenger RNA
Codons
9 8 7 6 5 4 3 2 1
Ribosome
Next amino acid
Transfer RNA
t RNA
Messenger RNA
9 8 7 6 5 4 3 2 1

Protein Synthesis
Figure 4.18

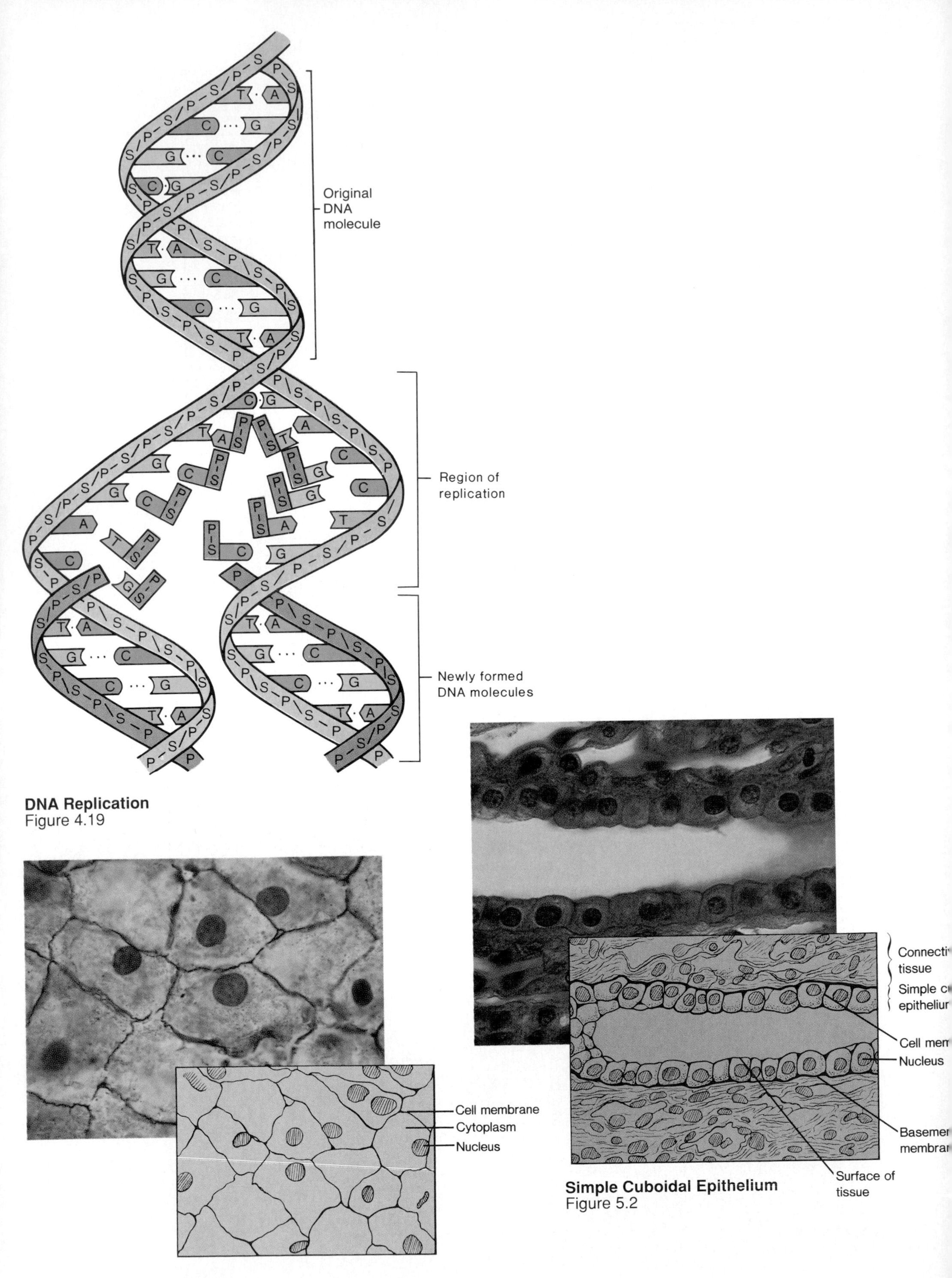

Original
DNA
molecule
Region of
replication
Newly formed
DNA molecules
Cell membrane
Cytoplasm
Nucleus
tissue
Nucleus
Surface of
tissue

DNA Replication
Figure 4.19

Simple Squamous Epithelium
Figure 5.1

Simple Cuboidal Epithelium
Figure 5.2

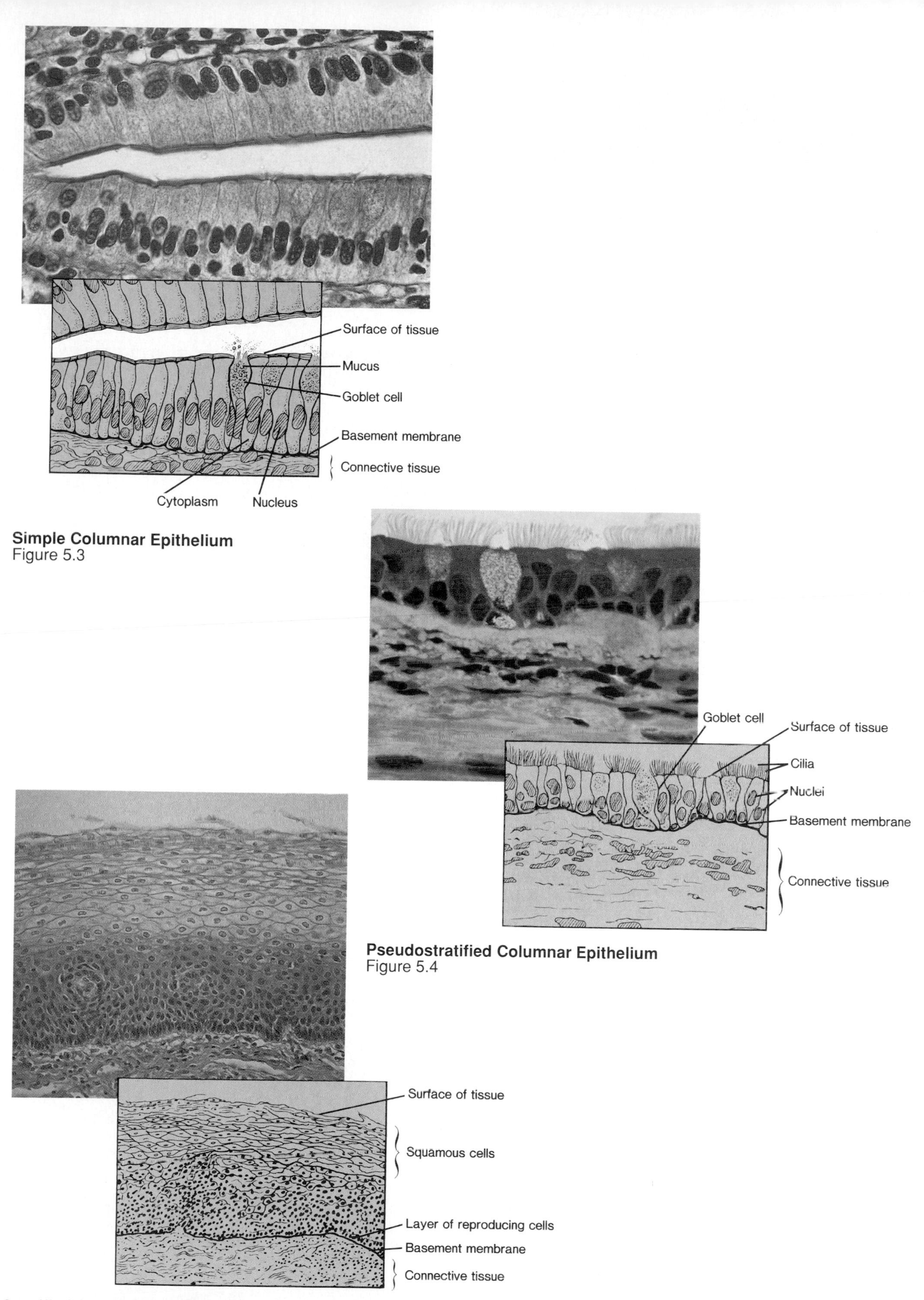
Surface of tissue
Mucus
Goblet cell
Basement membrane
Connective tissue
Cytoplasm
Nucleus
Goblet cell
Surface of tissue
Cilia
Nuclei
Basement membrane
Connective tissue
Surface of tissue
Squamous cells
Layer of reproducing cells
Basement membrane
Connective tissue

Simple Columnar Epithelium
Figure 5.3

Pseudostratified Columnar Epithelium
Figure 5.4

Stratified Squamous Epithelium
Figure 5.5

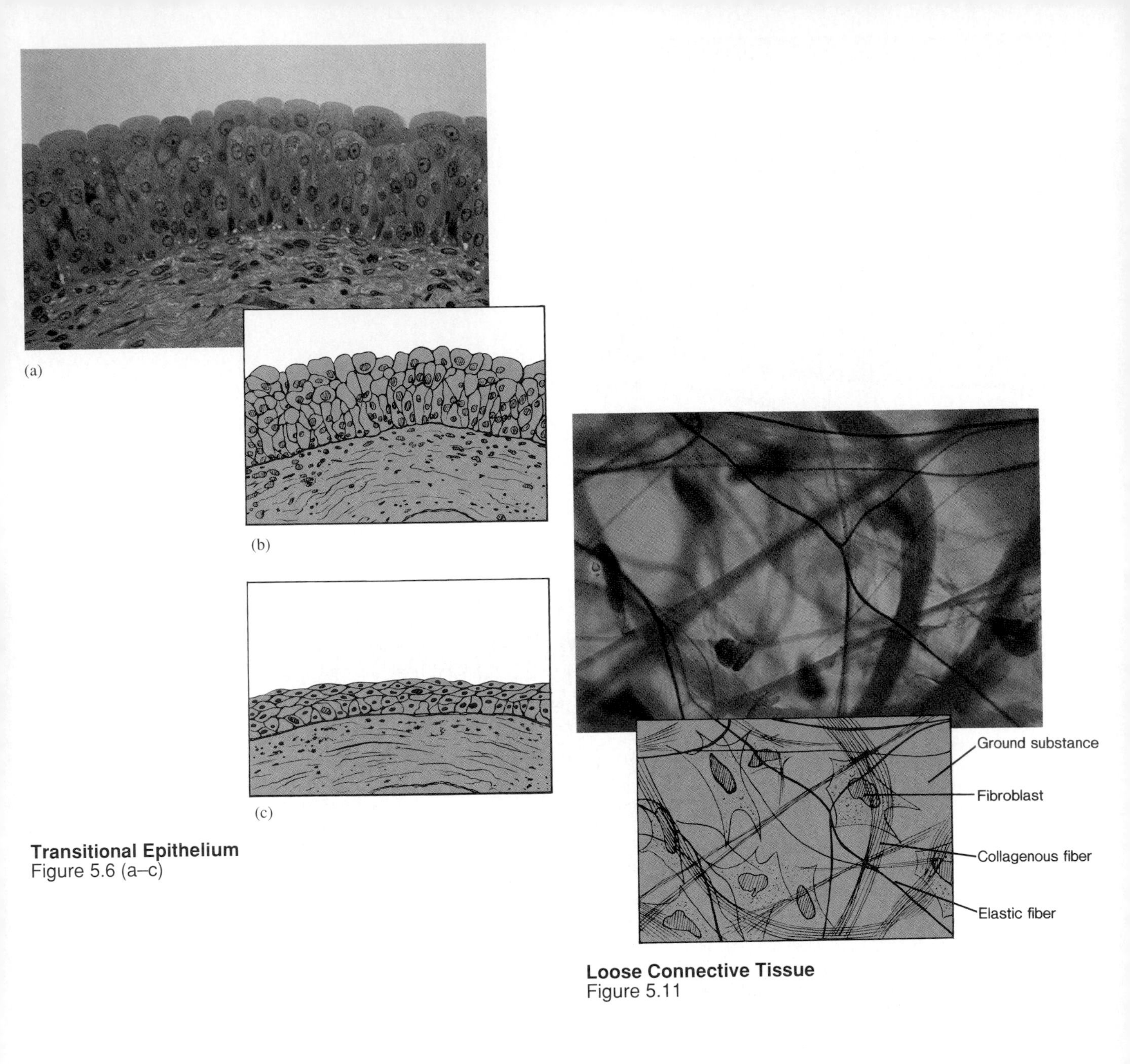
(a)
(b)
(c)
Ground substance
Fibroblast
Collagenous fiber
Elastic fiber

Transitional Epithelium
Figure 5.6 (a–c)

Loose Connective Tissue
Figure 5.11

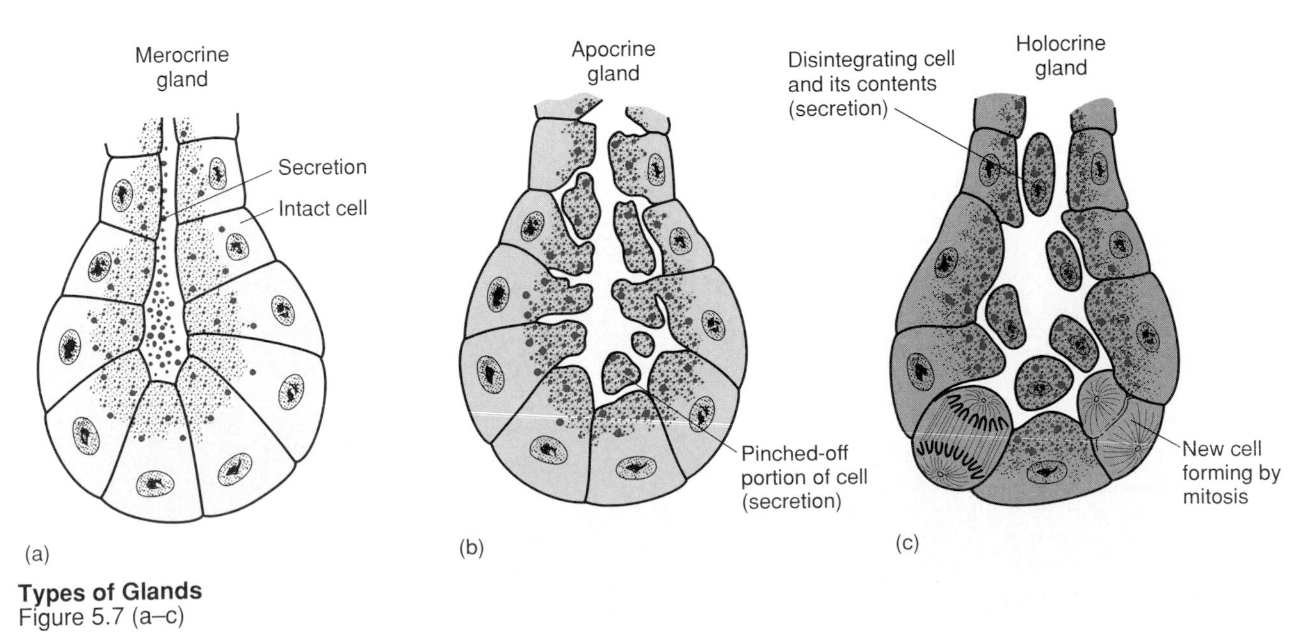
Merocrine
gland
Secretion
Intact cell
Apocrine
gland
Pinched-off
portion of cell
(secretion)
Disintegrating cell
and its contents
(secretion)
Holocrine
gland
New cell
forming by
mitosis
(a)
(b)
(c)

Types of Glands
Figure 5.7 (a–c)

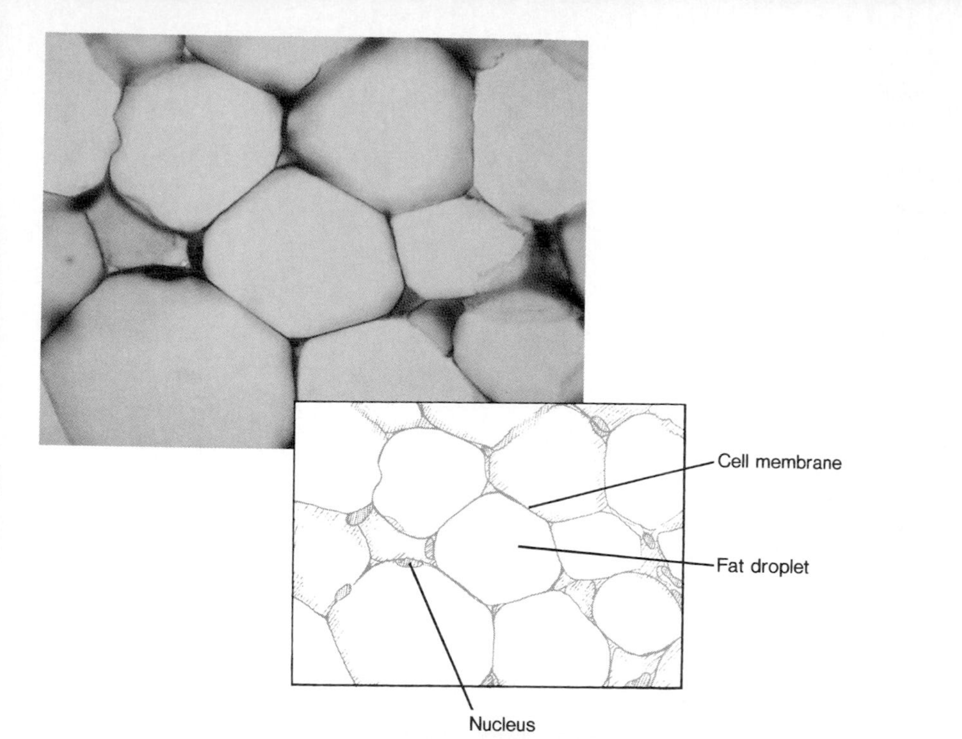
Cell membrane
Fat droplet
Nucleus

Adipose Tissue
Figure 5.12

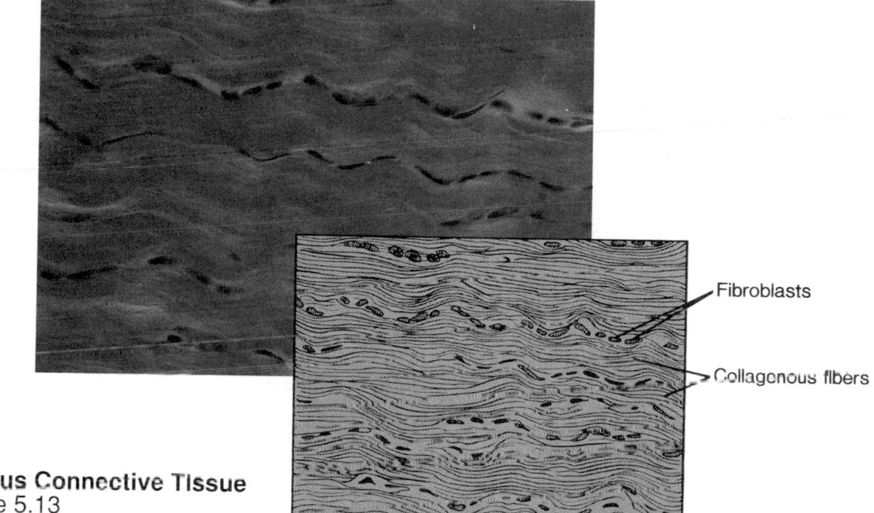
Fibroblasts
Collagenous fibers

Fibrous Connective Tissue
Figure 5.13

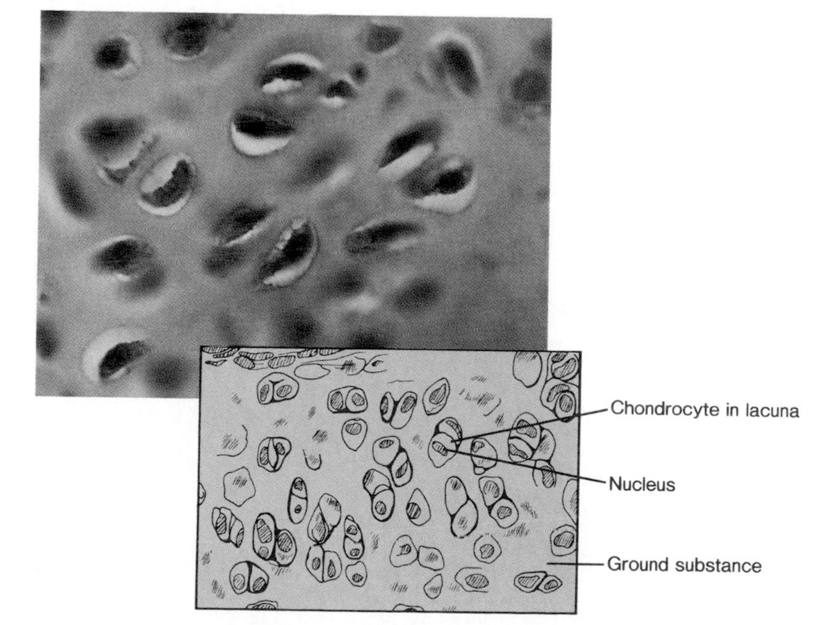
Chondrocyte in lacuna
Nucleus
Ground substance

Hyaline Cartilage
Figure 5.14

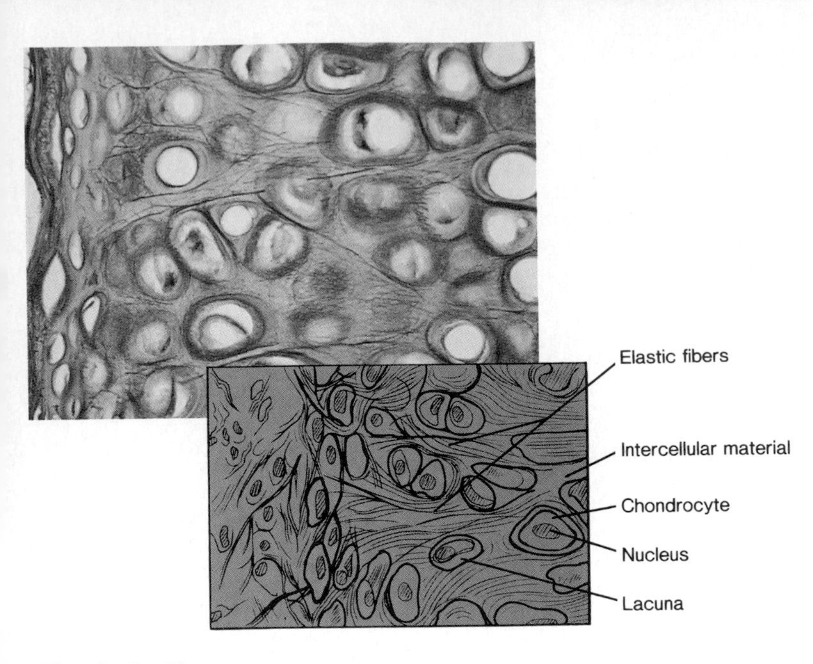
Elastic fibers
Intercellular material
Chondrocyte
Nucleus
Lacuna

Elastic Cartilage
Figure 5.15

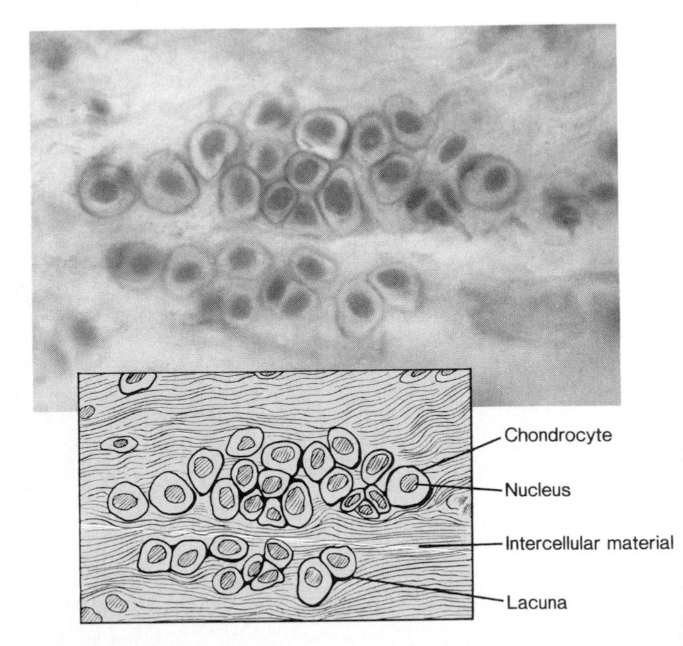
Chondrocyte
Nucleus
Intercellular material
Lacuna

Fibrocartilage
Figure 5.16

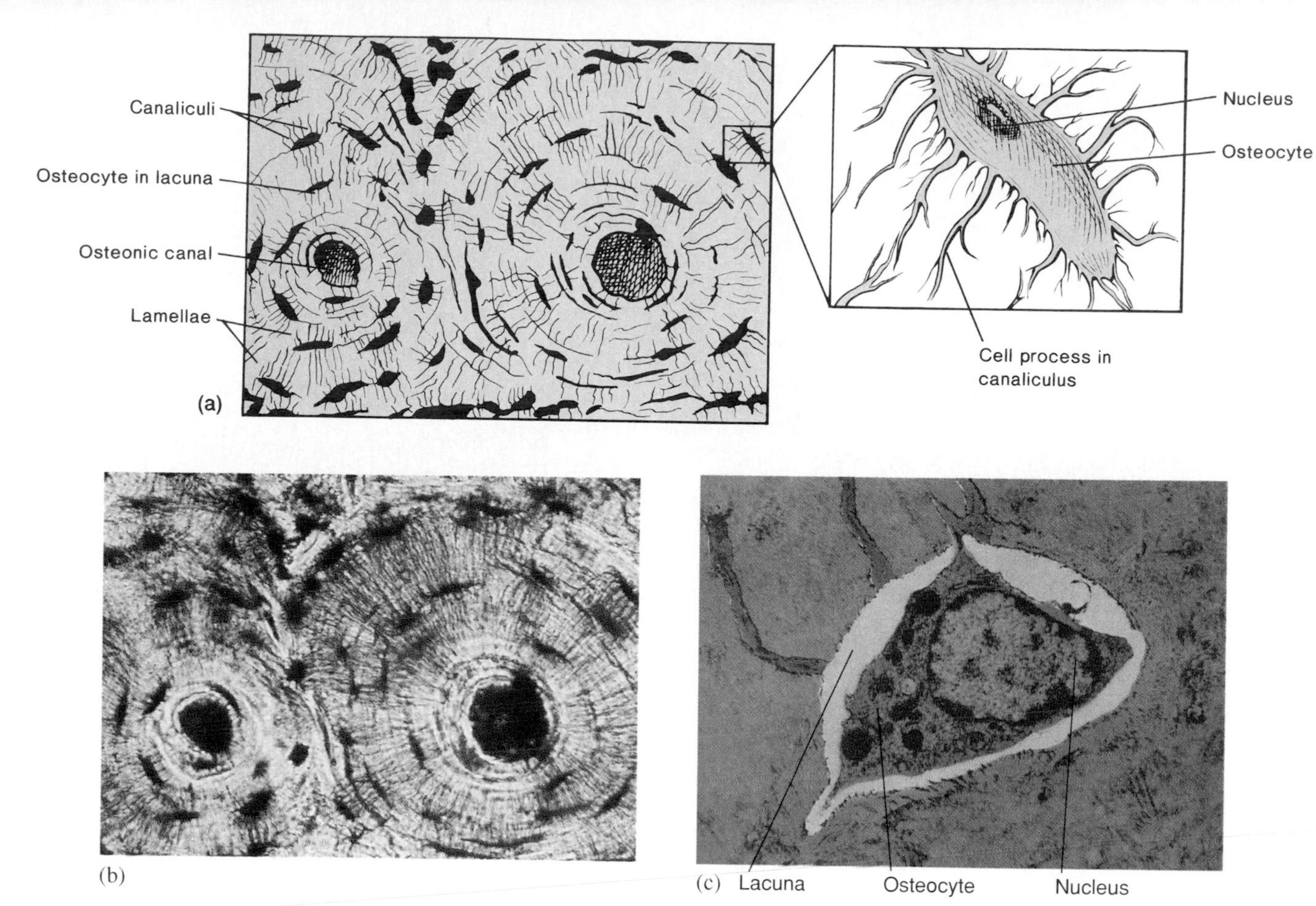
Canaliculi
Osteocyte in lacuna
Osteonic canal
Lamellae
(a)
Nucleus
Osteocyte
Cell process in canaliculus
(b)
(c) Lacuna
Osteocyte
Nucleus

Bone Tissue
Figure 5.17 (a–b)

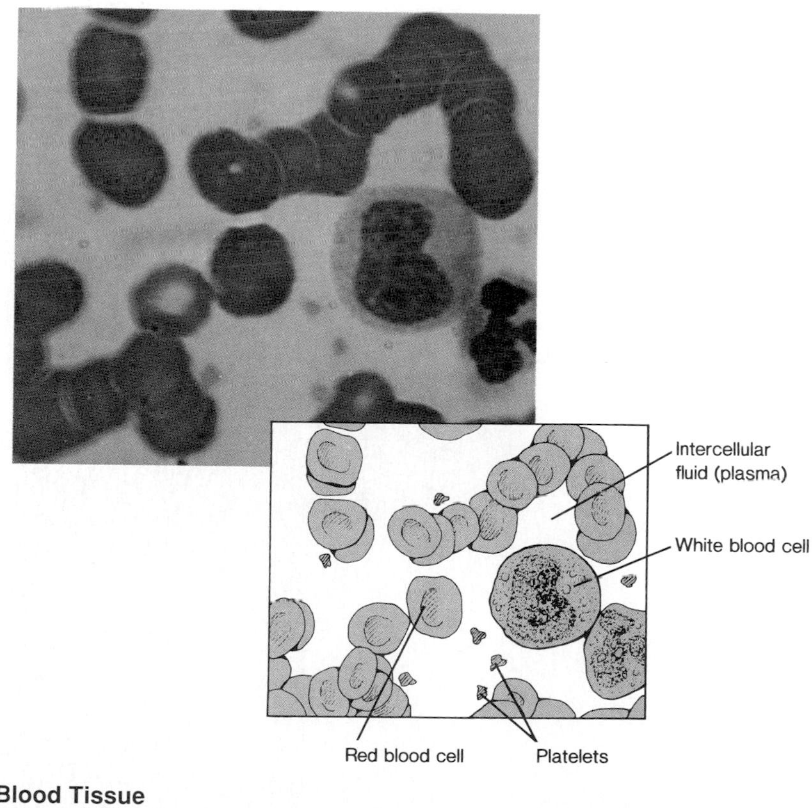
Intercellular fluid (plasma)
White blood cell
Red blood cell
Platelets

Blood Tissue
Figure 5.18

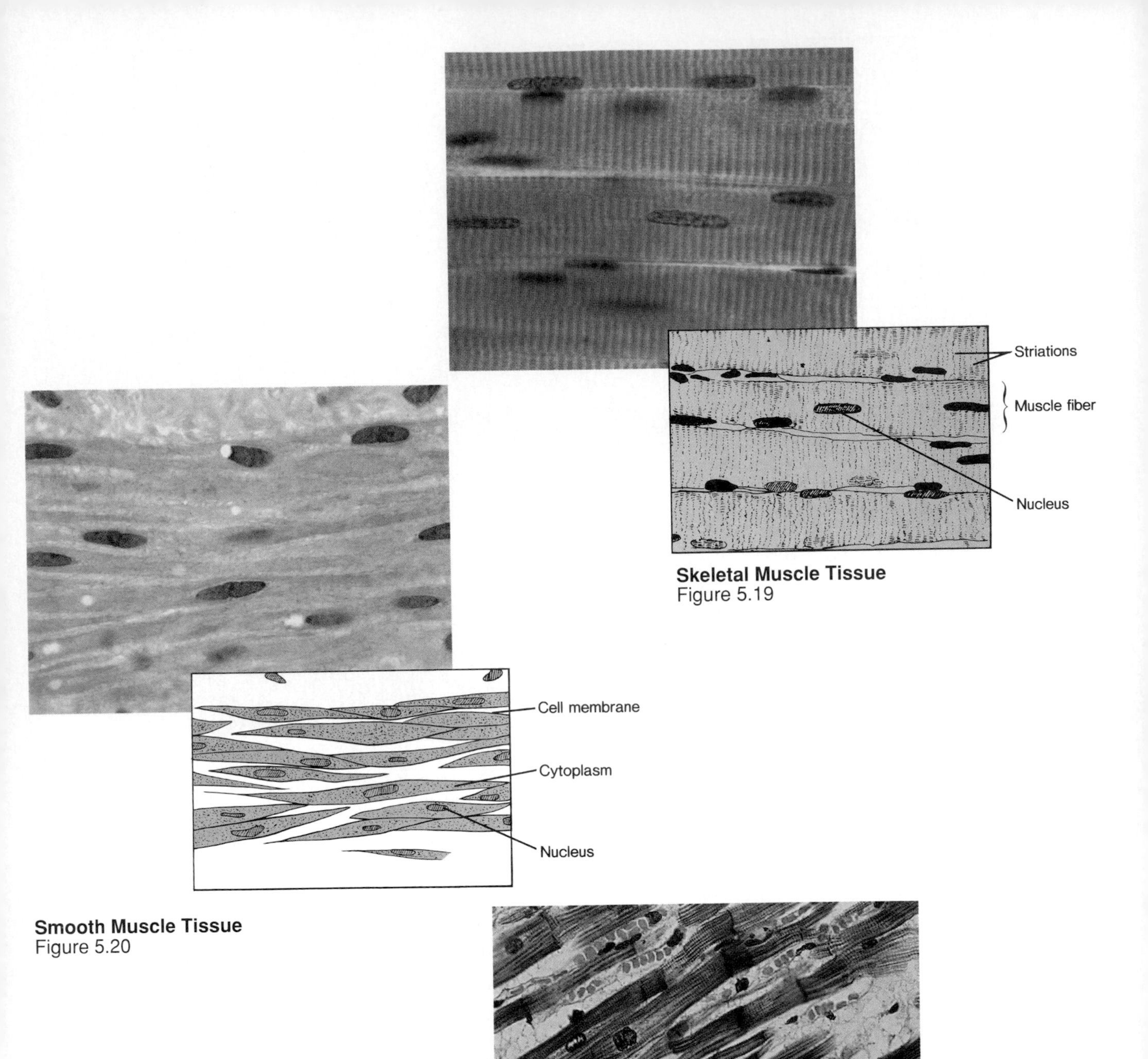
Striations
Muscle fiber
Nucleus
Cell membrane
Cytoplasm
Nucleus

Skeletal Muscle Tissue
Figure 5.19

Smooth Muscle Tissue
Figure 5.20

Striations
Nucleus
Intercalated disk

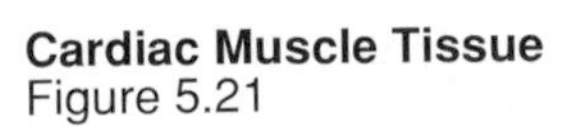

Cardiac Muscle Tissue
Figure 5.21

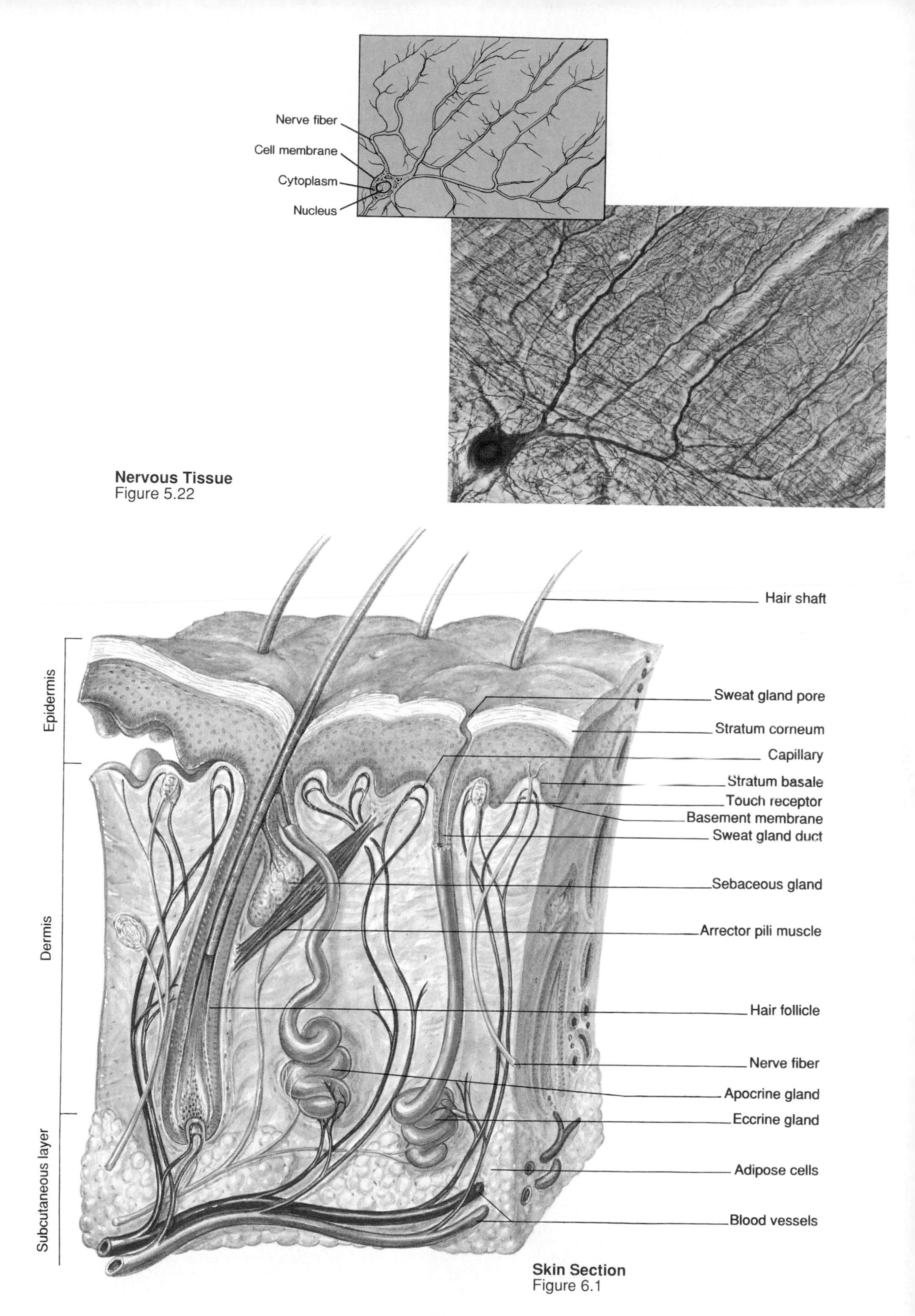

Nerve fiber
Cell membrane
Cytoplasm
Nucleus
Hair shaft
Sweat gland pore
Stratum corneum
Capillary
Stratum basale
Touch receptor
Basement membrane
Sweat gland duct
Sebaceous gland
Arrector pili muscle
Hair follicle
Nerve fiber
Apocrine gland
Eccrine gland
Adipose cells
Blood vessels
Epidermis
Dermis
Subcutaneous layer

Nervous Tissue
Figure 5.22

Skin Section
Figure 6.1

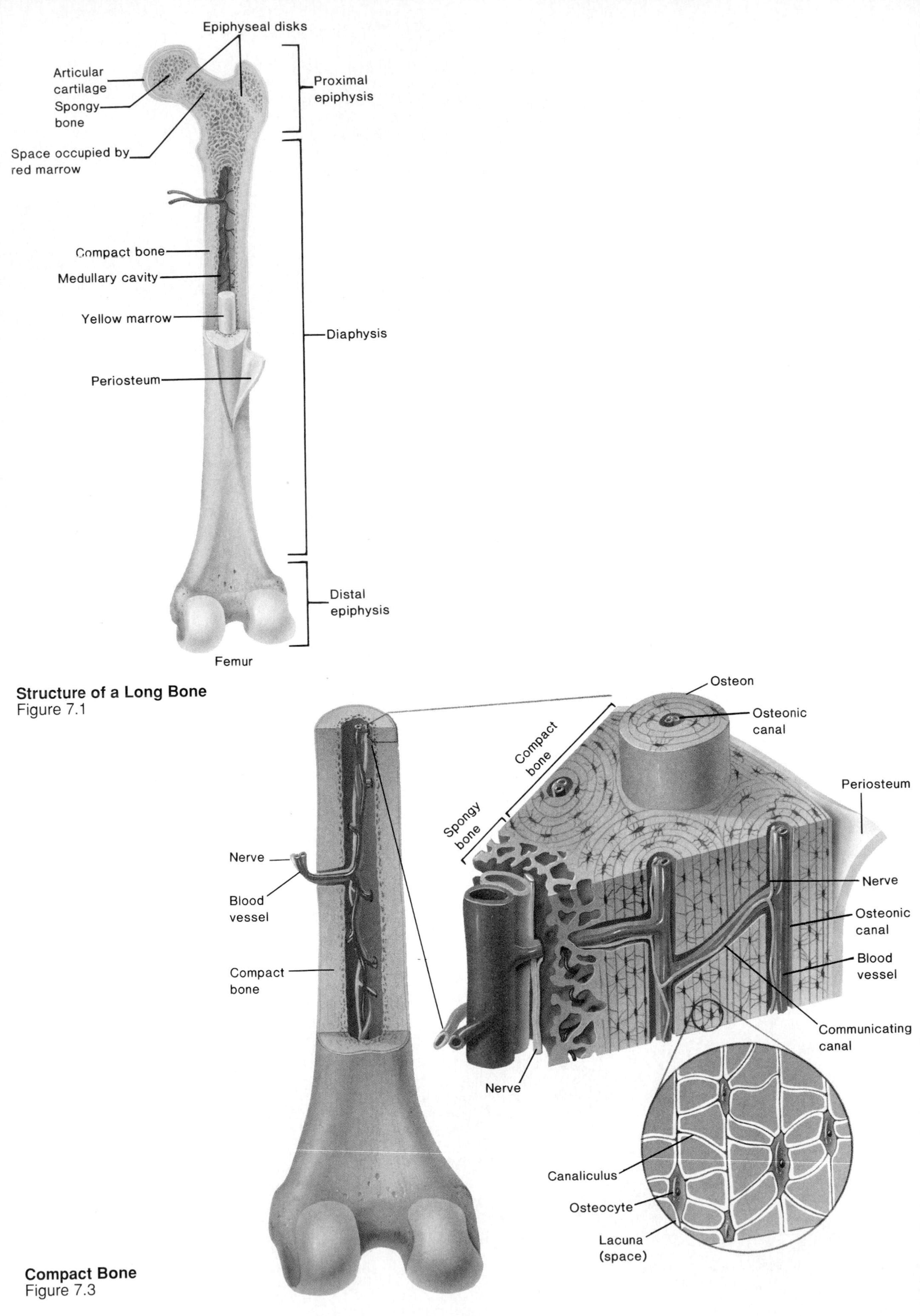
Epiphyseal disks
Articular cartilage
Spongy bone
Space occupied by red marrow
Proximal epiphysis
Compact bone
Medullary cavity
Yellow marrow
Diaphysis
Periosteum
Distal epiphysis
Femur
Osteon
Osteonic canal
Compact bone
Spongy bone
Periosteum
Nerve
Blood vessel
Compact bone
Nerve
Osteonic canal
Blood vessel
Communicating canal
Nerve
Canaliculus
Osteocyte
Lacuna (space)

Structure of a Long Bone
Figure 7.1

Compact Bone
Figure 7.3

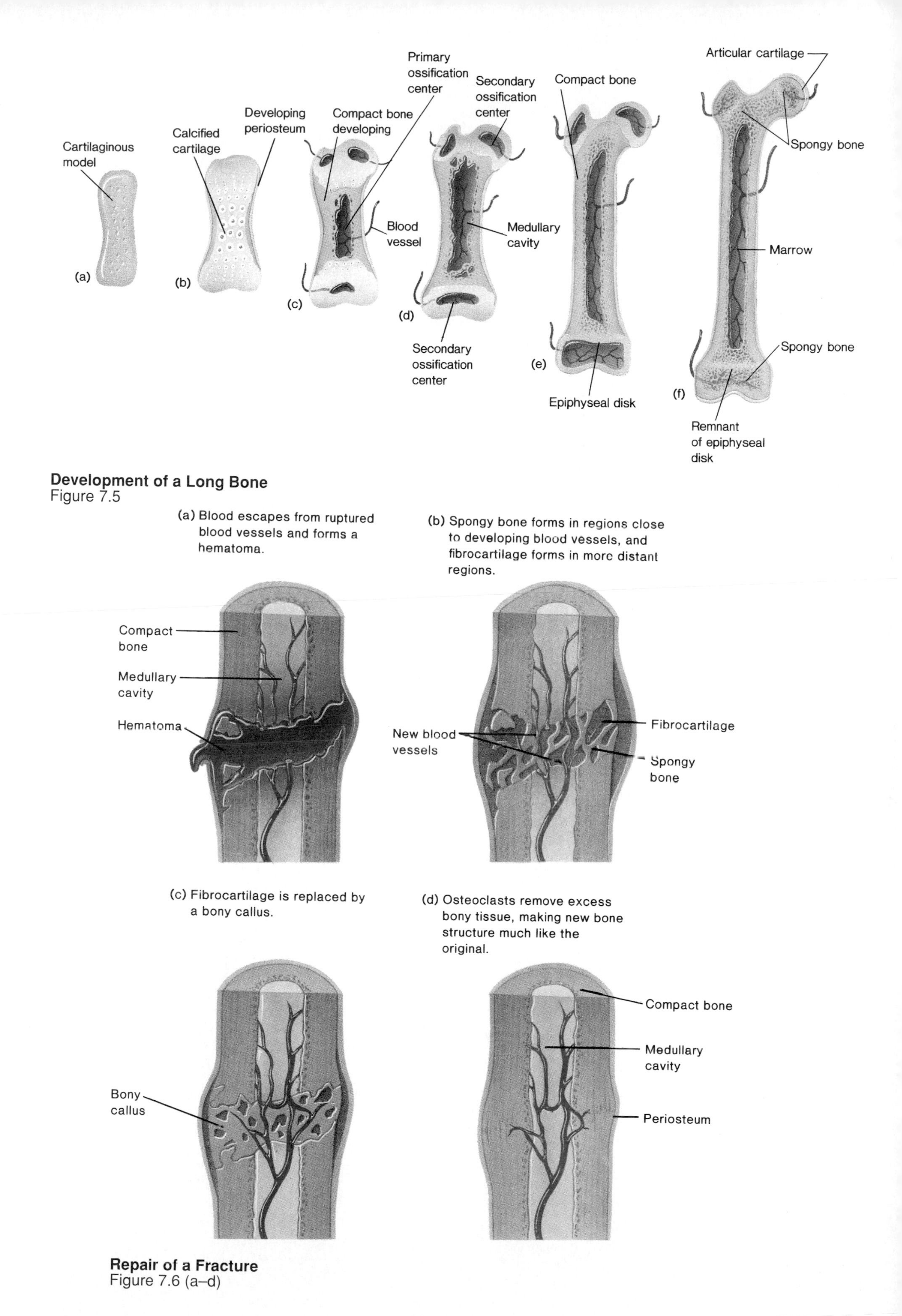
Cartilaginous model
Calcified cartilage
Developing periosteum
Compact bone developing
Primary ossification center
Secondary ossification center
Compact bone
Articular cartilage
Spongy bone
Blood vessel
Medullary cavity
Marrow
(a)
(b)
(c)
(d)
(e)
(f)
Secondary ossification center
Epiphyseal disk
Spongy bone
Remnant of epiphyseal disk

Development of a Long Bone
Figure 7.5

(a) Blood escapes from ruptured blood vessels and forms a hematoma.
(b) Spongy bone forms in regions close to developing blood vessels, and fibrocartilage forms in more distant regions.
Compact bone
Medullary cavity
Hematoma
New blood vessels
Fibrocartilage
Spongy bone
(c) Fibrocartilage is replaced by a bony callus.
(d) Osteoclasts remove excess bony tissue, making new bone structure much like the original.
Bony callus
Compact bone
Medullary cavity
Periosteum

Repair of a Fracture
Figure 7.6 (a–d)

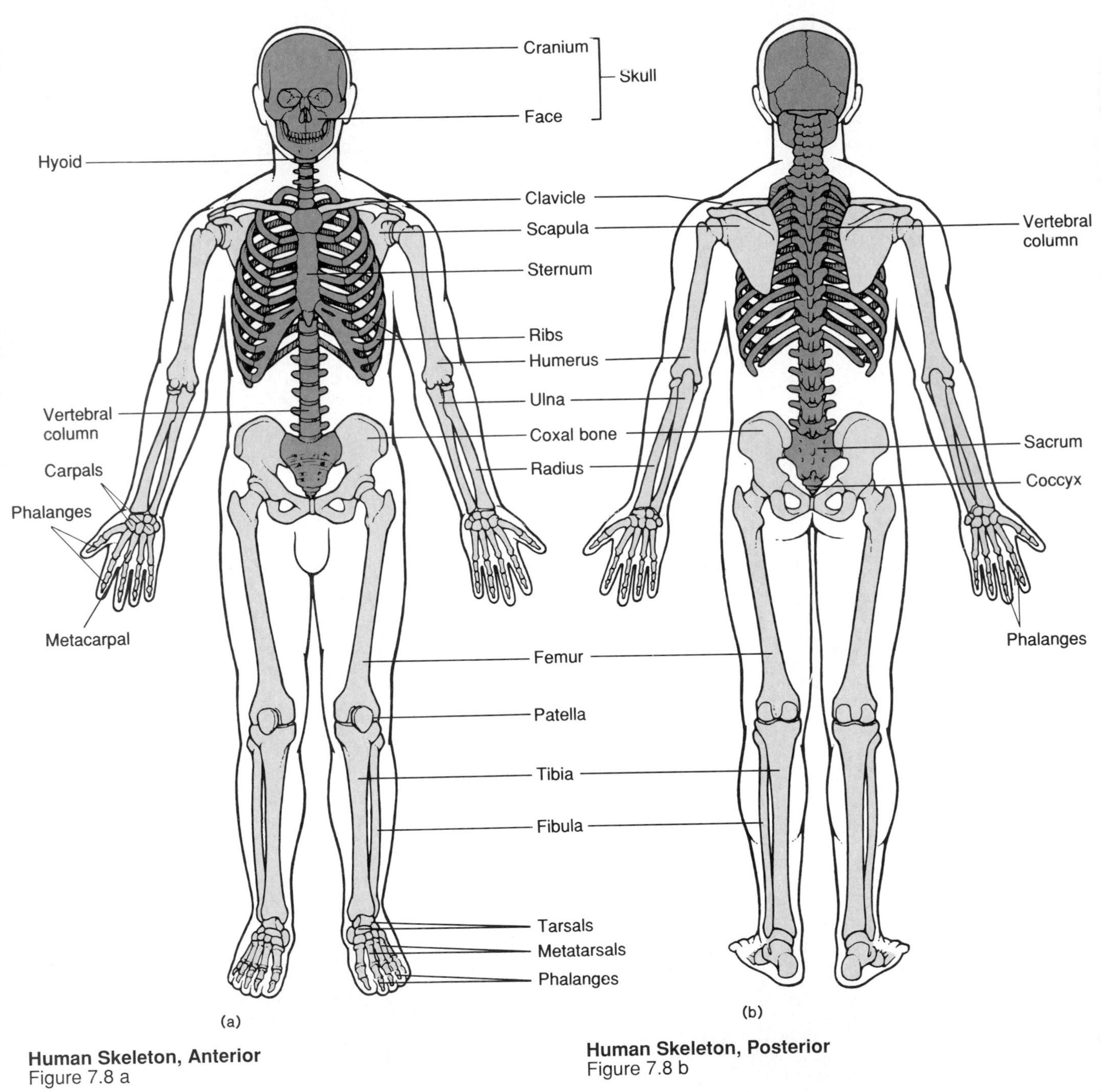

Cranium
Skull
Face
Hyoid
Clavicle
Scapula
Sternum
Ribs
Humerus
Ulna
Vertebral column
Coxal bone
Radius
Carpals
Phalanges
Metacarpal
Femur
Patella
Tibia
Fibula
Tarsals
Metatarsals
Phalanges
(a)
Vertebral column
Sacrum
Coccyx
Phalanges
(b)

Human Skeleton, Anterior
Figure 7.8 a

Human Skeleton, Posterior
Figure 7.8 b

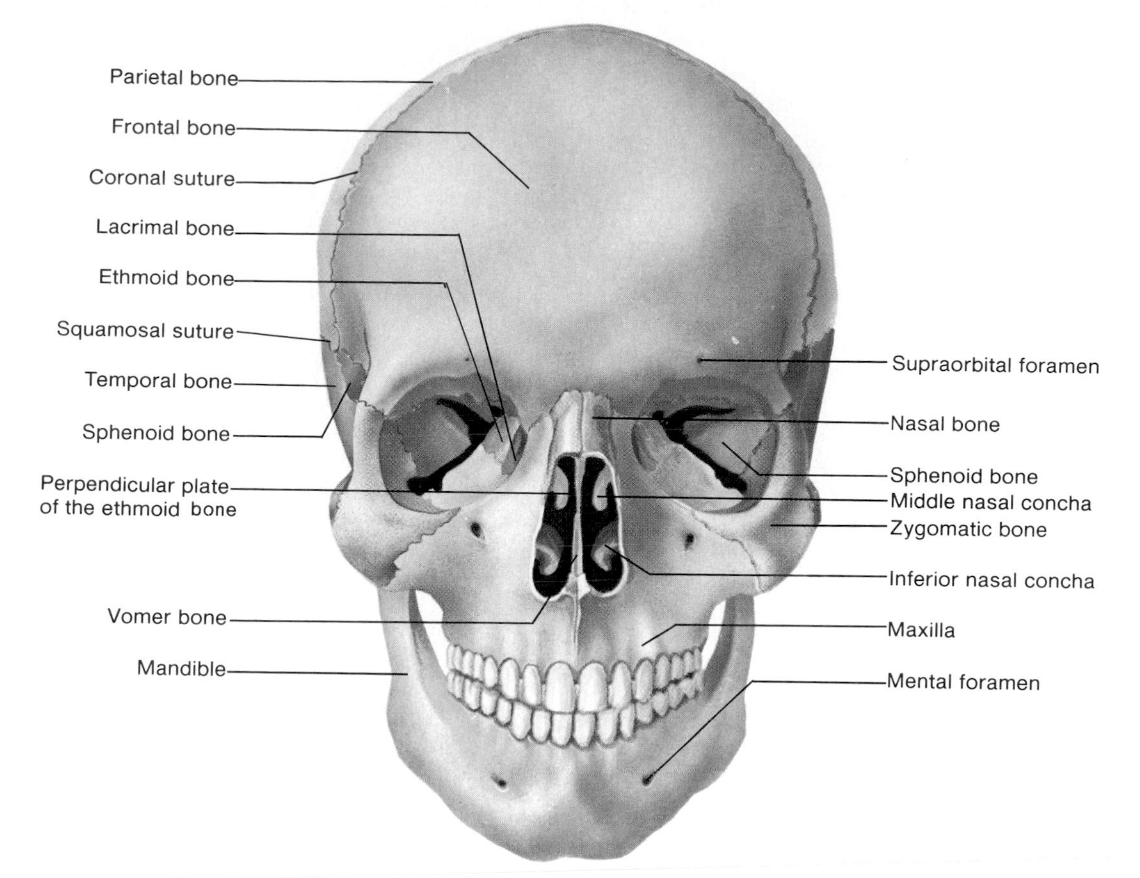
Parietal bone
Frontal bone
Coronal suture
Lacrimal bone
Ethmoid bone
Squamosal suture
Temporal bone
Sphenoid bone
Perpendicular plate
of the ethmoid bone
Vomer bone
Mandible
Supraorbital foramen
Nasal bone
Sphenoid bone
Middle nasal concha
Zygomatic bone
Inferior nasal concha
Maxilla
Mental foramen

Human Skull, Anterior
Figure 7.9

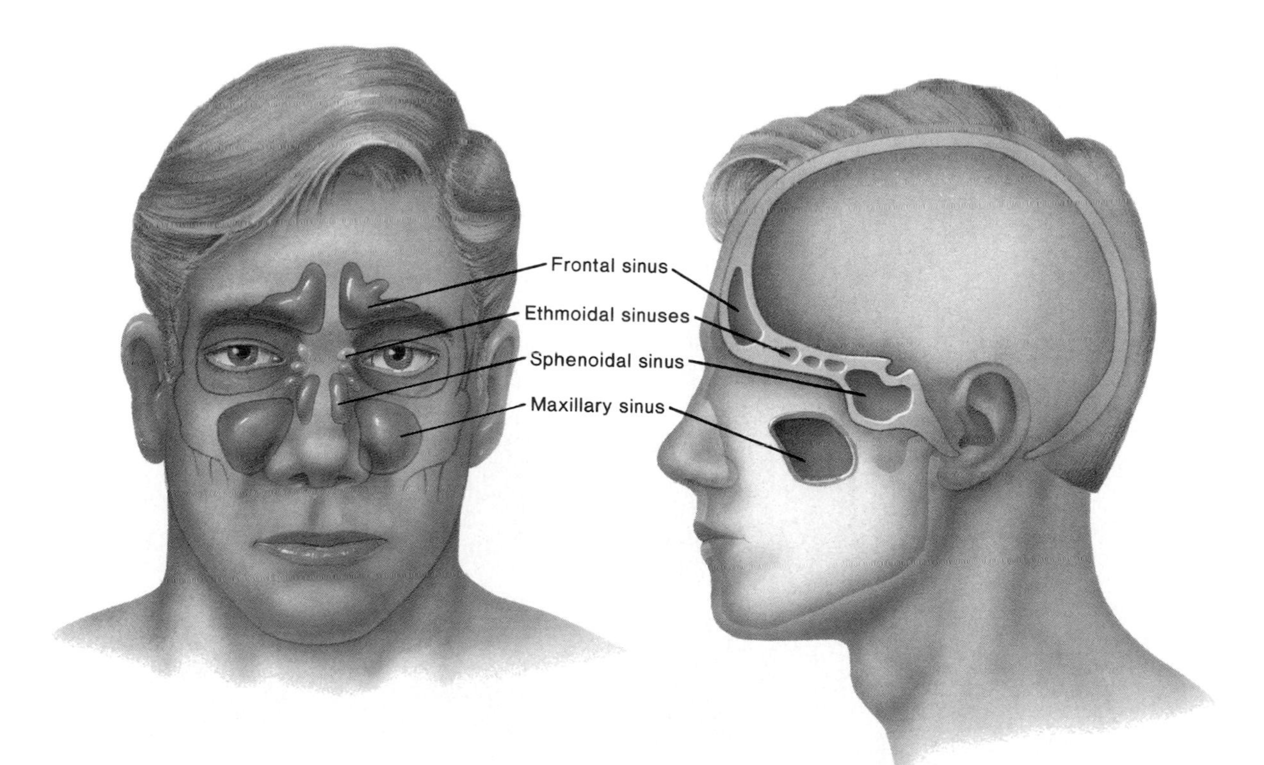
Frontal sinus
Ethmoidal sinuses
Sphenoidal sinus
Maxillary sinus

Paranasal Sinuses
Figure 7.10

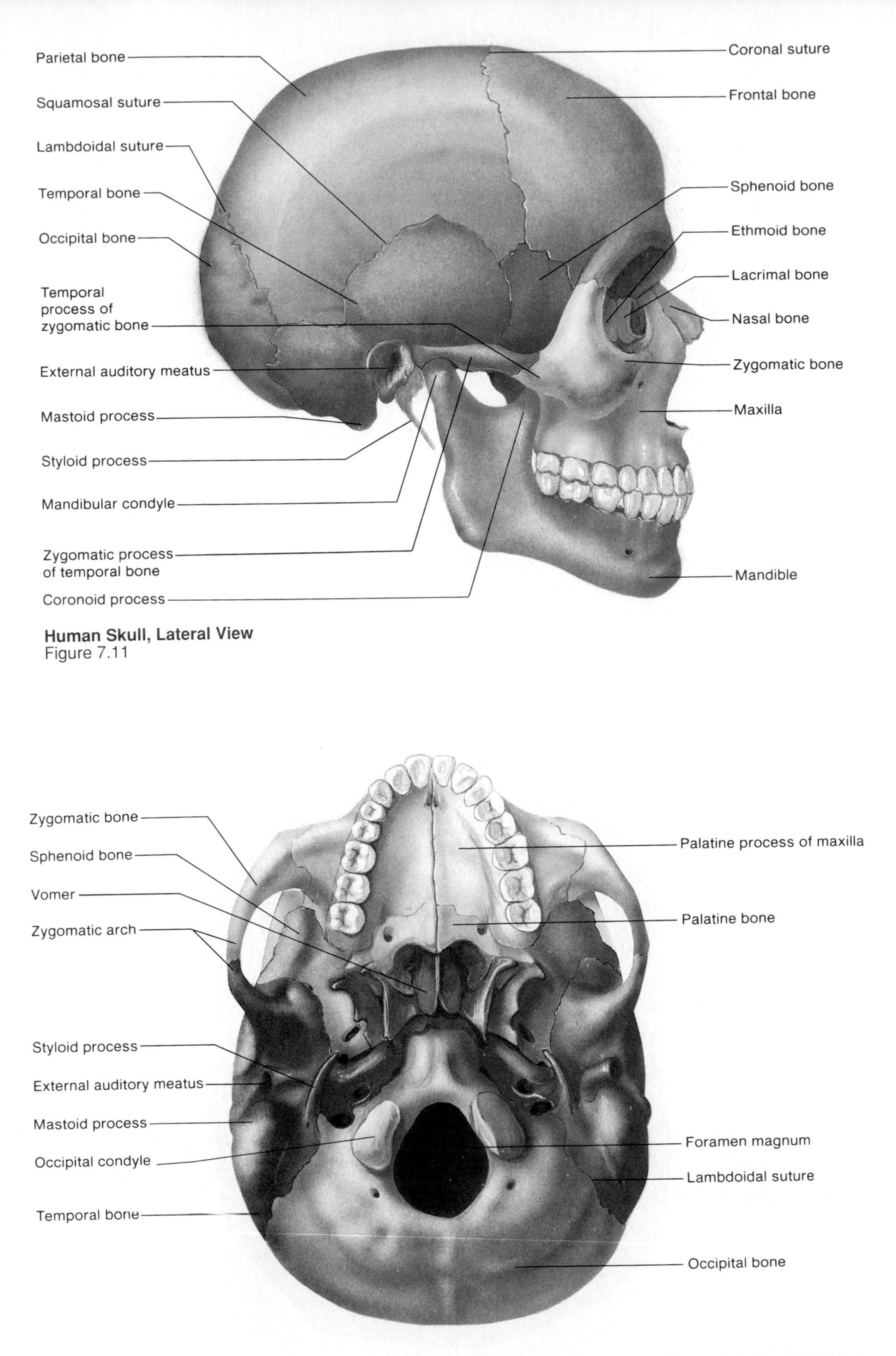
Parietal bone
Squamosal suture
Lambdoidal suture
Temporal bone
Occipital bone
Temporal process of zygomatic bone
External auditory meatus
Mastoid process
Styloid process
Mandibular condyle
Zygomatic process of temporal bone
Coronoid process
Coronal suture
Frontal bone
Sphenoid bone
Ethmoid bone
Lacrimal bone
Nasal bone
Zygomatic bone
Maxilla
Mandible
Zygomatic bone
Sphenoid bone
Vomer
Zygomatic arch
Styloid process
External auditory meatus
Mastoid process
Occipital condyle
Temporal bone
Palatine process of maxilla
Palatine bone
Foramen magnum
Lambdoidal suture
Occipital bone

Human Skull, Lateral View
Figure 7.11

Human Skull, Inferior
Figure 7.12

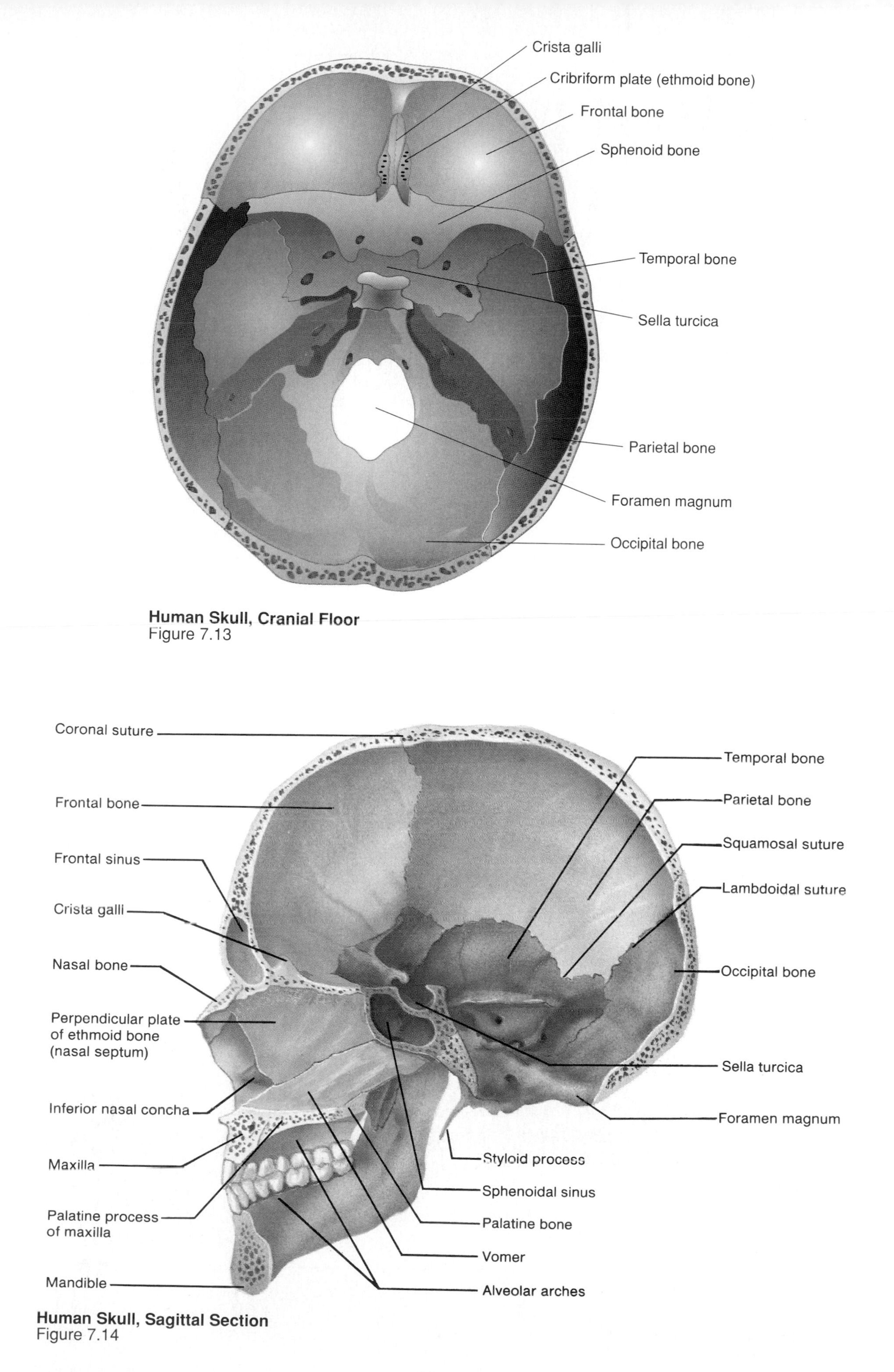
Crista galli
Cribriform plate (ethmoid bone)
Frontal bone
Sphenoid bone
Temporal bone
Sella turcica
Parietal bone
Foramen magnum
Occipital bone

Human Skull, Cranial Floor
Figure 7.13

Coronal suture
Frontal bone
Frontal sinus
Crista galli
Nasal bone
Perpendicular plate of ethmoid bone (nasal septum)
Inferior nasal concha
Maxilla
Palatine process of maxilla
Mandible
Temporal bone
Parietal bone
Squamosal suture
Lambdoidal suture
Occipital bone
Sella turcica
Foramen magnum
Styloid process
Sphenoidal sinus
Palatine bone
Vomer
Alveolar arches

Human Skull, Sagittal Section
Figure 7.14

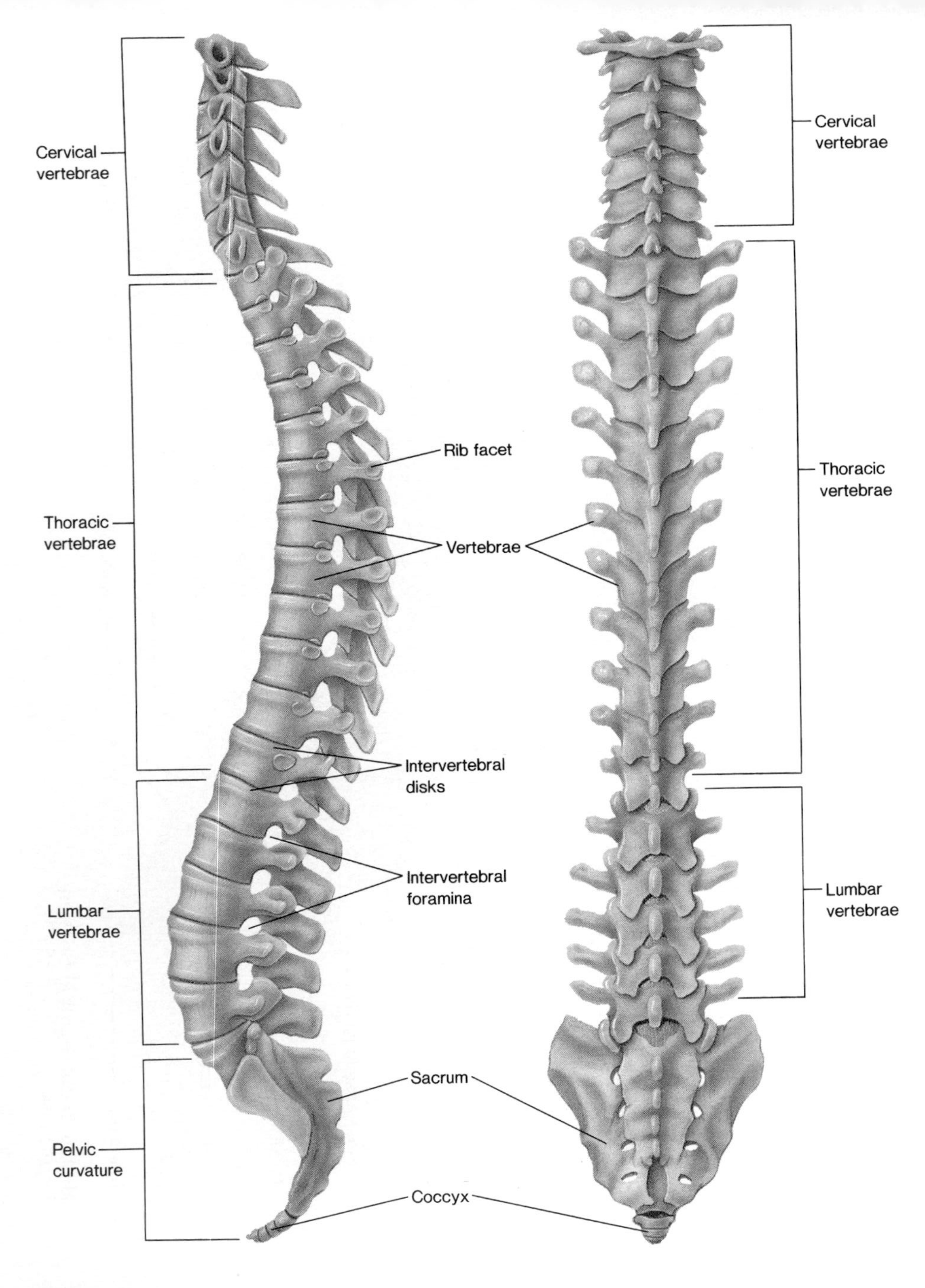
Cervical vertebrae
Thoracic vertebrae
Lumbar vertebrae
Pelvic curvature
Rib facet
Vertebrae
Intervertebral disks
Intervertebral foramina
Sacrum
Coccyx
Cervical vertebrae
Thoracic vertebrae
Lumbar vertebrae

Vertebral Column

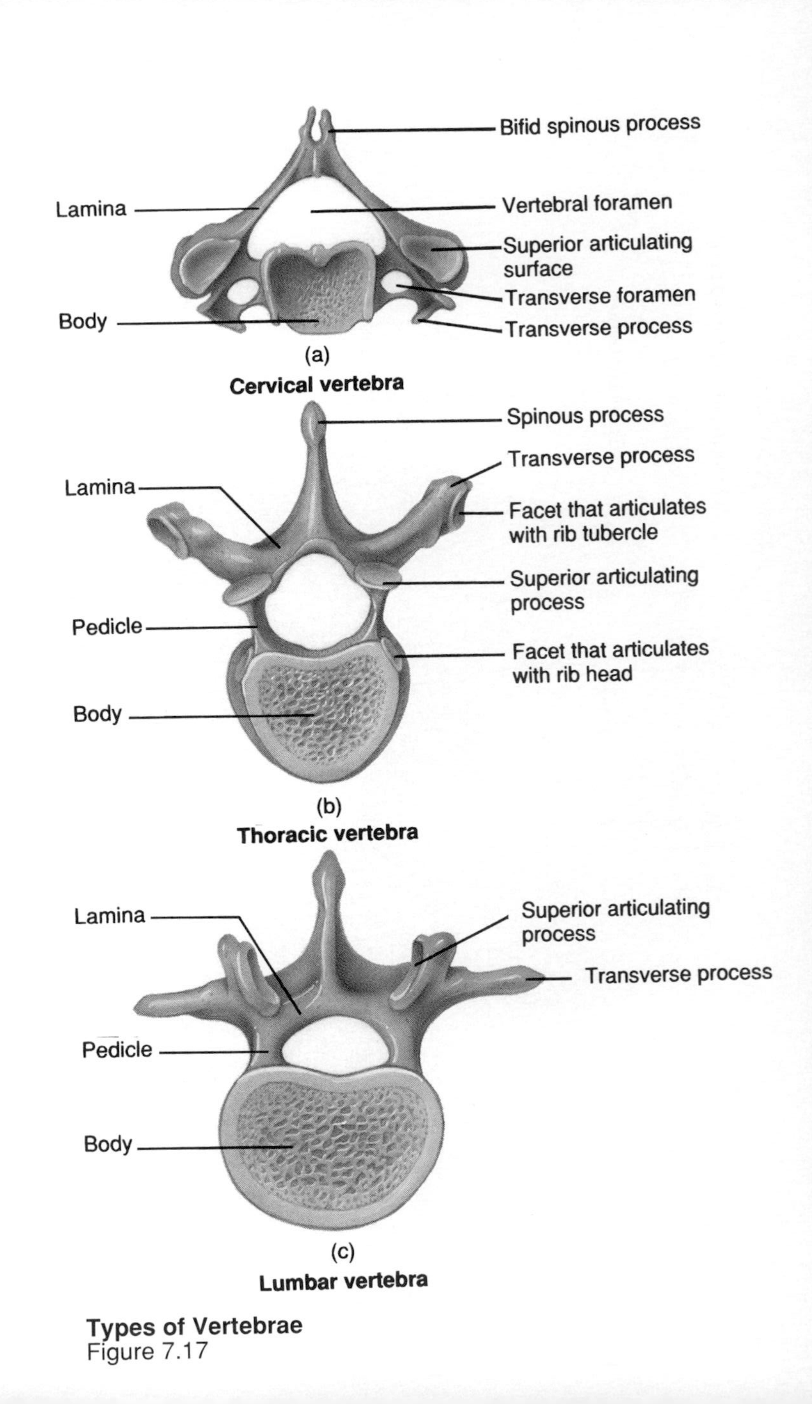
Lamina
Body
Bifid spinous process
Vertebral foramen
Superior articulating surface
Transverse foramen
Transverse process
(a)
Cervical vertebra
Lamina
Pedicle
Body
Spinous process
Transverse process
Facet that articulates with rib tubercle
Superior articulating process
Facet that articulates with rib head
(b)
Thoracic vertebra
Lamina
Pedicle
Body
Superior articulating process
Transverse process
(c)
Lumbar vertebra

Types of Vertebrae
Figure 7.17

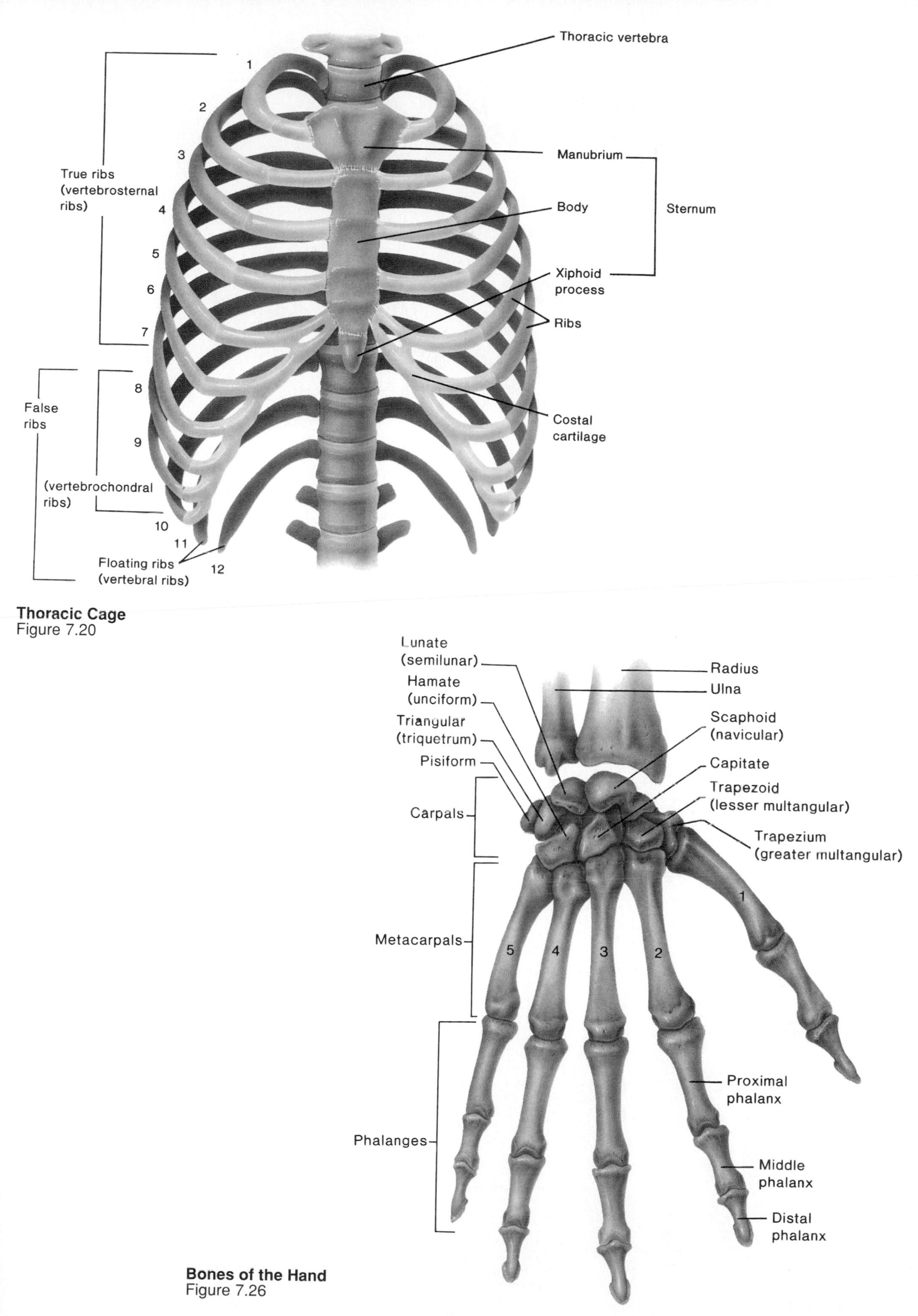
Thoracic vertebra
1
2
3
4
5
6
7
8
9
10
11
12
True ribs
(vertebrosternal
ribs)
False
ribs
(vertebrochondral
ribs)
Floating ribs
(vertebral ribs)
Manubrium
Body
Sternum
Xiphoid
process
Ribs
Costal
cartilage
Lunate
(semilunar)
Hamate
(unciform)
Triangular
(triquetrum)
Pisiform
Carpals
Metacarpals
Phalanges
Radius
Ulna
Scaphoid
(navicular)
Capitate
Trapezoid
(lesser multangular)
Trapezium
(greater multangular)
1
2
3
4
5
Proximal
phalanx
Middle
phalanx
Distal
phalanx

Thoracic Cage
Figure 7.20

Bones of the Hand
Figure 7.26

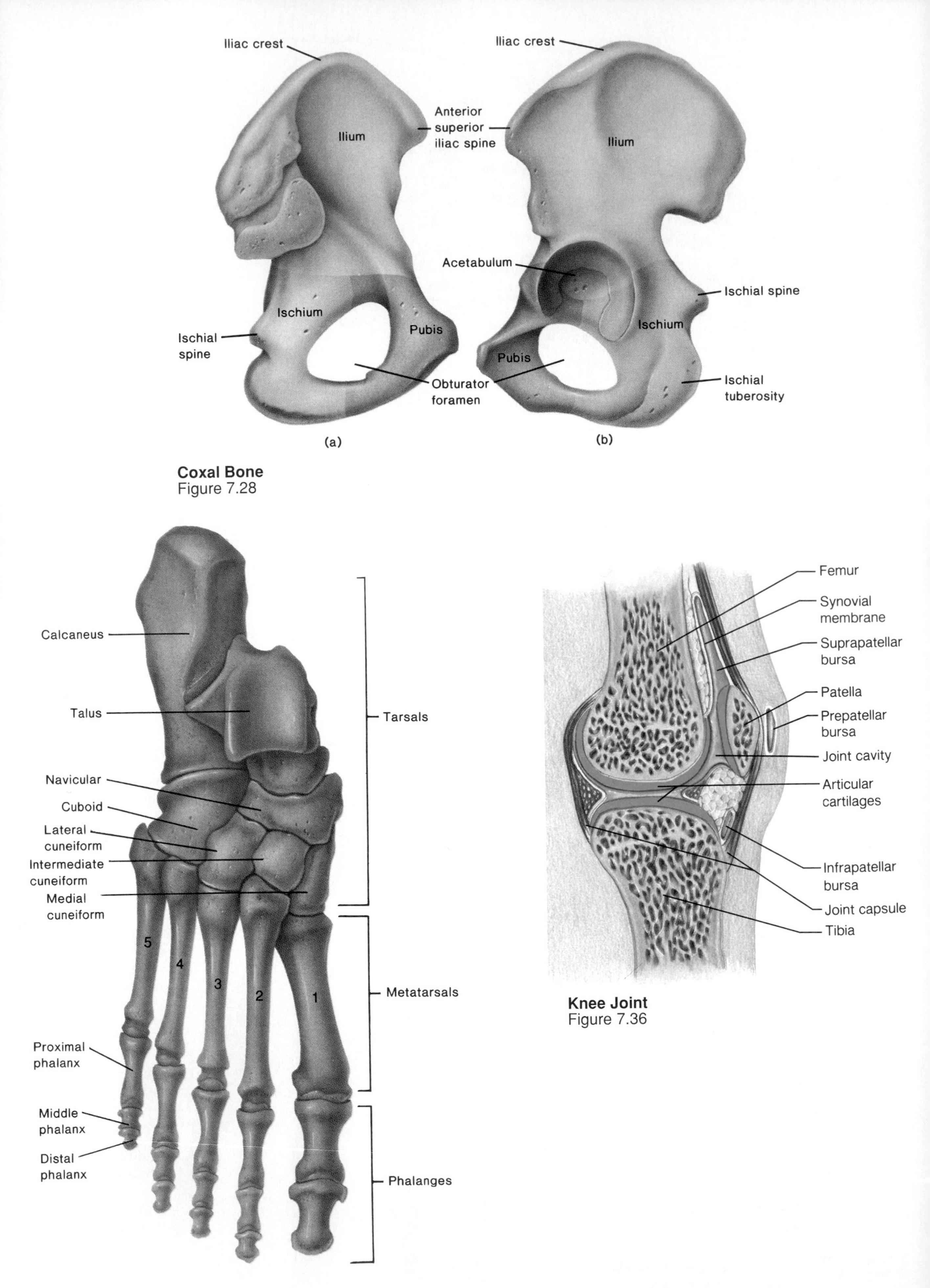

Iliac crest
Anterior
superior
iliac spine
Ilium
Ischium
Ischial
spine
Pubis
Obturator
foramen
(a)
Iliac crest
Ilium
Acetabulum
Ischial spine
Ischium
Pubis
Ischial
tuberosity
(b)
Calcaneus
Talus
Navicular
Cuboid
Lateral
cuneiform
Intermediate
cuneiform
Medial
cuneiform
Tarsals
5
4
3
2
1
Metatarsals
Proximal
phalanx
Middle
phalanx
Distal
phalanx
Phalanges
Femur
Synovial
membrane
Suprapatellar
bursa
Patella
Prepatellar
bursa
Joint cavity
Articular
cartilages
Infrapatellar
bursa
Joint capsule
Tibia

Coxal Bone
Figure 7.28

Knee Joint
Figure 7.36

Bones of the Foot
Figure 7.33

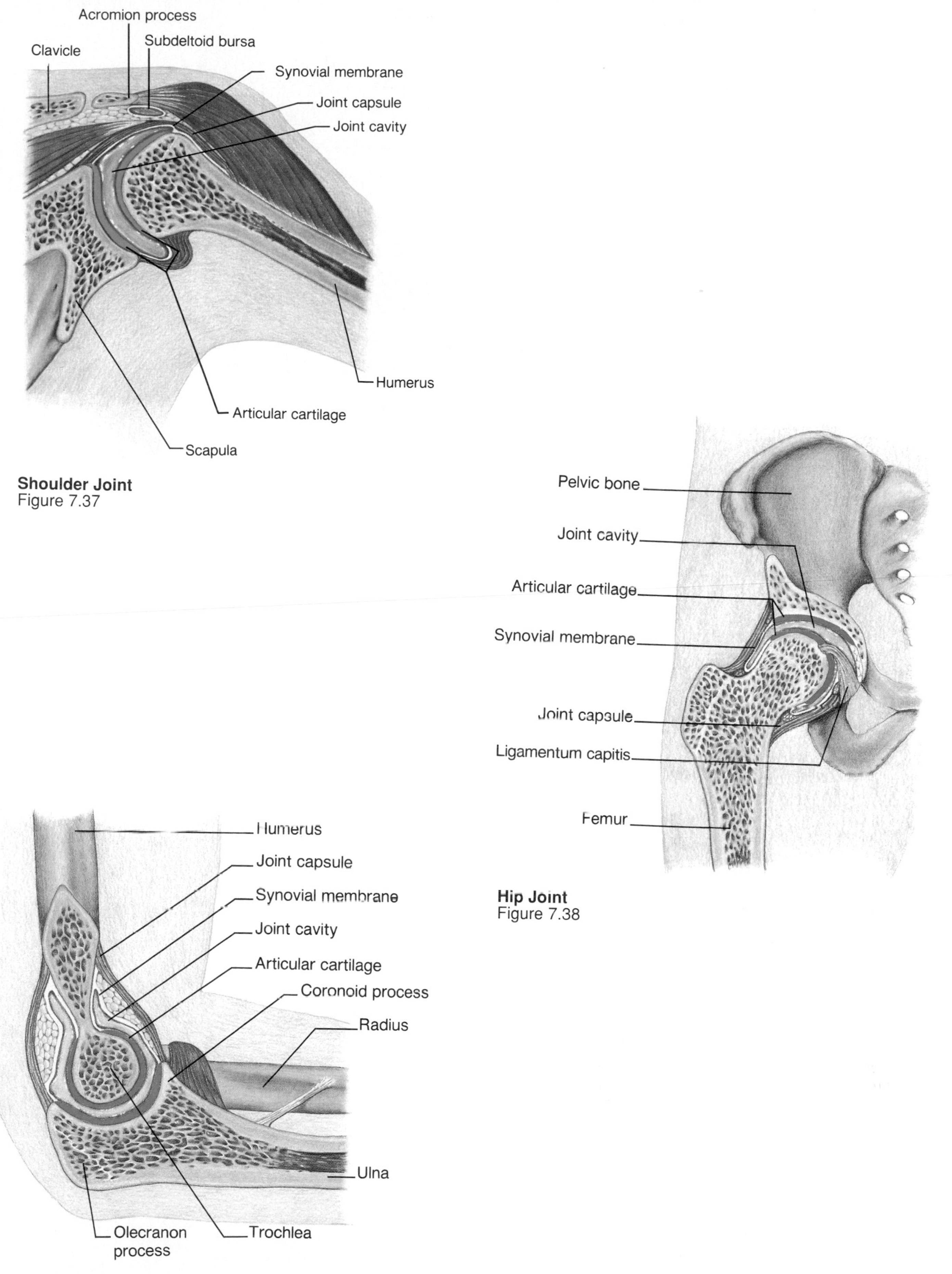
Acromion process
Clavicle
Subdeltoid bursa
Synovial membrane
Joint capsule
Joint cavity
Humerus
Articular cartilage
Scapula
Pelvic bone
Joint cavity
Articular cartilage
Synovial membrane
Joint capsule
Ligamentum capitis
Femur
Humerus
Joint capsule
Synovial membrane
Joint cavity
Articular cartilage
Coronoid process
Radius
Ulna
Olecranon process
Trochlea

Shoulder Joint
Figure 7.37

Hip Joint
Figure 7.38

Elbow Joint
Figure 7.39

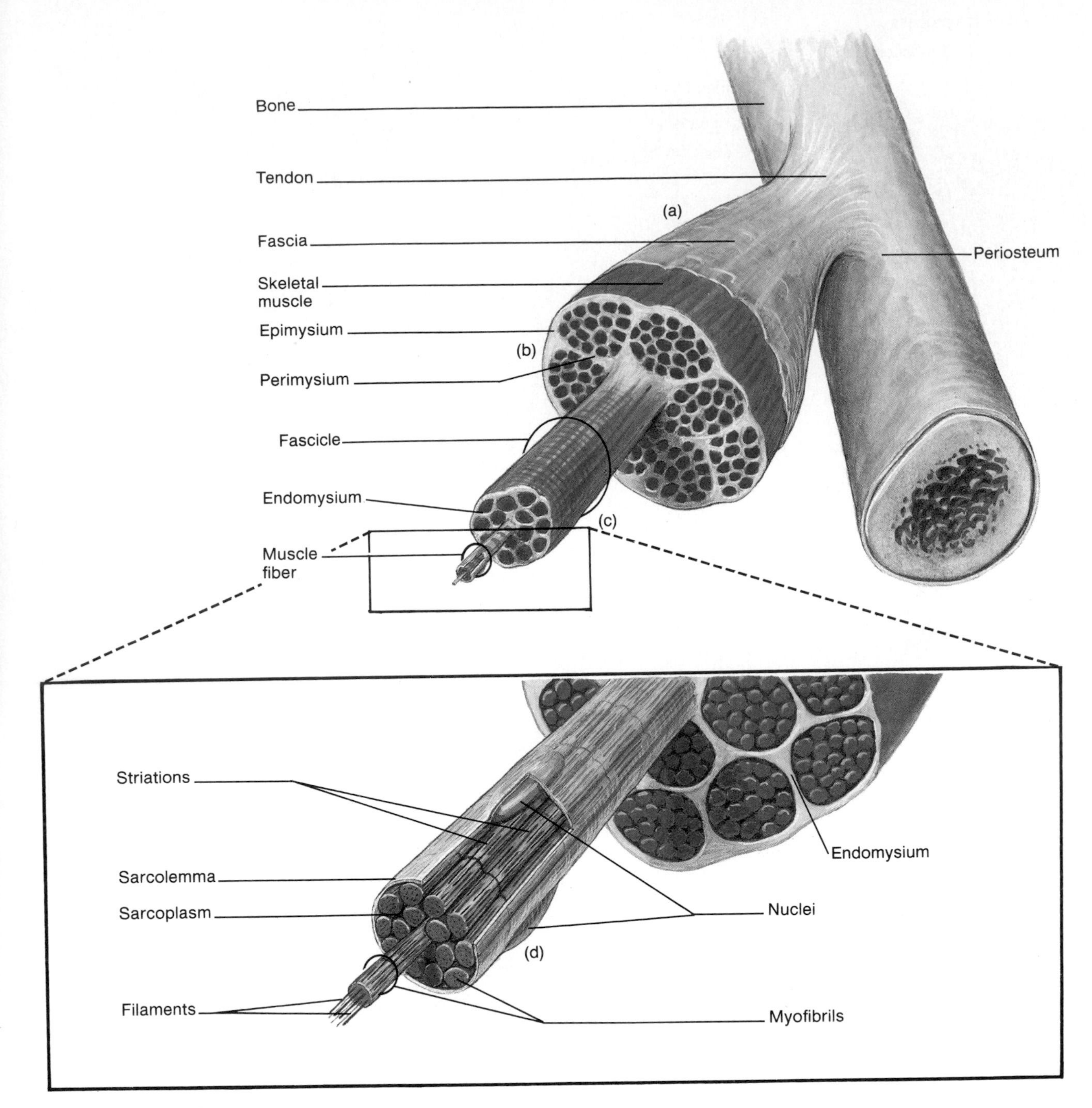
Bone
Tendon
(a)
Fascia
Periosteum
Skeletal muscle
Epimysium
(b)
Perimysium
Fascicle
Endomysium
(c)
Muscle fiber
Striations
Endomysium
Sarcolemma
Nuclei
Sarcoplasm
(d)
Filaments
Myofibrils

Skeletal Muscle Structure
Figure 8.1

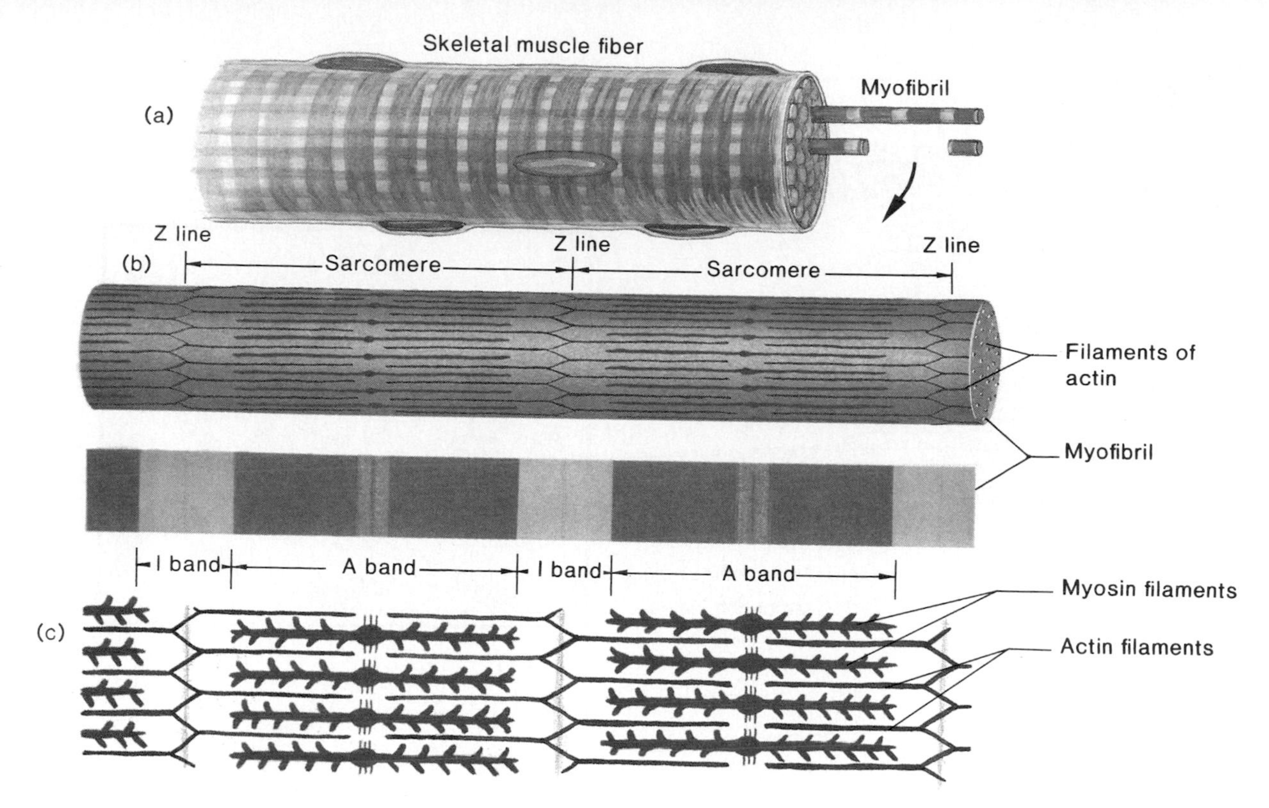
Skeletal muscle fiber
(a)
Myofibril
Z line
(b)
Sarcomere
Z line
Sarcomere
Z line
Filaments of actin
Myofibril
I band
A band
I band
A band
(c)
Myosin filaments
Actin filaments

Skeletal Muscle Fiber I
Figure 8.2

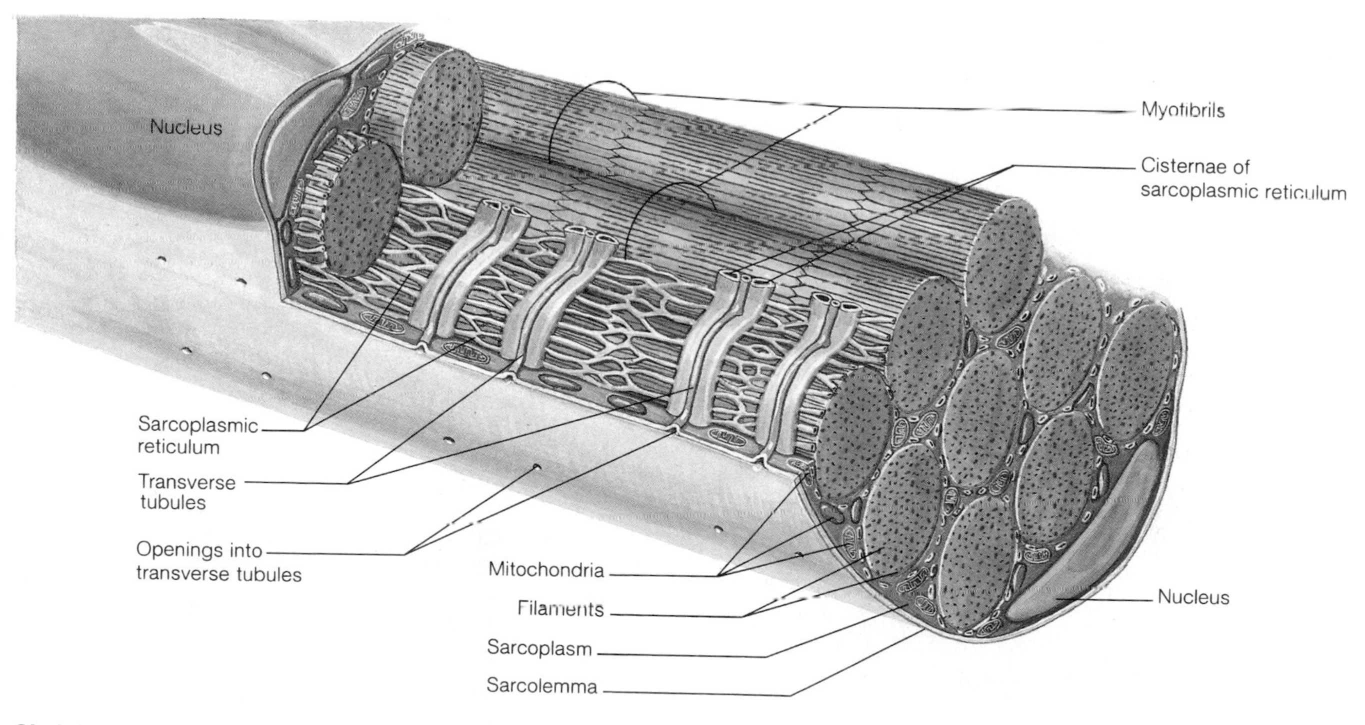
Nucleus
Myofibrils
Cisternae of sarcoplasmic reticulum
Sarcoplasmic reticulum
Transverse tubules
Openings into transverse tubules
Mitochondria
Filaments
Sarcoplasm
Sarcolemma
Nucleus

Skeletal Muscle Fiber II
Figure 8.4

Motor neuron fiber
Nerve fiber branches
Muscle fiber nucleus
Motor end plate
Myofibril of muscle fiber
Mitochondria
Synaptic vesicles
Synaptic cleft
Folded sarcolemma
Motor end plate

Neuromuscular Junction
Figure 8.5

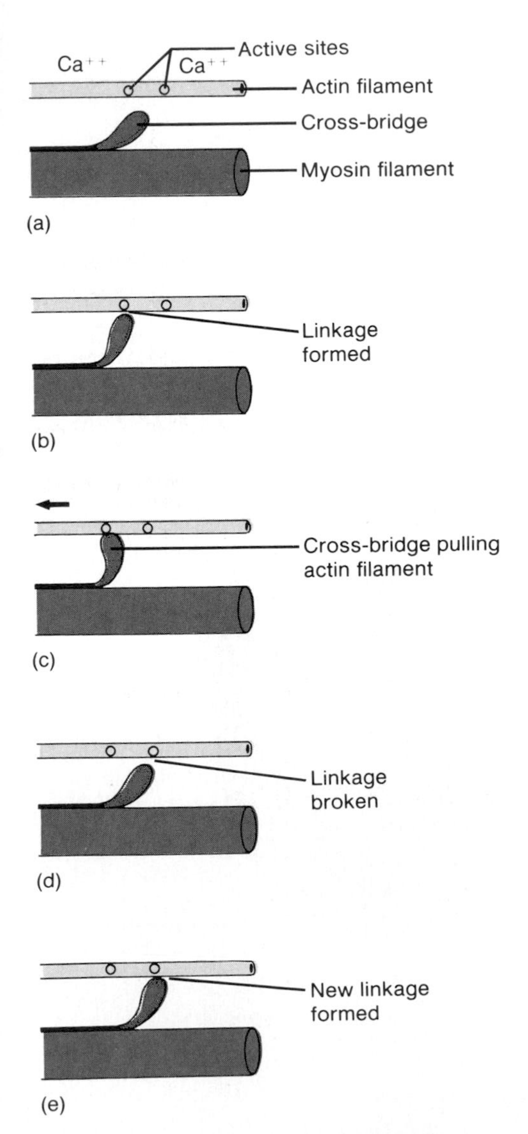

Ca++
Ca++
Active sites
Actin filament
Cross-bridge
Myosin filament
(a)
Linkage formed
(b)
Cross-bridge pulling actin filament
(c)
Linkage broken
(d)
New linkage formed
(e)

Ratchet Theory of Muscle Contraction
Figure 8.7

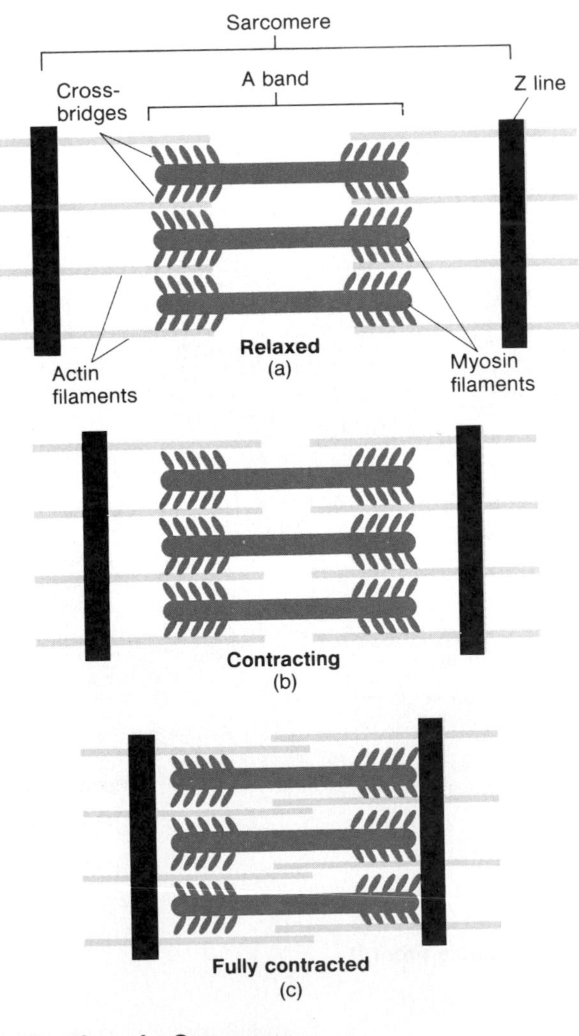

Sarcomere
Cross-bridges
A band
Z line
Relaxed
(a)
Actin filaments
Myosin filaments
Contracting
(b)
Fully contracted
(c)

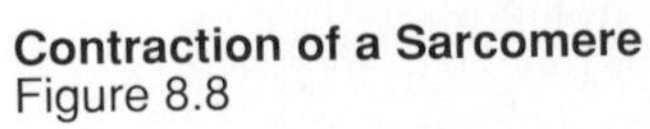

Contraction of a Sarcomere
Figure 8.8

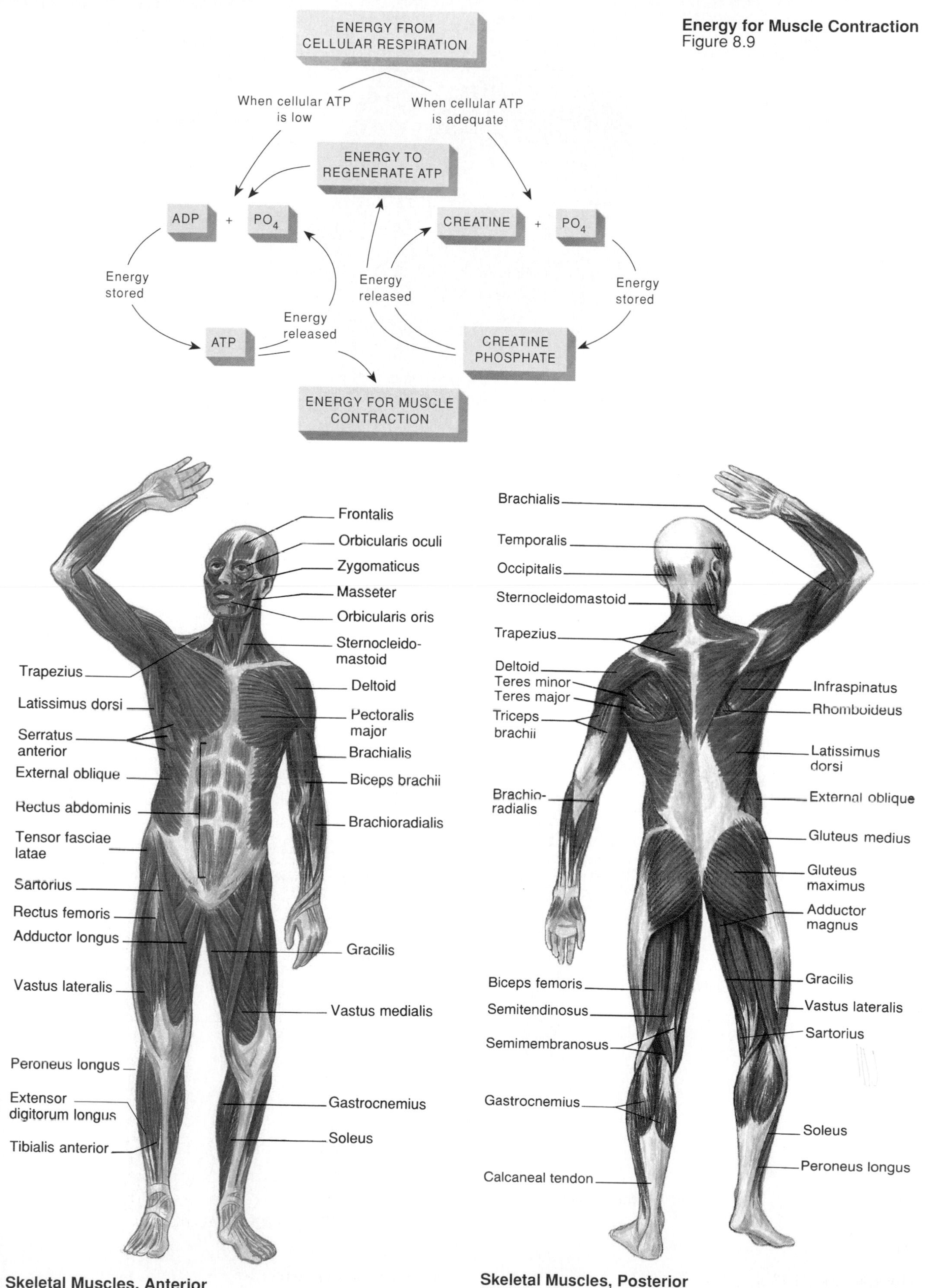
ENERGY FROM
CELLULAR RESPIRATION
When cellular ATP
is low
When cellular ATP
is adequate
ENERGY TO
REGENERATE ATP
ADP
+
PO_4
CREATINE
+
PO_4
Energy
stored
Energy
released
Energy
stored
ATP
Energy
released
CREATINE
PHOSPHATE
ENERGY FOR MUSCLE
CONTRACTION
Frontalis
Orbicularis oculi
Zygomaticus
Masseter
Orbicularis oris
Sternocleido-
mastoid
Trapezius
Deltoid
Latissimus dorsi
Pectoralis
major
Serratus
anterior
Brachialis
External oblique
Biceps brachii
Rectus abdominis
Brachioradialis
Tensor fasciae
latae
Sartorius
Rectus femoris
Adductor longus
Gracilis
Vastus lateralis
Vastus medialis
Peroneus longus
Extensor
digitorum longus
Gastrocnemius
Soleus
Tibialis anterior
Brachialis
Temporalis
Occipitalis
Sternocleidomastoid
Trapezius
Deltoid
Teres minor
Teres major
Infraspinatus
Rhomboideus
Triceps
brachii
Latissimus
dorsi
Brachio-
radialis
External oblique
Gluteus medius
Gluteus
maximus
Adductor
magnus
Biceps femoris
Gracilis
Semitendinosus
Vastus lateralis
Sartorius
Semimembranosus
Gastrocnemius
Soleus
Peroneus longus
Calcaneal tendon

Energy for Muscle Contraction
Figure 8.9

Skeletal Muscles, Anterior
Figure 8.14

Skeletal Muscles, Posterior
Figure 8.15

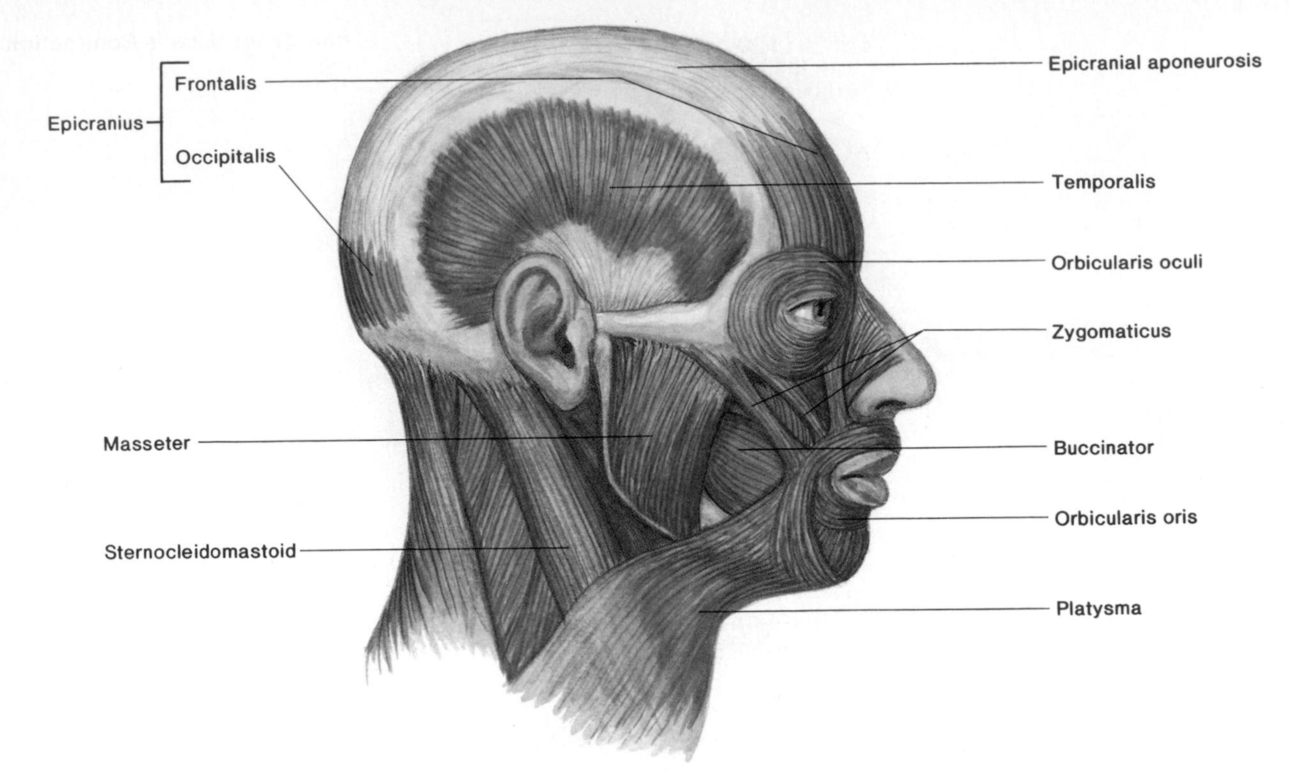

Epicranial aponeurosis
Frontalis
Epicranius
Occipitalis
Temporalis
Orbicularis oculi
Zygomaticus
Masseter
Buccinator
Orbicularis oris
Sternocleidomastoid
Platysma

Muscles of Expression and Mastication
Figure 8.16 a

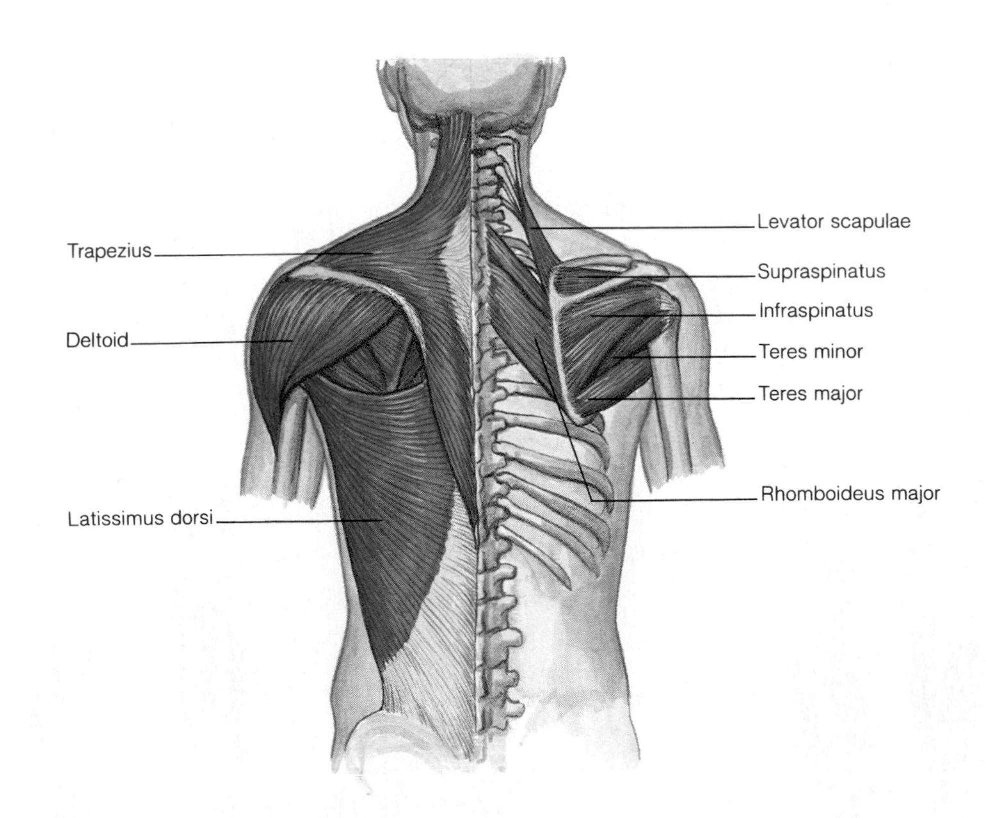

Levator scapulae
Trapezius
Supraspinatus
Infraspinatus
Deltoid
Teres minor
Teres major
Rhomboideus major
Latissimus dorsi

Muscles of Posterior Shoulder
Figure 8.17

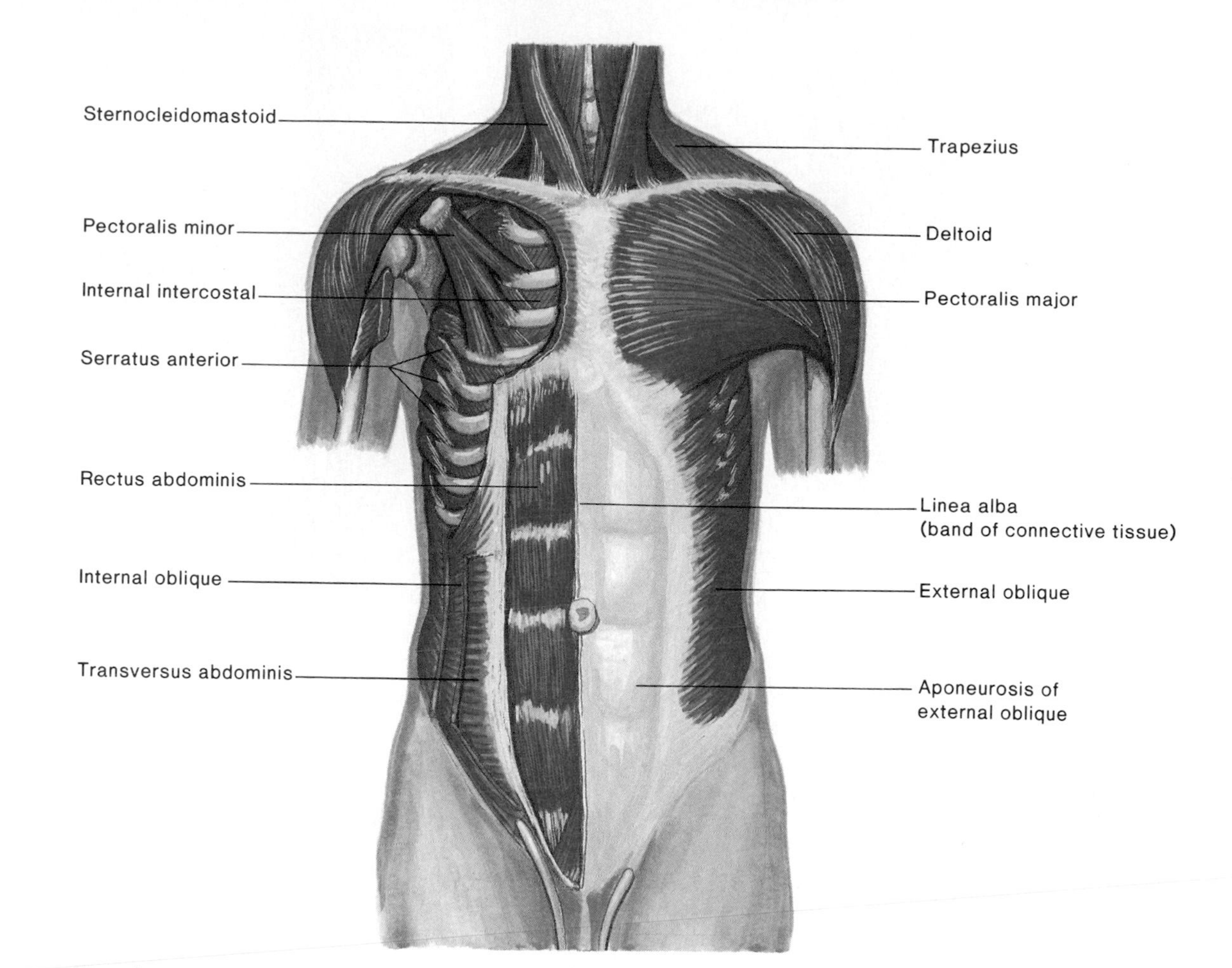
Sternocleidomastoid
Pectoralis minor
Internal intercostal
Serratus anterior
Rectus abdominis
Internal oblique
Transversus abdominis
Trapezius
Deltoid
Pectoralis major
Linea alba
(band of connective tissue)
External oblique
Aponeurosis of
external oblique

Muscles of the Anterior Chest and Abdominal Wall
Figure 8.18

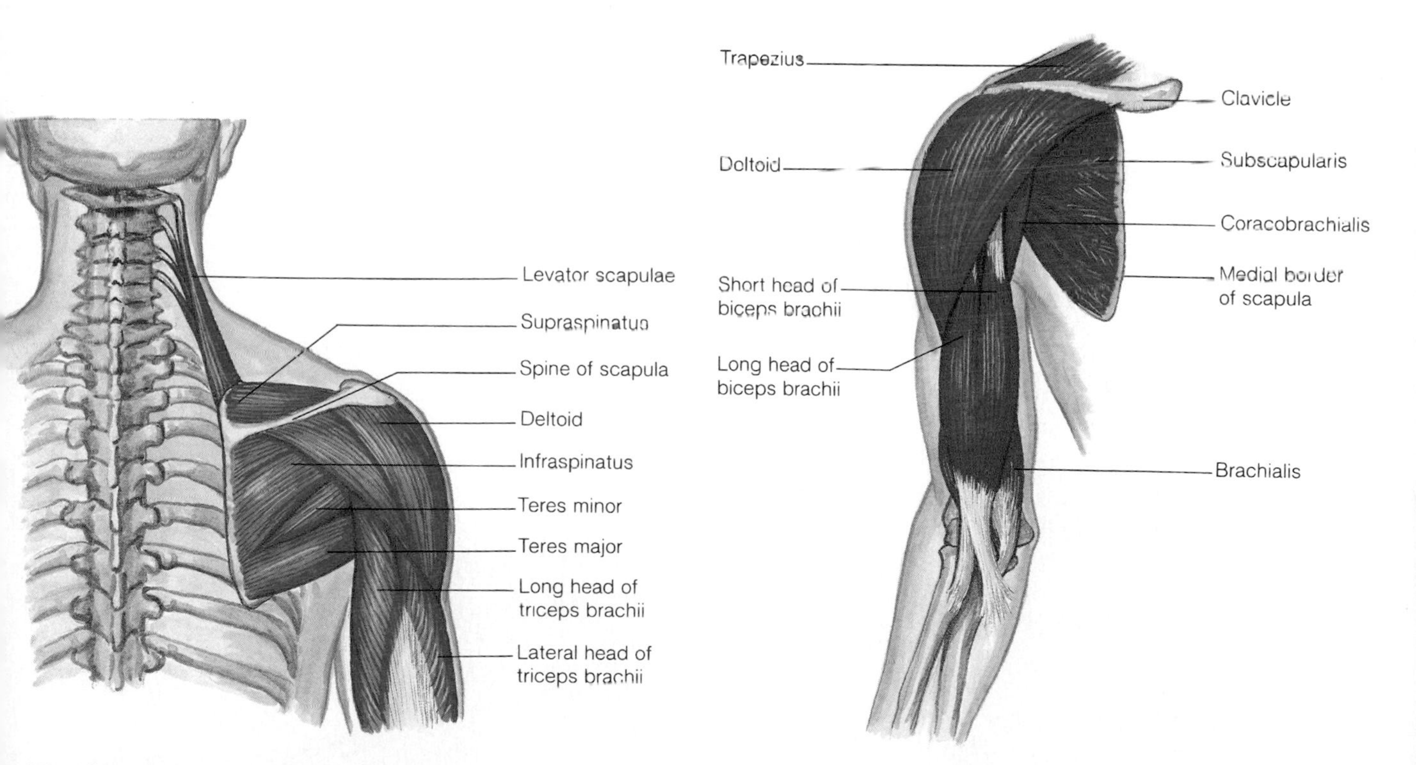
Levator scapulae
Supraspinatus
Spine of scapula
Deltoid
Infraspinatus
Teres minor
Teres major
Long head of
triceps brachii
Lateral head of
triceps brachii
Trapezius
Clavicle
Deltoid
Subscapularis
Coracobrachialis
Short head of
biceps brachii
Medial border
of scapula
Long head of
biceps brachii
Brachialis

Muscles of the Scapula and Upper Arm
Figure 8.19

Muscles of the Anterior Shoulder and Upper Arm
Figure 8.20

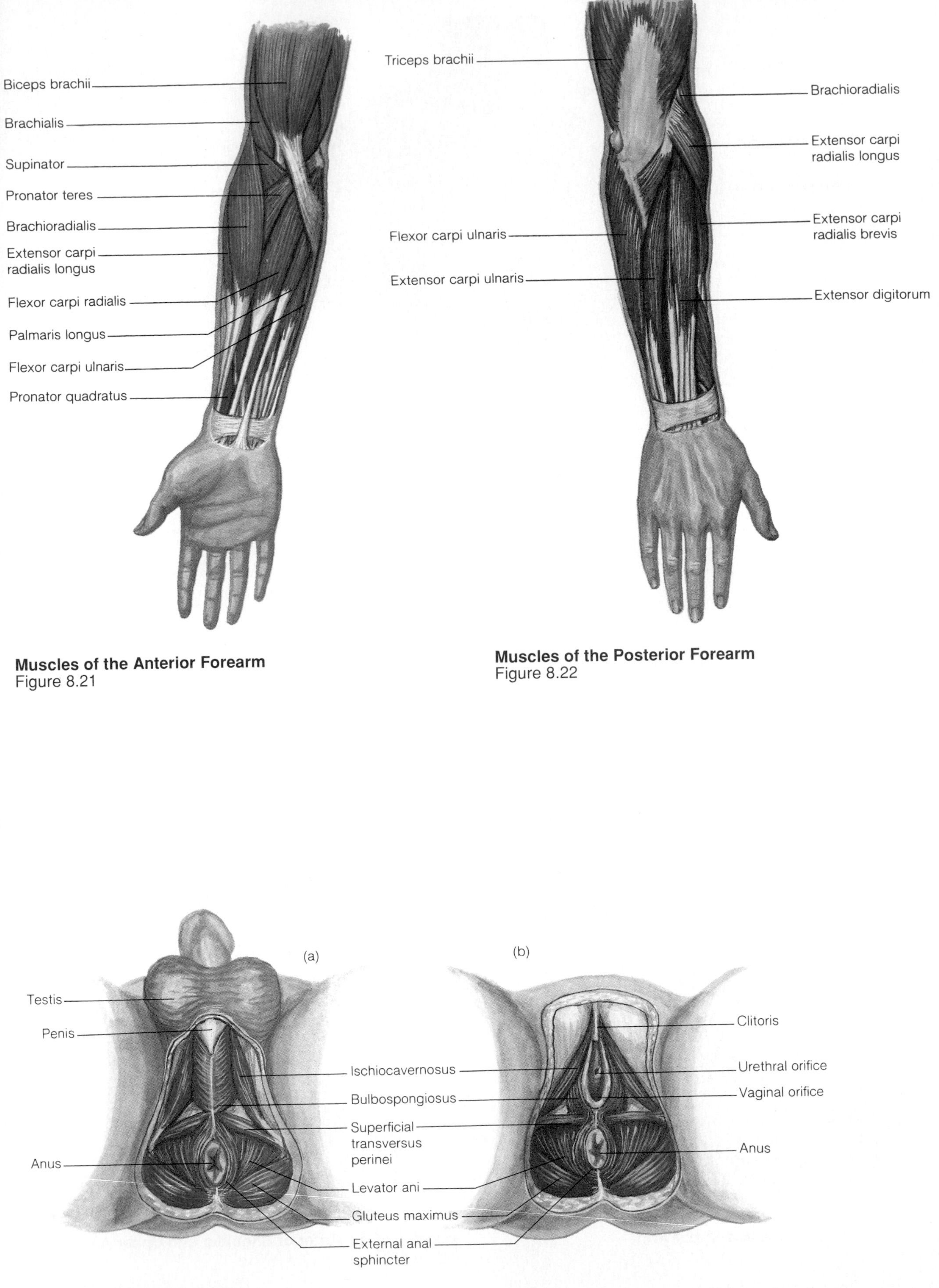
Biceps brachii
Brachialis
Supinator
Pronator teres
Brachioradialis
Extensor carpi
radialis longus
Flexor carpi radialis
Palmaris longus
Flexor carpi ulnaris
Pronator quadratus
Triceps brachii
Brachioradialis
Extensor carpi
radialis longus
Extensor carpi
radialis brevis
Flexor carpi ulnaris
Extensor carpi ulnaris
Extensor digitorum
(a)
(b)
Testis
Penis
Anus
Ischiocavernosus
Bulbospongiosus
Superficial
transversus
perinei
Levator ani
Gluteus maximus
External anal
sphincter
Clitoris
Urethral orifice
Vaginal orifice
Anus

Muscles of the Anterior Forearm
Figure 8.21

Muscles of the Posterior Forearm
Figure 8.22

Muscles of the Pelvic Outlet
Figure 8.23

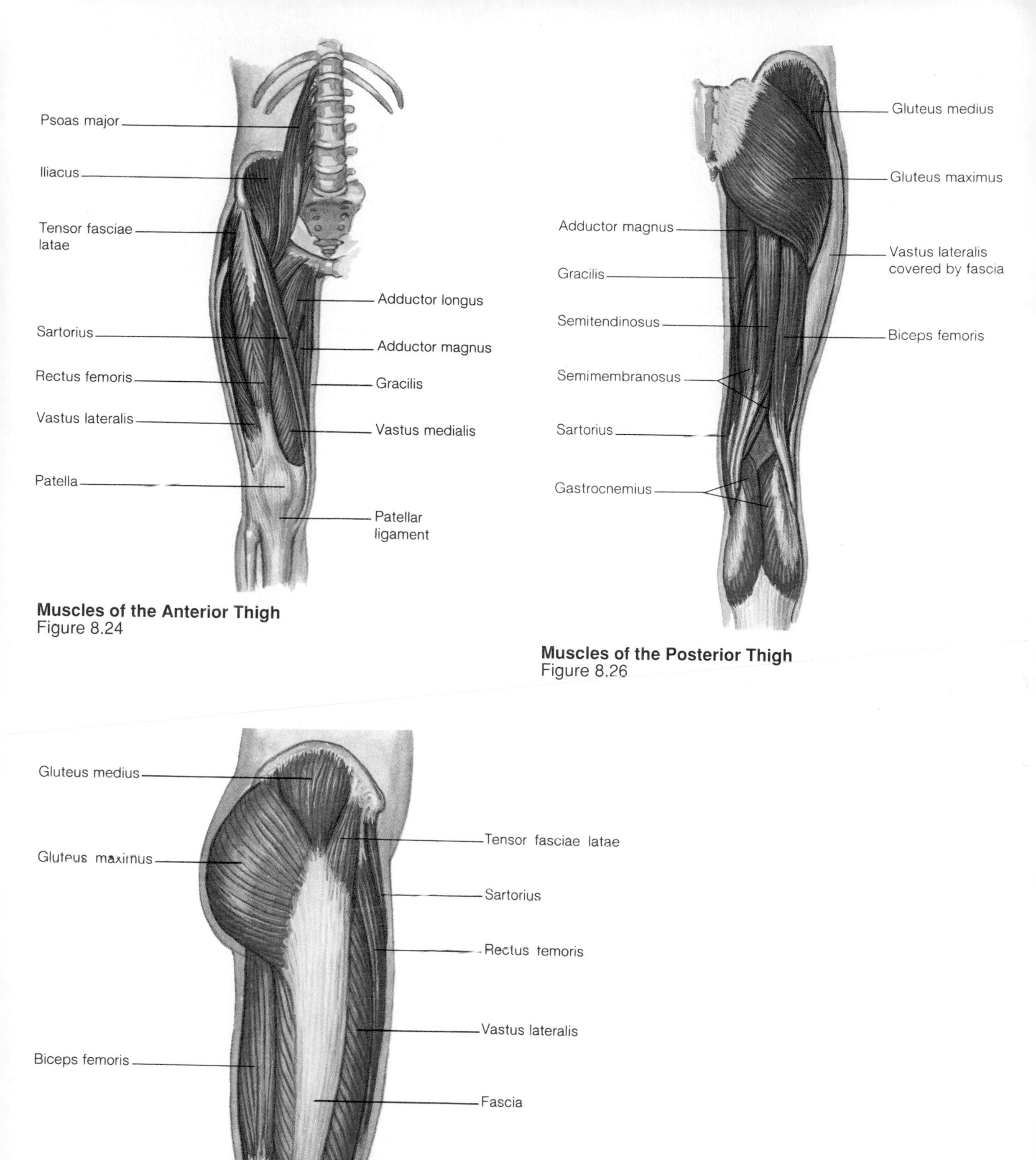
Psoas major
Iliacus
Tensor fasciae latae
Sartorius
Rectus femoris
Vastus lateralis
Patella
Adductor longus
Adductor magnus
Gracilis
Vastus medialis
Patellar ligament
Gluteus medius
Gluteus maximus
Vastus lateralis covered by fascia
Biceps femoris
Adductor magnus
Gracilis
Semitendinosus
Semimembranosus
Sartorius
Gastrocnemius
Gluteus medius
Gluteus maximus
Tensor fasciae latae
Sartorius
Rectus femoris
Vastus lateralis
Biceps femoris
Fascia
Patella

Muscles of the Anterior Thigh
Figure 8.24

Muscles of the Posterior Thigh
Figure 8.26

Muscles of the Lateral Thigh
Figure 8.25

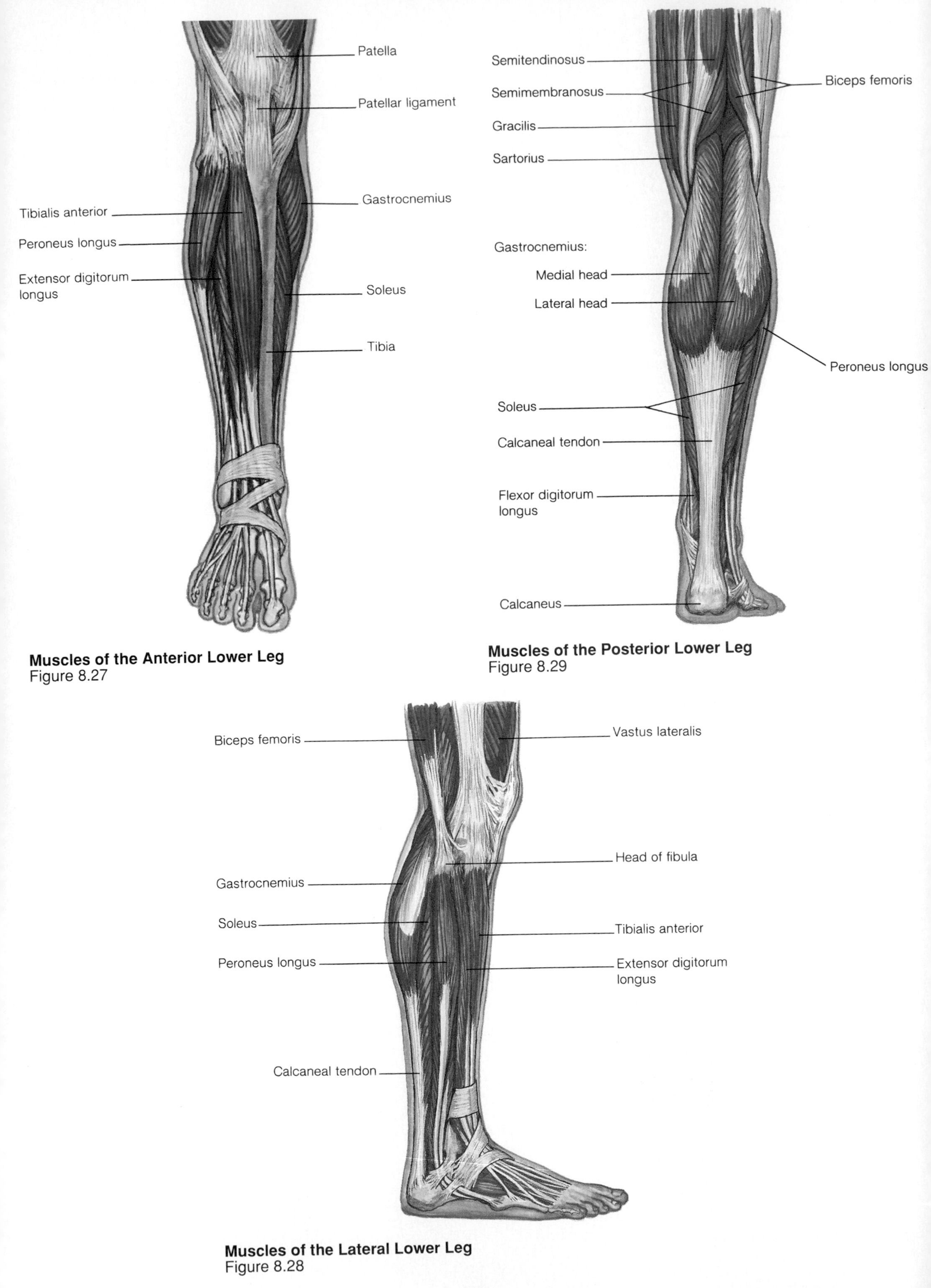
Patella
Patellar ligament
Tibialis anterior
Peroneus longus
Extensor digitorum longus
Gastrocnemius
Soleus
Tibia
Semitendinosus
Semimembranosus
Gracilis
Sartorius
Biceps femoris
Gastrocnemius:
Medial head
Lateral head
Peroneus longus
Soleus
Calcaneal tendon
Flexor digitorum longus
Calcaneus
Biceps femoris
Vastus lateralis
Head of fibula
Gastrocnemius
Soleus
Tibialis anterior
Peroneus longus
Extensor digitorum longus
Calcaneal tendon

Muscles of the Anterior Lower Leg
Figure 8.27

Muscles of the Posterior Lower Leg
Figure 8.29

Muscles of the Lateral Lower Leg
Figure 8.28

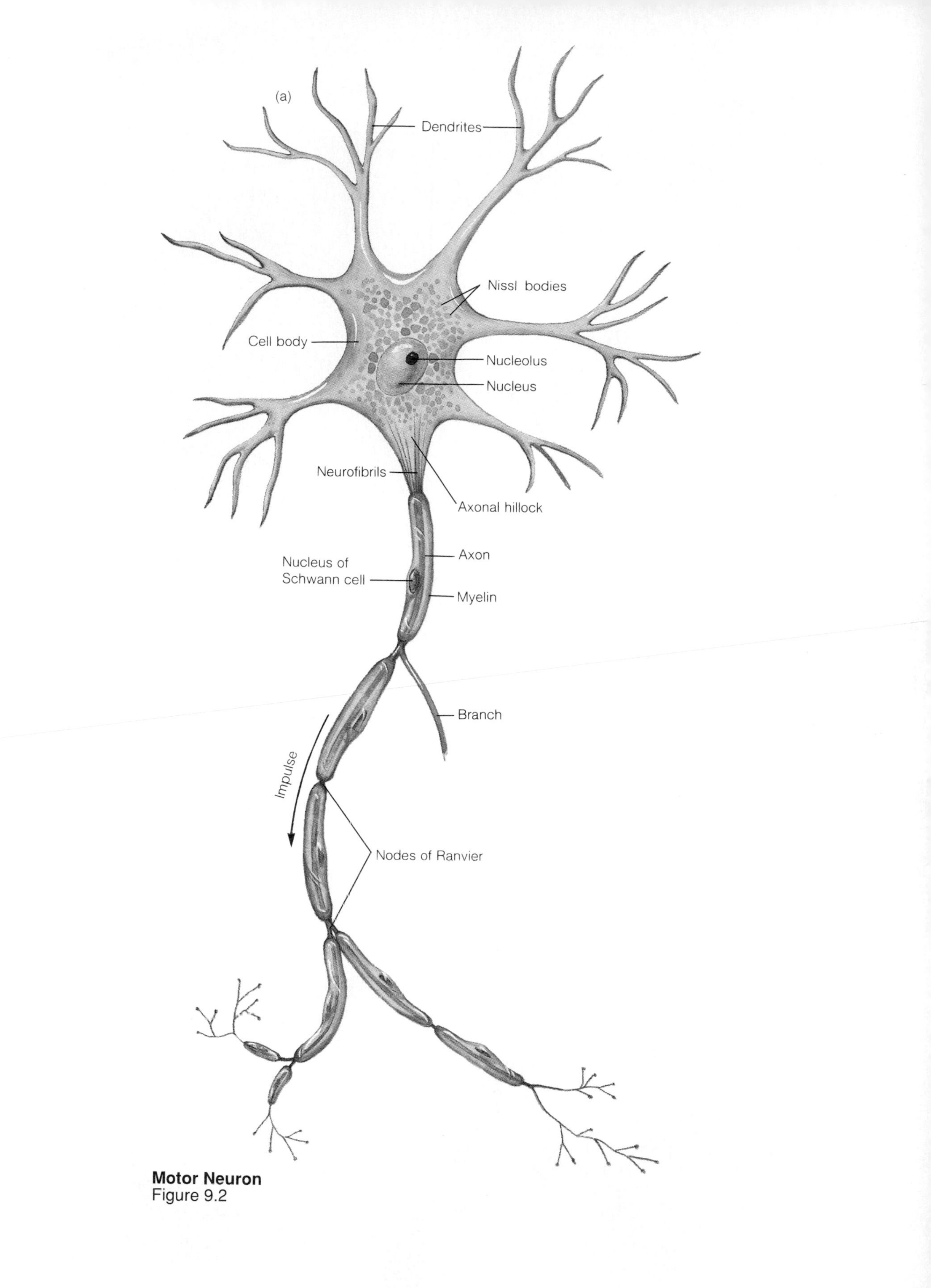

(a)
Dendrites
Nissl bodies
Cell body
Nucleolus
Nucleus
Neurofibrils
Axonal hillock
Axon
Nucleus of
Schwann cell
Myelin
Branch
Impulse
Nodes of Ranvier

Motor Neuron
Figure 9.2

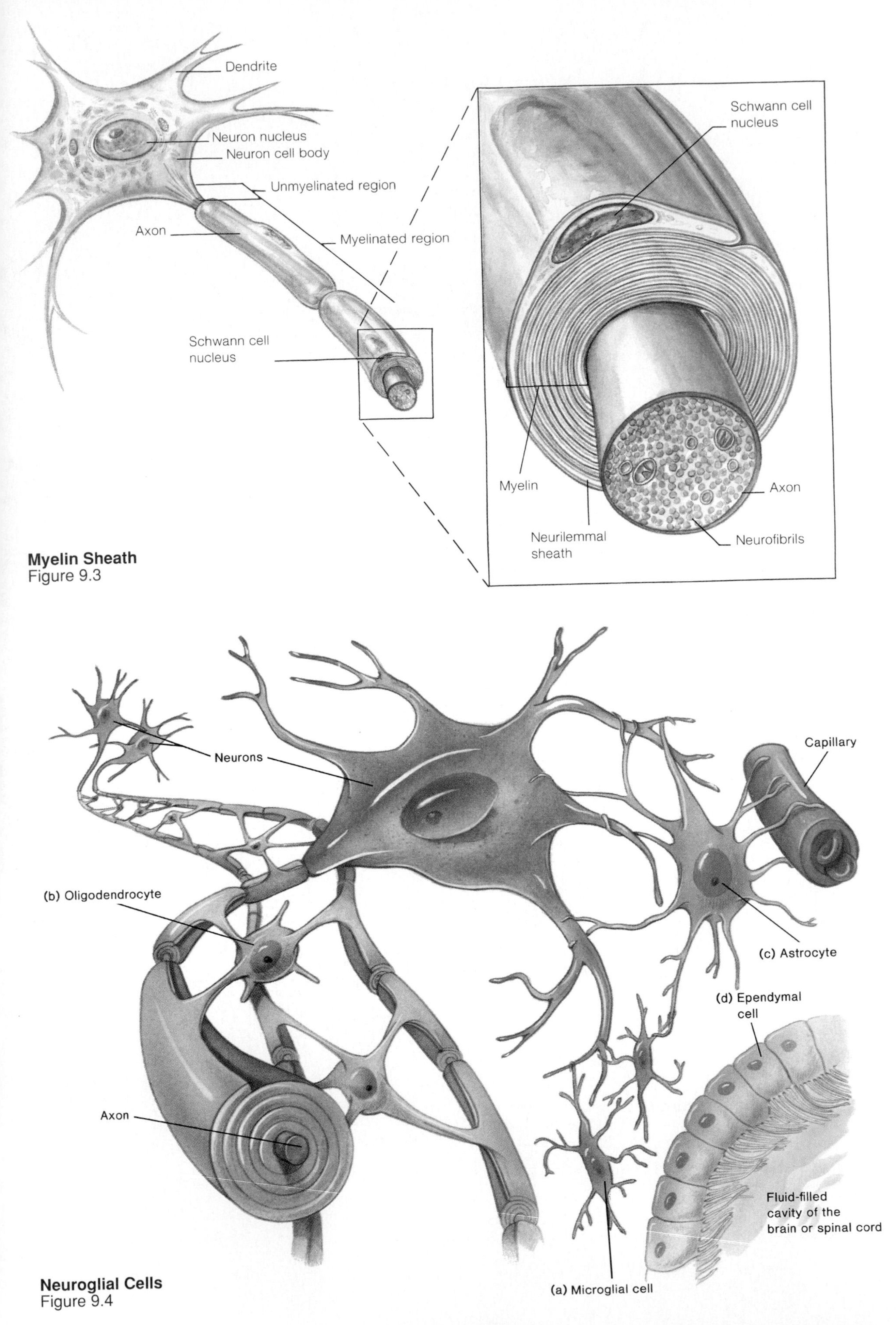
Dendrite
Neuron nucleus
Neuron cell body
Unmyelinated region
Axon
Myelinated region
Schwann cell
nucleus
Schwann cell
nucleus
Myelin
Axon
Neurilemmal
sheath
Neurofibrils

Myelin Sheath
Figure 9.3

Neurons
Capillary
(b) Oligodendrocyte
(c) Astrocyte
(d) Ependymal
cell
Axon
Fluid-filled
cavity of the
brain or spinal cord
(a) Microglial cell

Neuroglial Cells
Figure 9.4

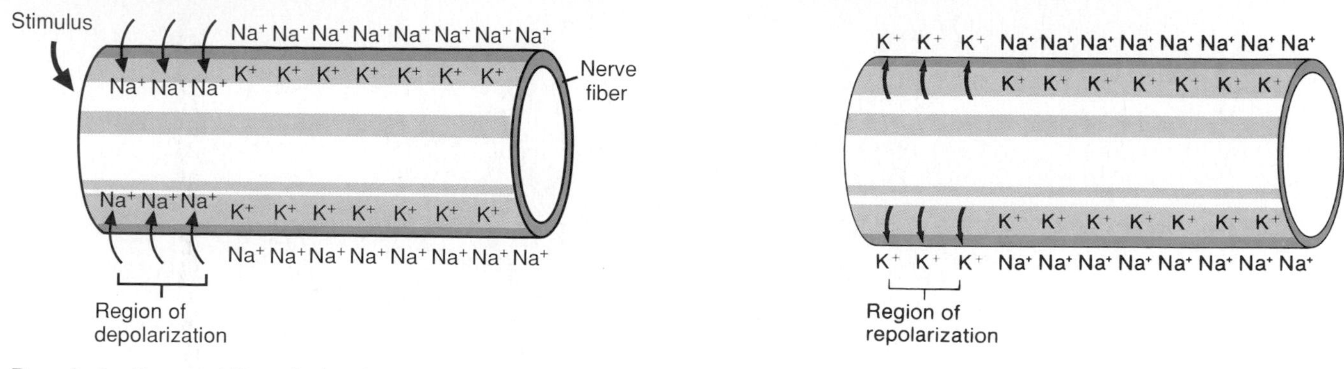
Stimulus
Na+ Na+ Na+ Na+ Na+ Na+ Na+ Na+
Na+ Na+ Na+
K+ K+ K+ K+ K+ K+ K+
Nerve fiber
Region of depolarization
K+ K+ K+
Region of repolarization

Depolarization and Repolarization
Figures 9.8 and 9.9

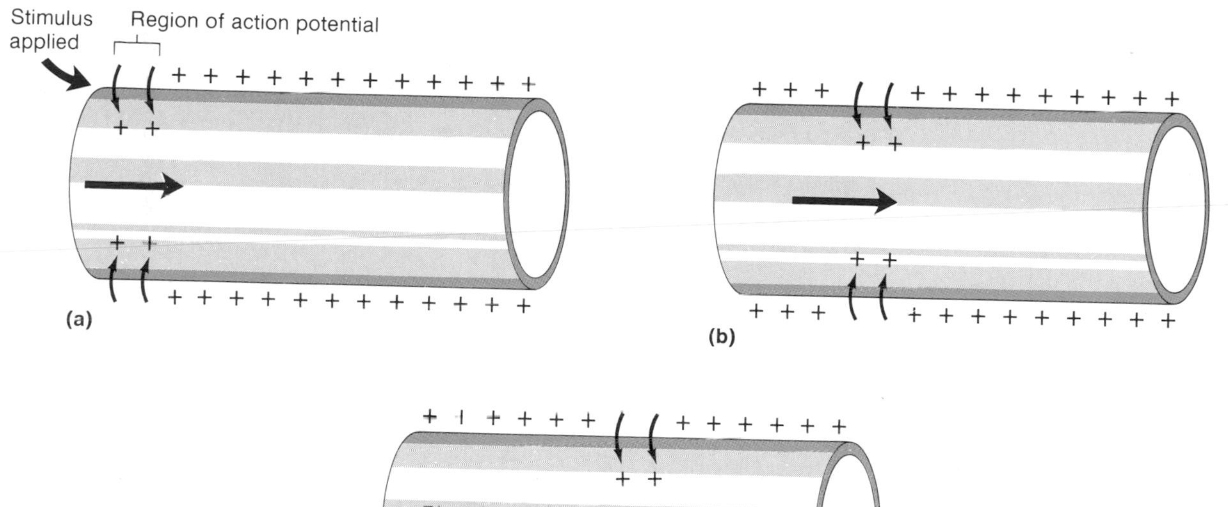
Stimulus applied
Region of action potential
(a)
(b)

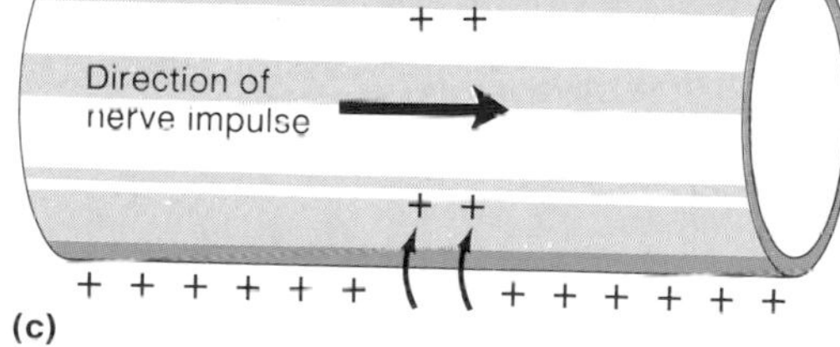
Direction of nerve impulse
(c)

Action Potential
Figure 9.10

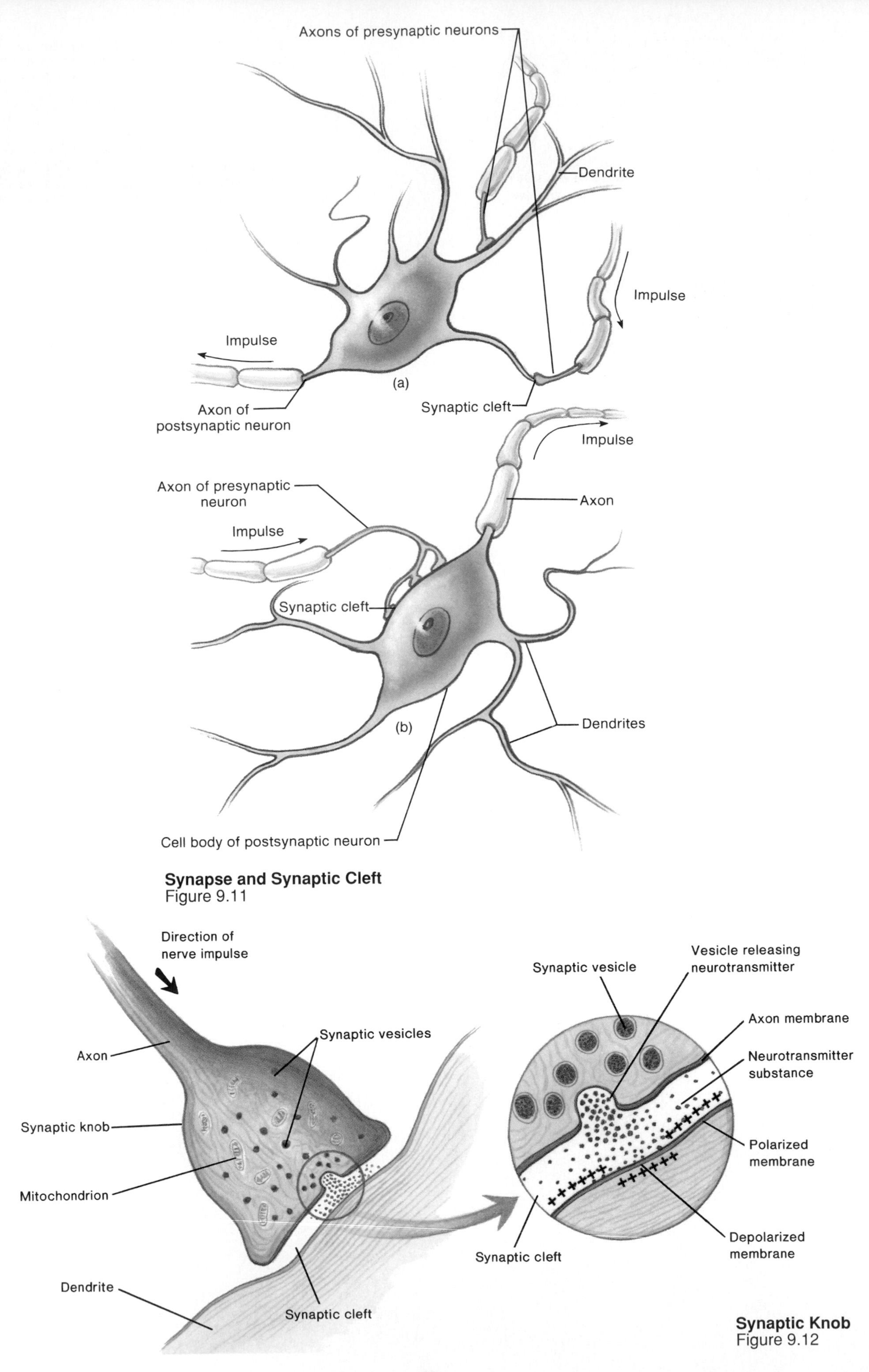
Axons of presynaptic neurons
Dendrite
Impulse
Impulse
(a)
Synaptic cleft
Axon of
postsynaptic neuron
Impulse
Axon of presynaptic
neuron
Axon
Impulse
Synaptic cleft
(b)
Dendrites
Cell body of postsynaptic neuron

Synapse and Synaptic Cleft
Figure 9.11

Direction of
nerve impulse
Axon
Synaptic vesicles
Synaptic knob
Mitochondrion
Dendrite
Synaptic cleft
Synaptic vesicle
Vesicle releasing
neurotransmitter
Axon membrane
Neurotransmitter
substance
Polarized
membrane
Depolarized
membrane
Synaptic cleft

Synaptic Knob
Figure 9.12

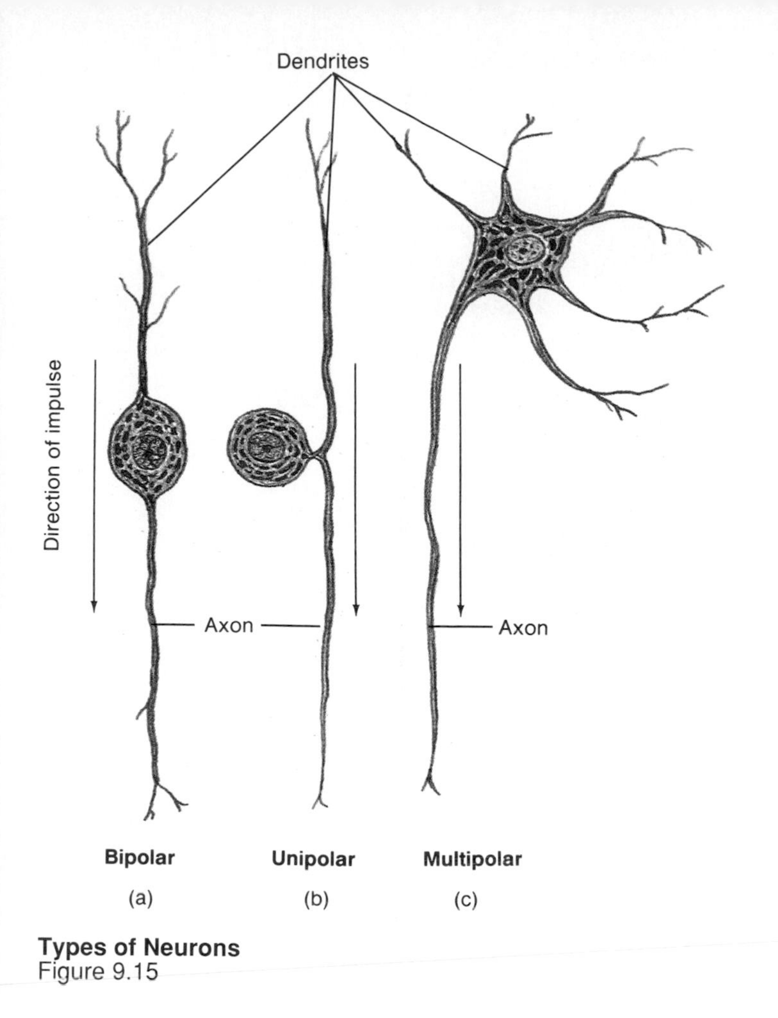
Dendrites
Direction of impulse
Axon
Axon
Bipolar
(a)
Unipolar
(b)
Multipolar
(c)

Types of Neurons
Figure 9.15

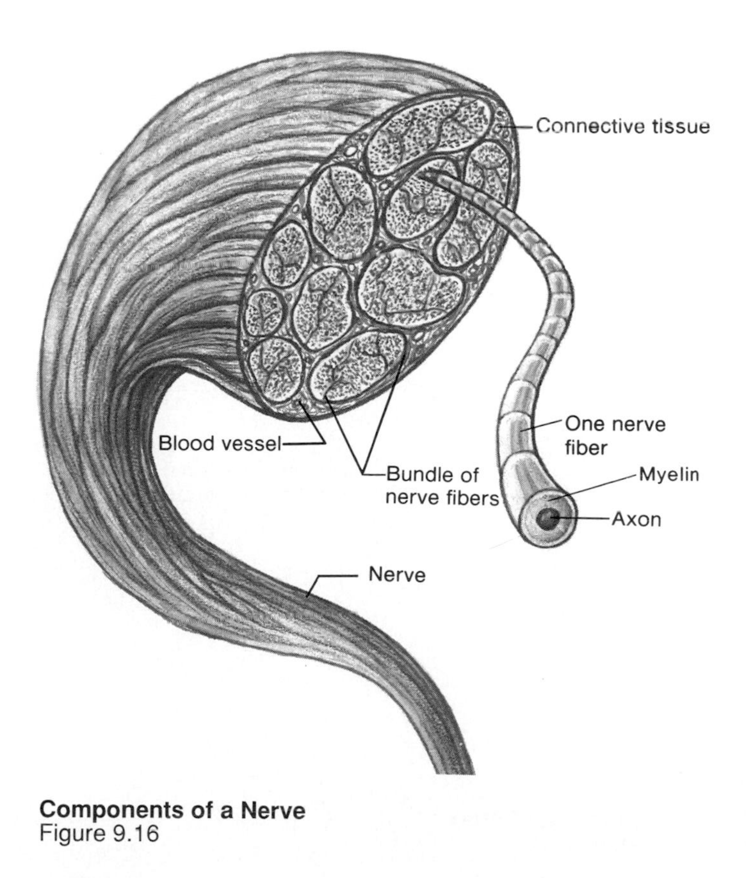
Connective tissue
Blood vessel
Bundle of nerve fibers
One nerve fiber
Myelin
Axon
Nerve

Components of a Nerve
Figure 9.16

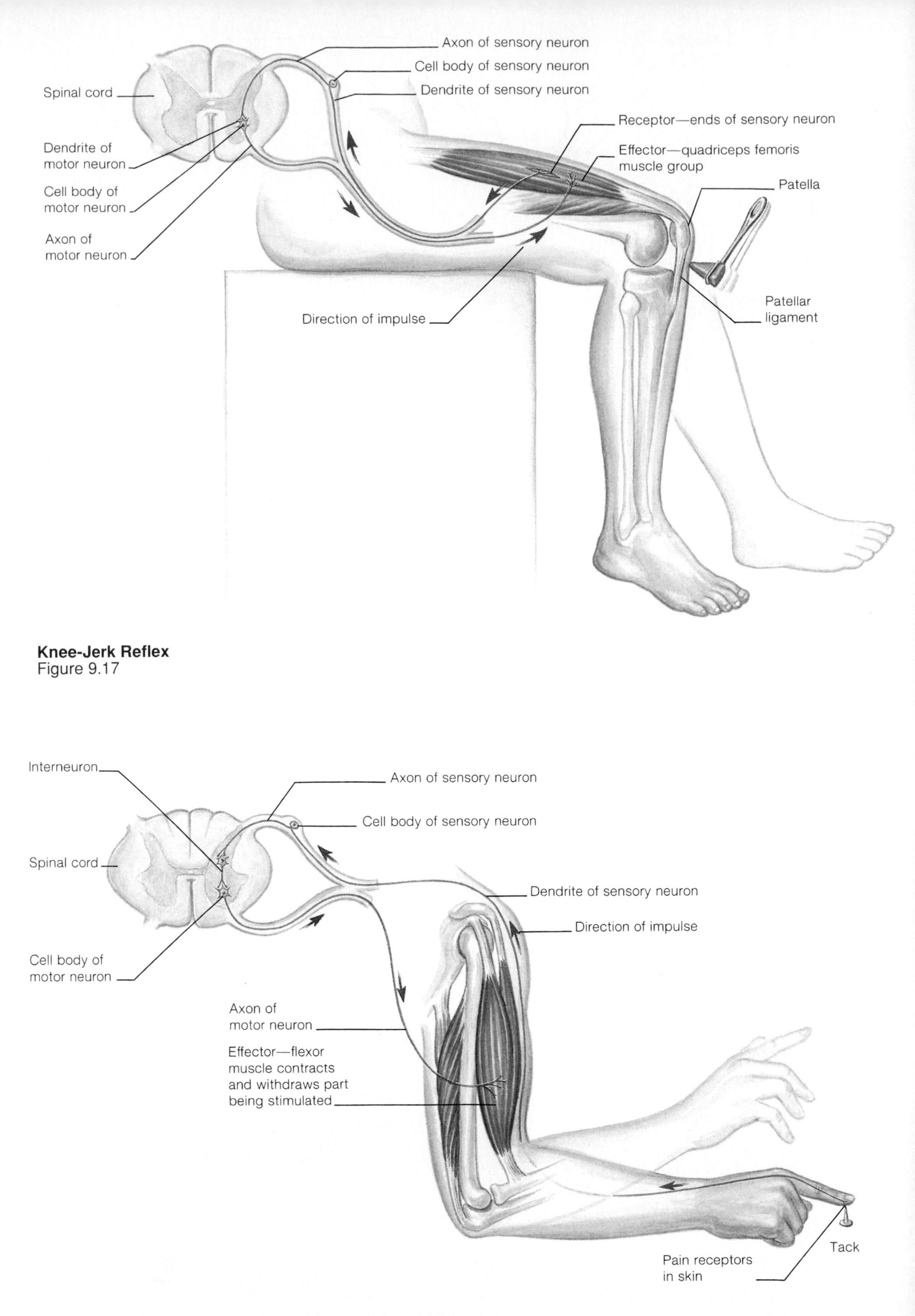
Axon of sensory neuron
Cell body of sensory neuron
Dendrite of sensory neuron
Spinal cord
Receptor—ends of sensory neuron
Dendrite of motor neuron
Effector—quadriceps femoris muscle group
Cell body of motor neuron
Patella
Axon of motor neuron
Patellar ligament
Direction of impulse

Knee-Jerk Reflex
Figure 9.17

Interneuron
Axon of sensory neuron
Cell body of sensory neuron
Spinal cord
Dendrite of sensory neuron
Direction of impulse
Cell body of motor neuron
Axon of motor neuron
Effector—flexor muscle contracts and withdraws part being stimulated
Tack
Pain receptors in skin

Withdrawal Reflex
Figure 9.18

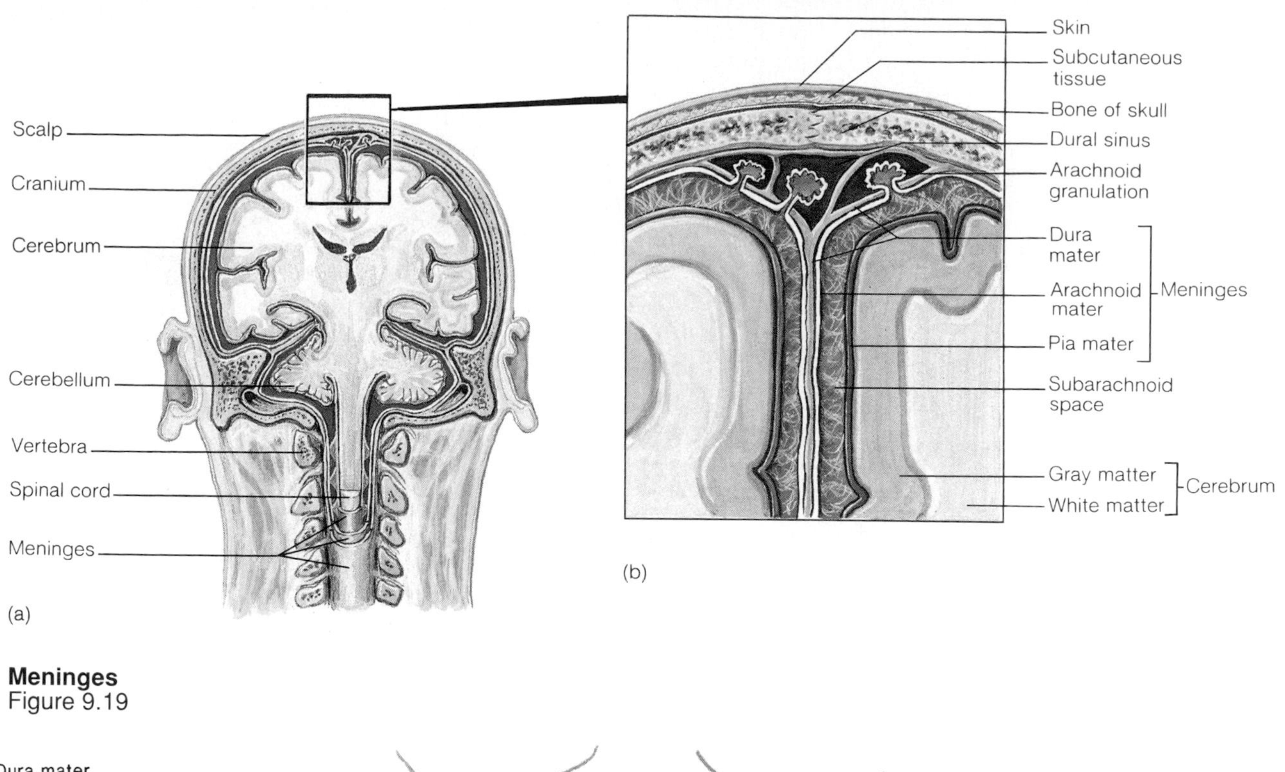
Scalp
Cranium
Cerebrum
Cerebellum
Vertebra
Spinal cord
Meninges
(a)
Skin
Subcutaneous tissue
Bone of skull
Dural sinus
Arachnoid granulation
Dura mater
Arachnoid mater
Pia mater
Meninges
Subarachnoid space
Gray matter
White matter
Cerebrum
(b)

Meninges
Figure 9.19

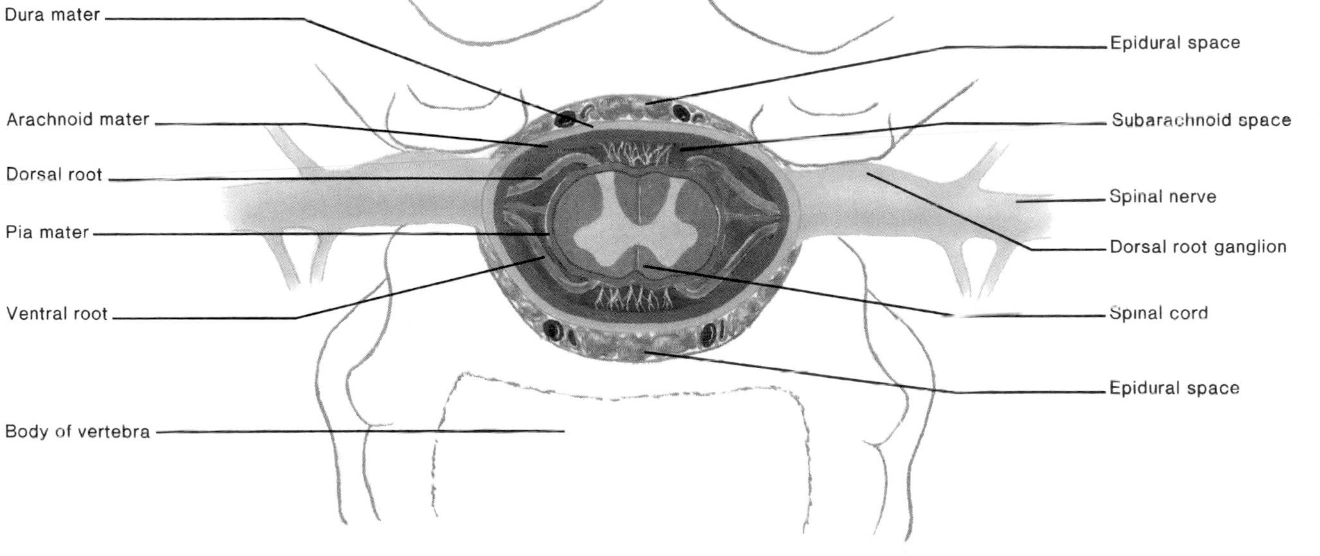
Dura mater
Arachnoid mater
Dorsal root
Pia mater
Ventral root
Body of vertebra
Epidural space
Subarachnoid space
Spinal nerve
Dorsal root ganglion
Spinal cord
Epidural space

Spinal Cord Section
Figure 9.20

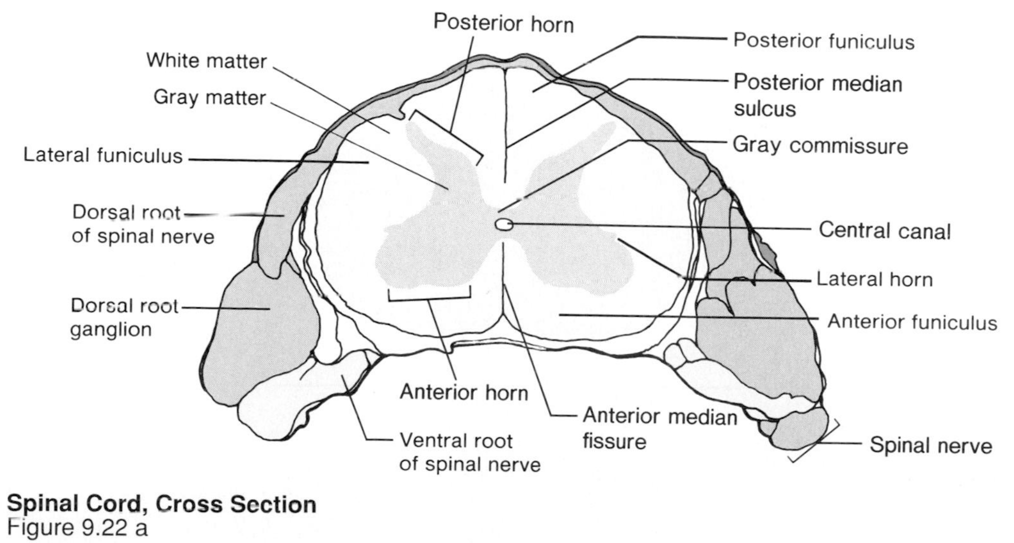
Posterior horn
White matter
Gray matter
Lateral funiculus
Dorsal root of spinal nerve
Dorsal root ganglion
Posterior funiculus
Posterior median sulcus
Gray commissure
Central canal
Lateral horn
Anterior funiculus
Anterior horn
Ventral root of spinal nerve
Anterior median fissure
Spinal nerve

Spinal Cord, Cross Section
Figure 9.22 a

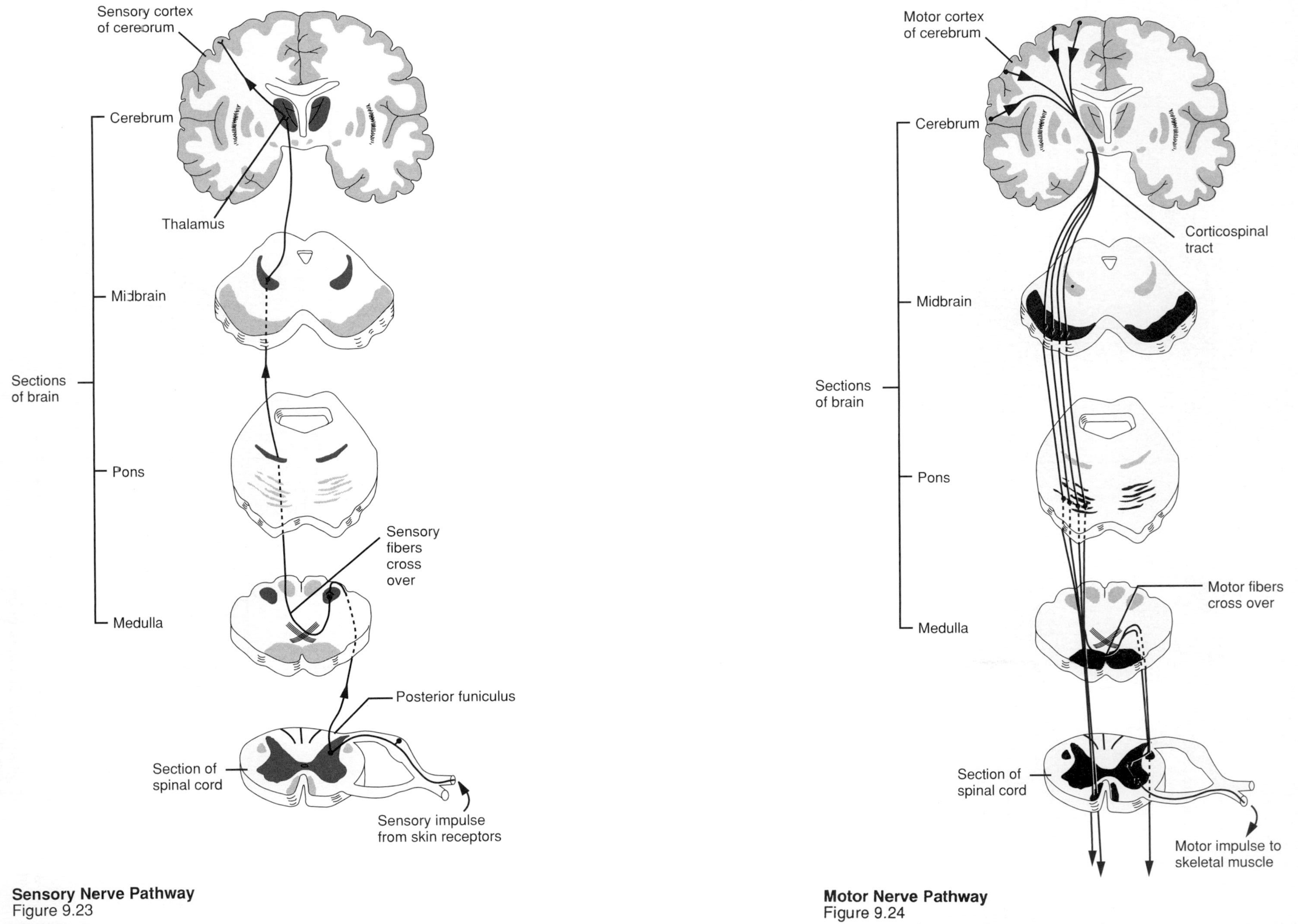
Sensory cortex of cerebrum
Cerebrum
Thalamus
Midbrain
Sections of brain
Pons
Sensory fibers cross over
Medulla
Posterior funiculus
Section of spinal cord
Sensory impulse from skin receptors
Motor cortex of cerebrum
Cerebrum
Corticospinal tract
Midbrain
Sections of brain
Pons
Motor fibers cross over
Medulla
Section of spinal cord
Motor impulse to skeletal muscle

Sensory Nerve Pathway
Figure 9.23

Motor Nerve Pathway
Figure 9.24

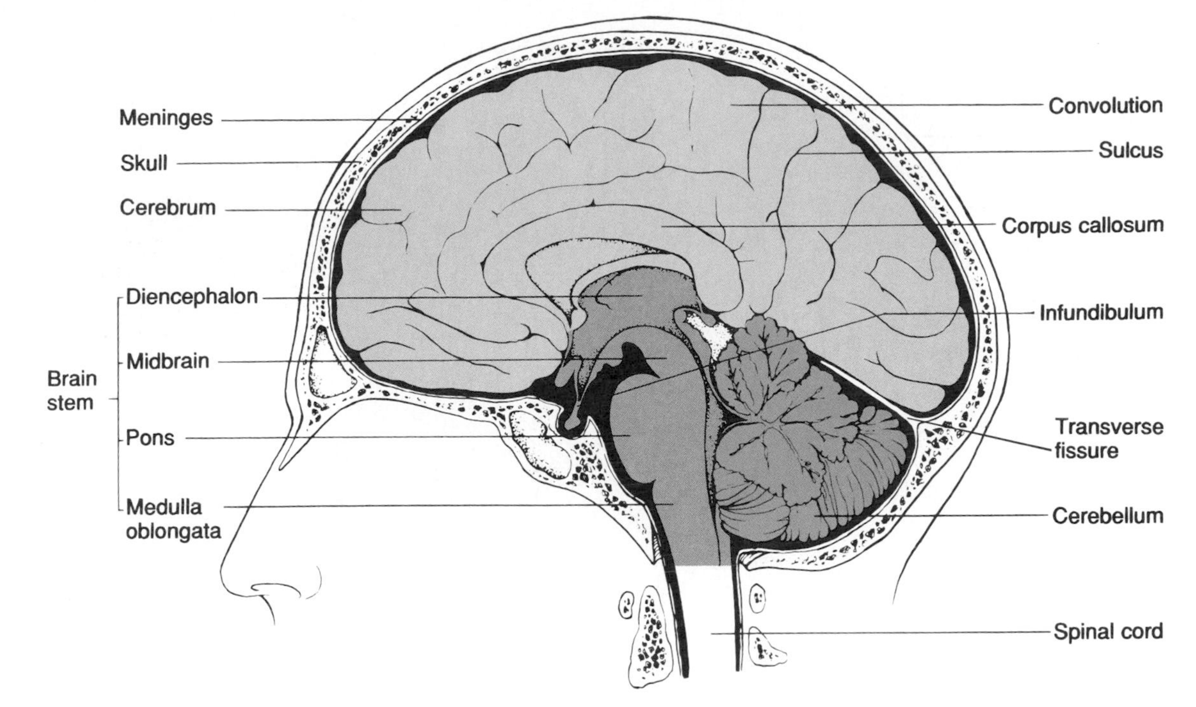
Meninges
Skull
Cerebrum
Diencephalon
Midbrain
Brain stem
Pons
Medulla oblongata
Convolution
Sulcus
Corpus callosum
Infundibulum
Transverse fissure
Cerebellum
Spinal cord

Brain, Sagittal Section
Figure 9.25

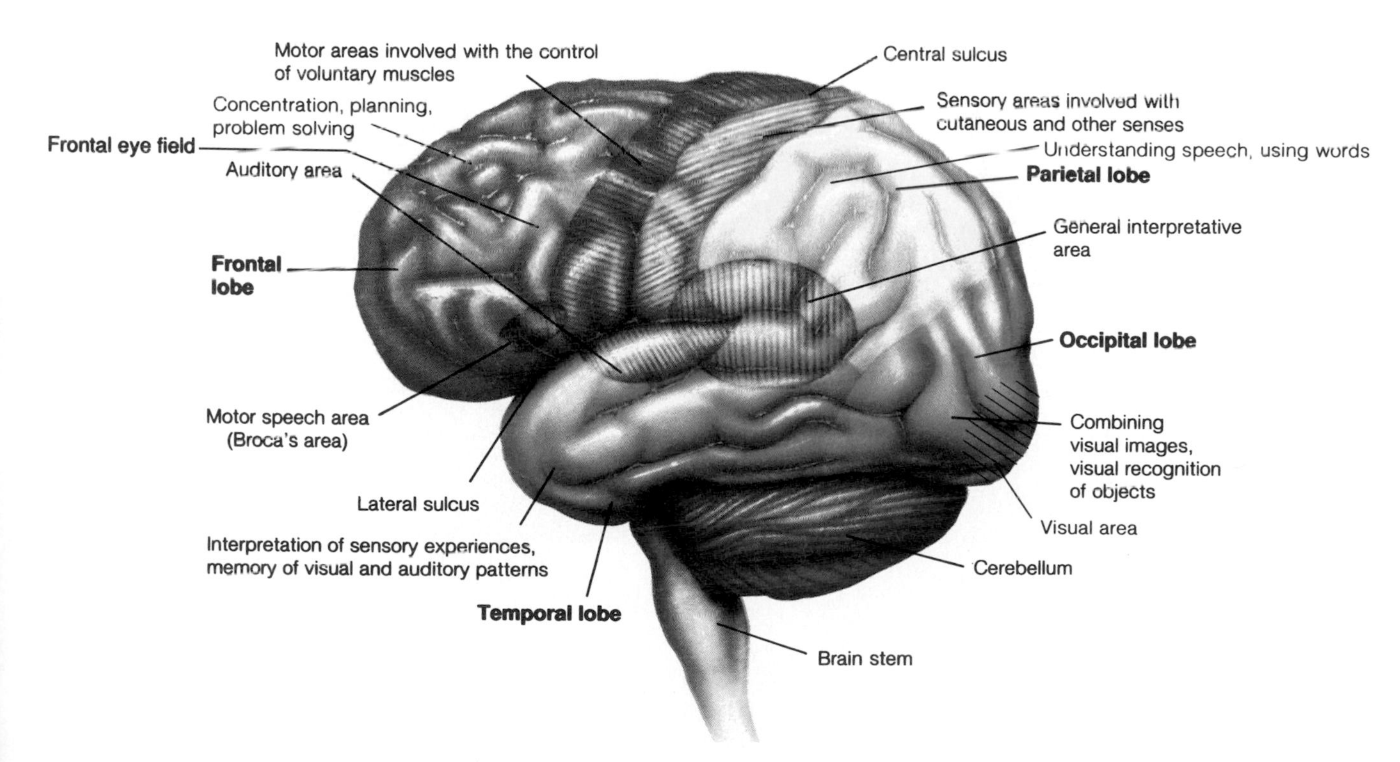
Motor areas involved with the control of voluntary muscles
Concentration, planning, problem solving
Frontal eye field
Auditory area
Frontal lobe
Motor speech area (Broca's area)
Lateral sulcus
Interpretation of sensory experiences, memory of visual and auditory patterns
Temporal lobe
Central sulcus
Sensory areas involved with cutaneous and other senses
Understanding speech, using words
Parietal lobe
General interpretative area
Occipital lobe
Combining visual images, visual recognition of objects
Visual area
Cerebellum
Brain stem

Sensory, Motor, and Association Areas
Figure 9.26

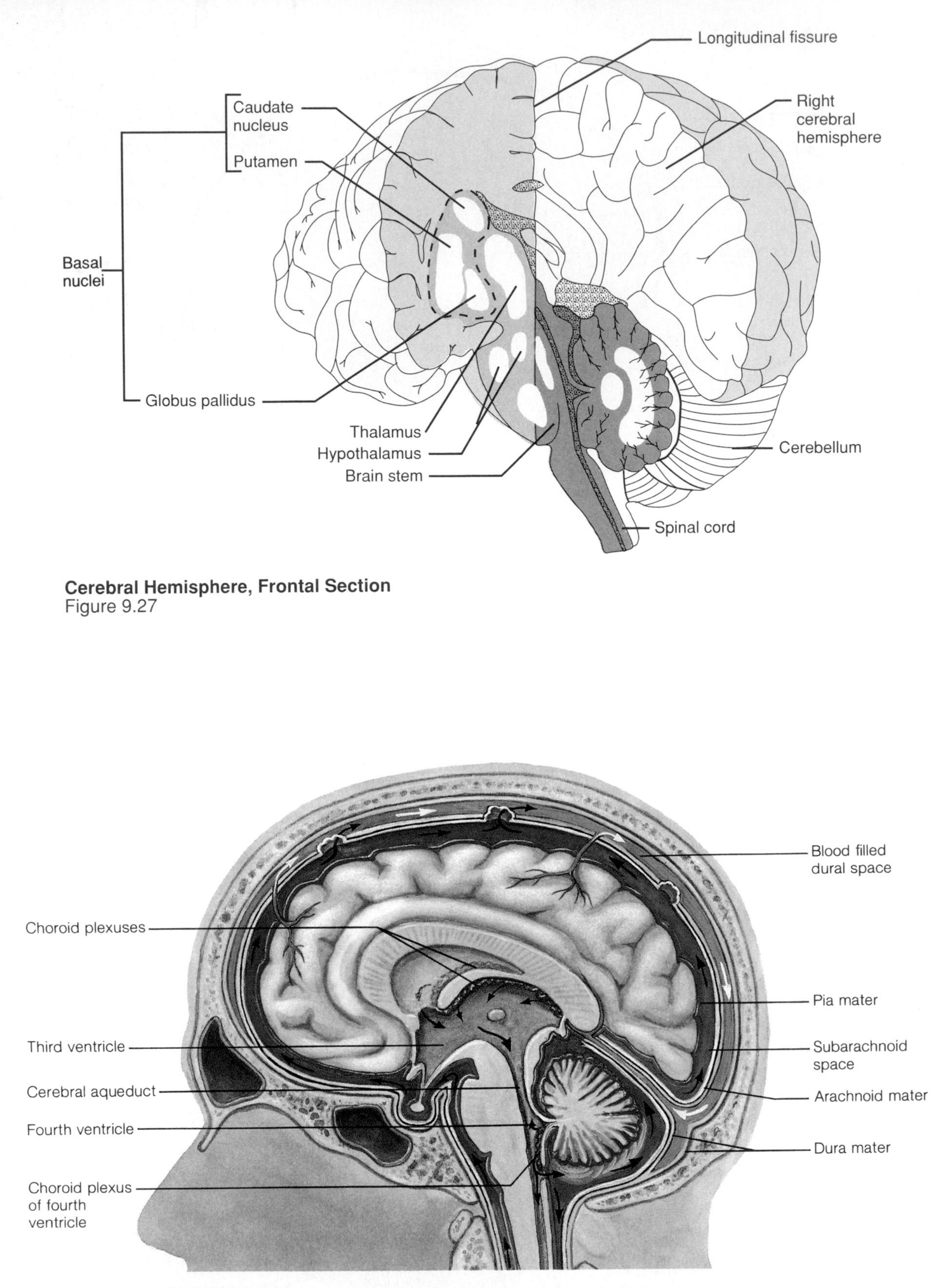
Longitudinal fissure
Caudate nucleus
Putamen
Right cerebral hemisphere
Basal nuclei
Globus pallidus
Thalamus
Hypothalamus
Brain stem
Cerebellum
Spinal cord
Blood filled dural space
Choroid plexuses
Pia mater
Third ventricle
Subarachnoid space
Cerebral aqueduct
Arachnoid mater
Fourth ventricle
Dura mater
Choroid plexus of fourth ventricle

Cerebral Hemisphere, Frontal Section
Figure 9.27

Cerebrospinal Fluid Circulation
Figure 9.29

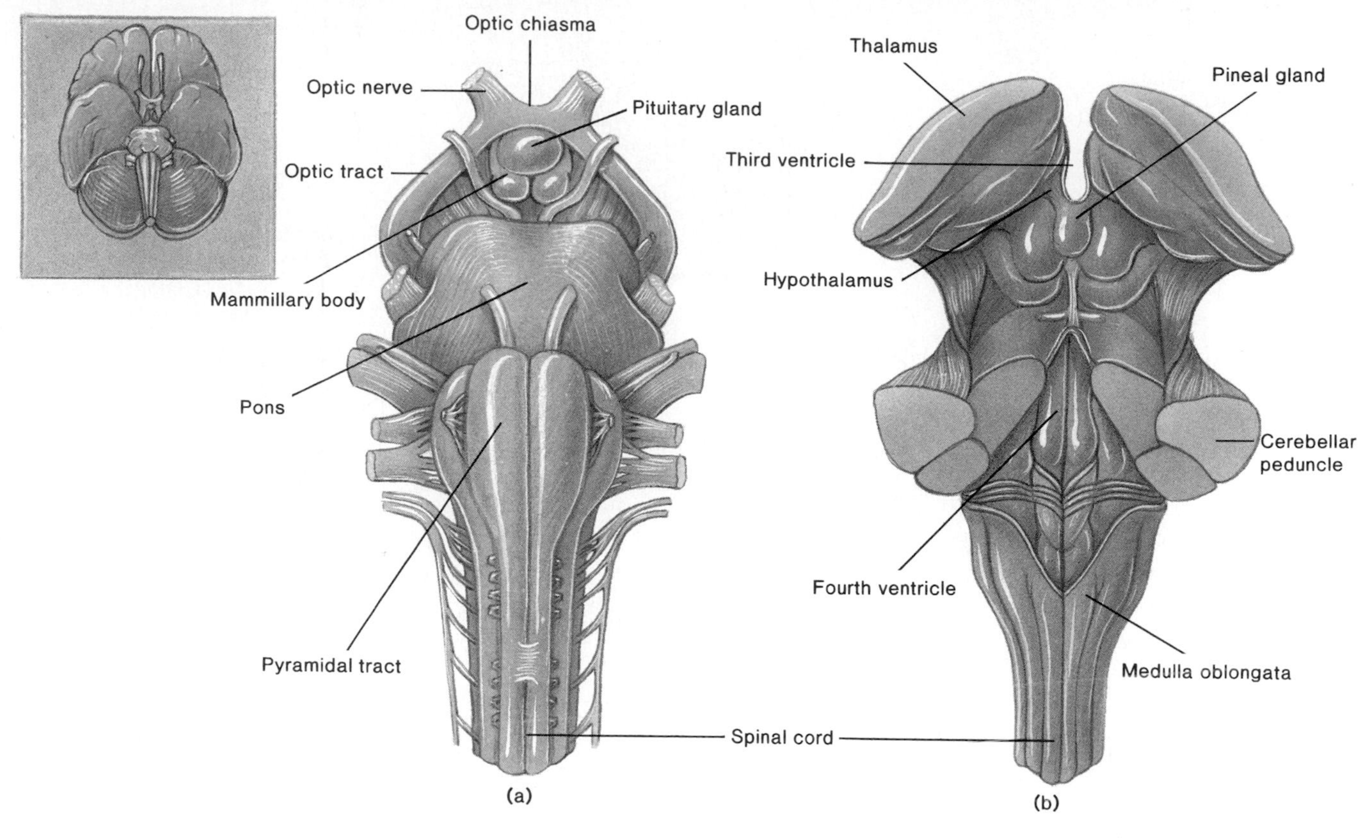

Optic chiasma
Optic nerve
Pituitary gland
Optic tract
Mammillary body
Pons
Pyramidal tract
Spinal cord
(a)
Thalamus
Pineal gland
Third ventricle
Hypothalamus
Cerebellar peduncle
Fourth ventricle
Medulla oblongata
(b)

Brain Stem
Figure 9.30

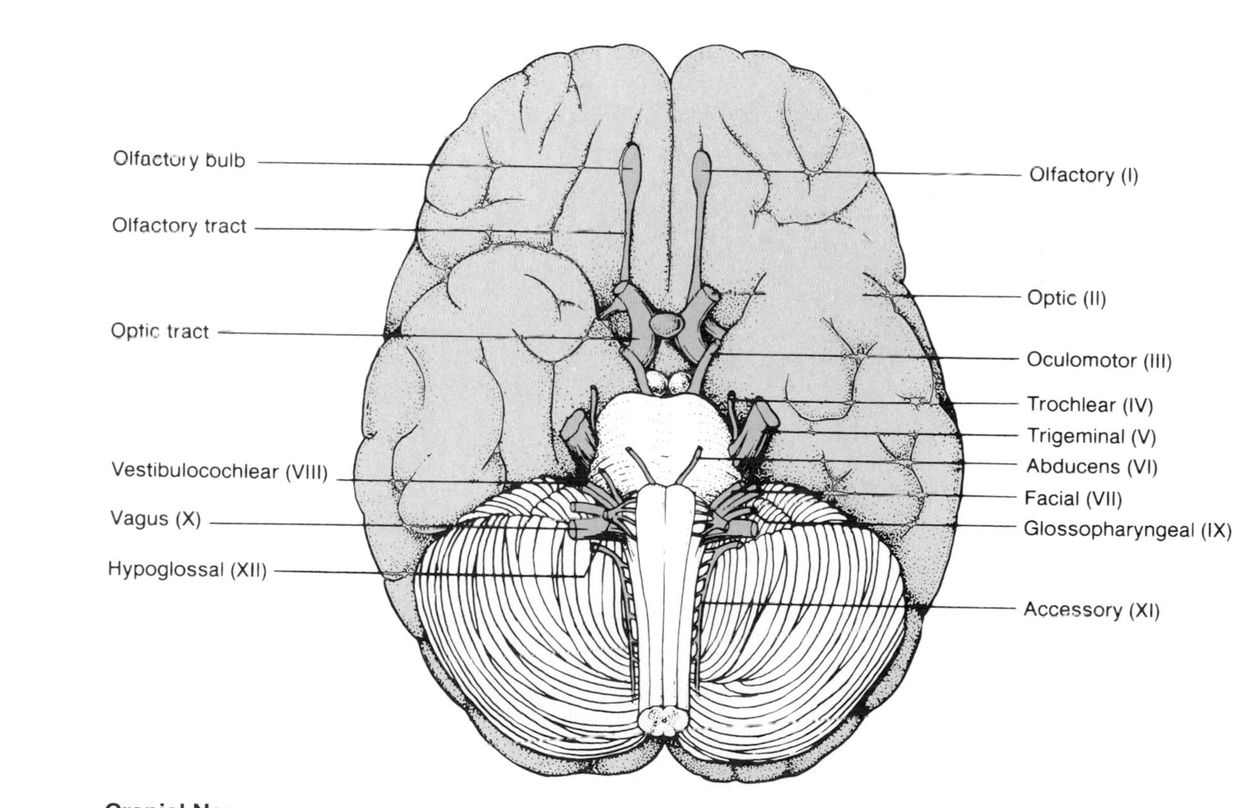

Olfactory bulb
Olfactory tract
Optic tract
Vestibulocochlear (VIII)
Vagus (X)
Hypoglossal (XII)
Olfactory (I)
Optic (II)
Oculomotor (III)
Trochlear (IV)
Trigeminal (V)
Abducens (VI)
Facial (VII)
Glossopharyngeal (IX)
Accessory (XI)

Cranial Nerves
Figure 9.31

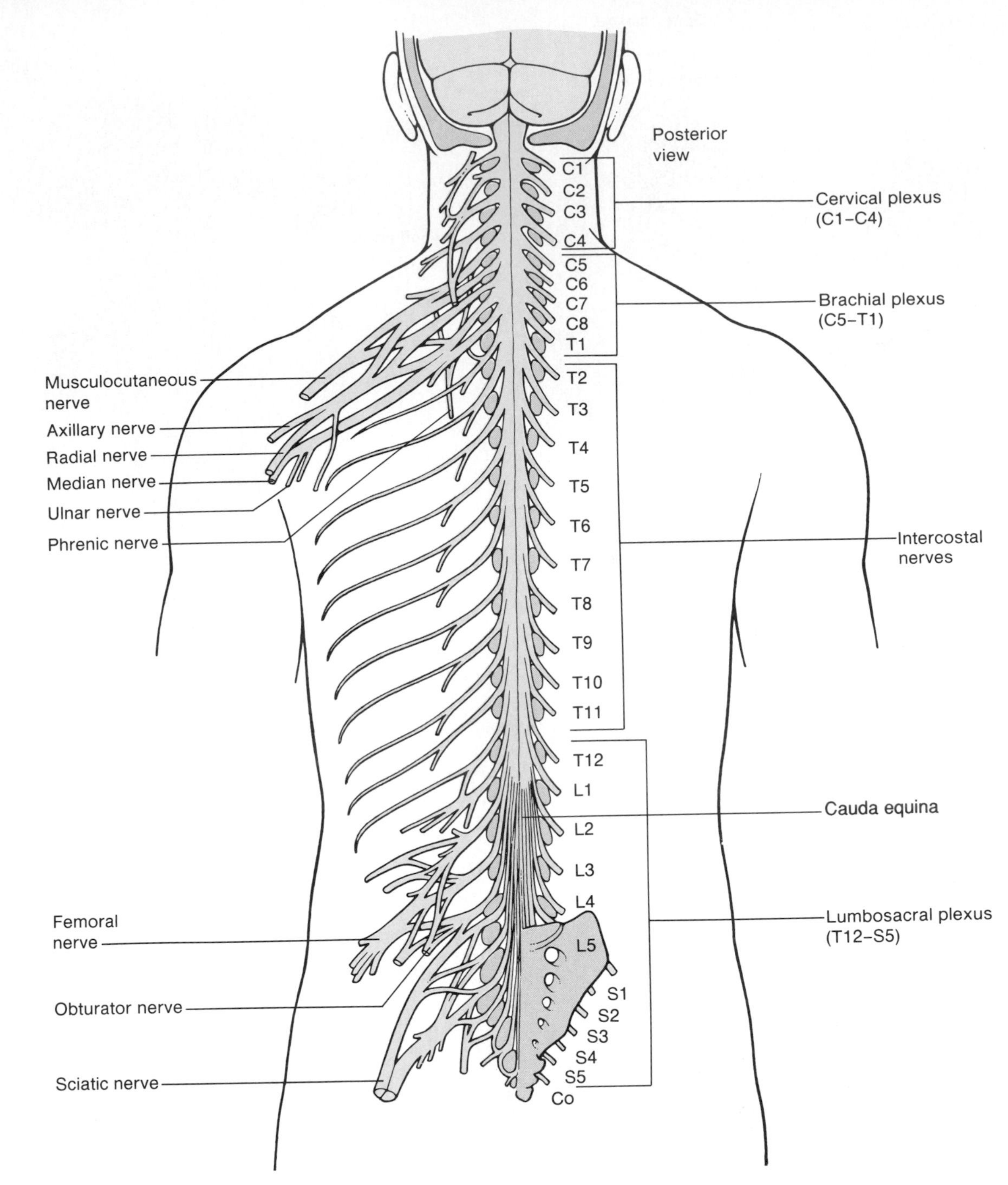
Posterior
view
C1
C2
C3
C4
C5
C6
C7
C8
T1
T2
T3
T4
T5
T6
T7
T8
T9
T10
T11
T12
L1
L2
L3
L4
L5
S1
S2
S3
S4
S5
Co
Cervical plexus
(C1–C4)
Brachial plexus
(C5–T1)
Musculocutaneous
nerve
Axillary nerve
Radial nerve
Median nerve
Ulnar nerve
Phrenic nerve
Intercostal
nerves
Cauda equina
Lumbosacral plexus
(T12–S5)
Femoral
nerve
Obturator nerve
Sciatic nerve

Spinal Nerves
Figure 9.32

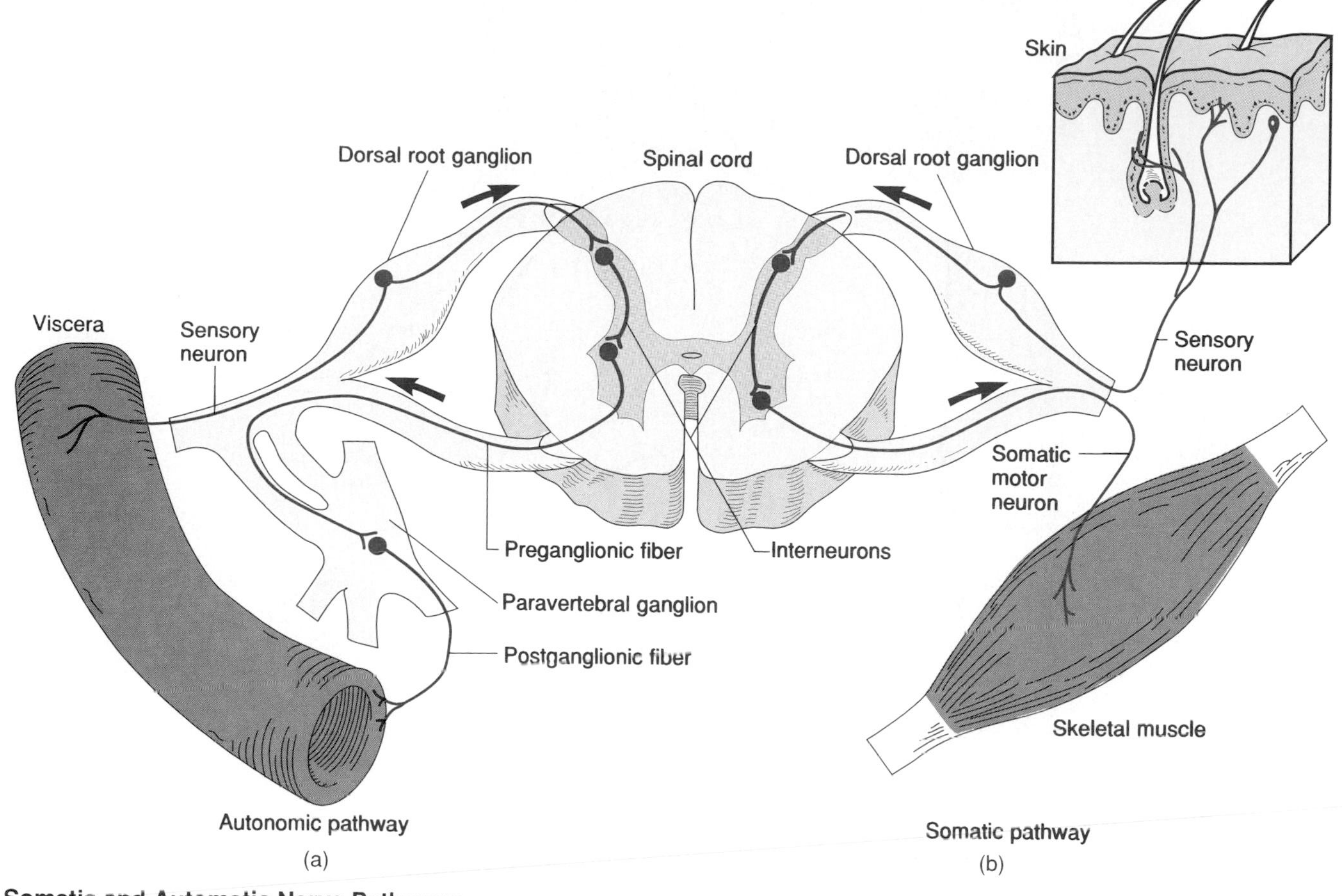

Skin
Dorsal root ganglion
Spinal cord
Dorsal root ganglion
Viscera
Sensory neuron
Sensory neuron
Somatic motor neuron
Preganglionic fiber
Interneurons
Paravertebral ganglion
Postganglionic fiber
Skeletal muscle
Autonomic pathway
(a)
Somatic pathway
(b)

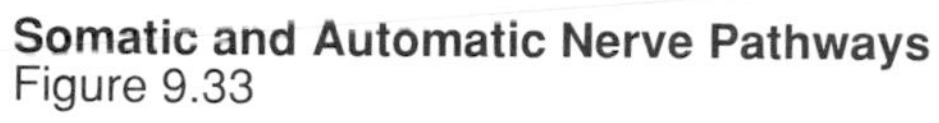

Somatic and Automatic Nerve Pathways
Figure 9.33

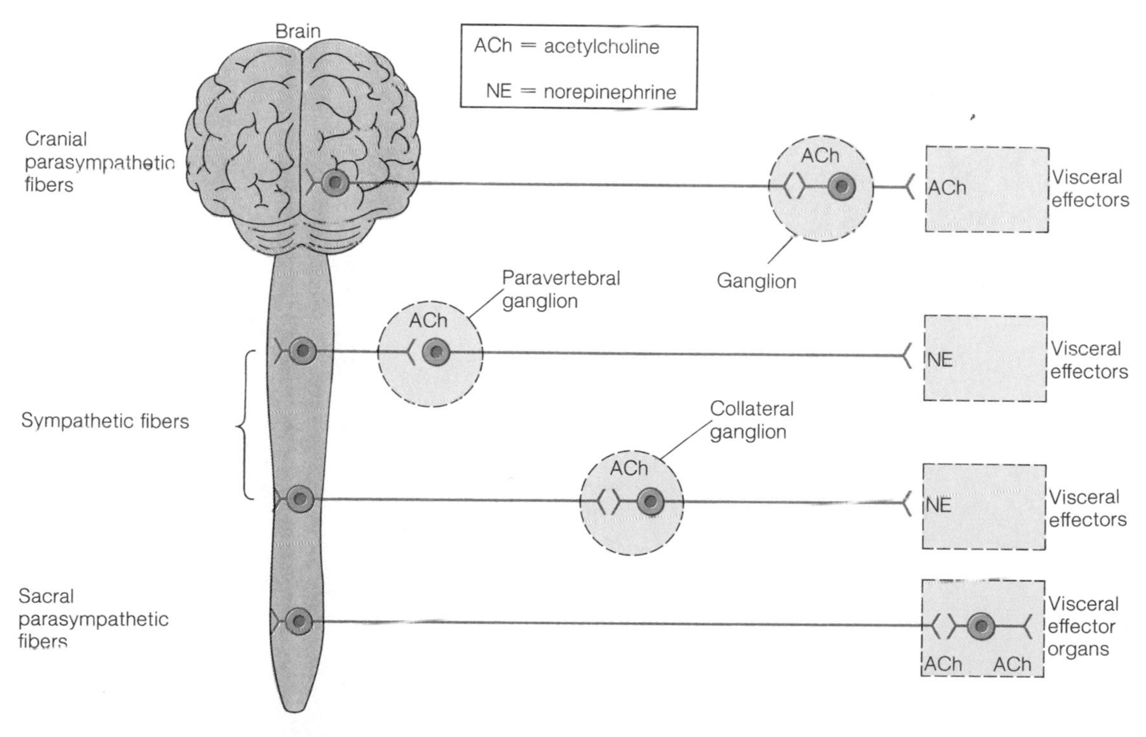

Brain
ACh = acetylcholine
NE = norepinephrine
Cranial parasympathetic fibers
ACh
ACh
Visceral effectors
Ganglion
Paravertebral ganglion
ACh
NE
Visceral effectors
Sympathetic fibers
Collateral ganglion
ACh
NE
Visceral effectors
Sacral parasympathetic fibers
ACh ACh
Visceral effector organs

Autonomic Neurotransmitters
Figure 9.36

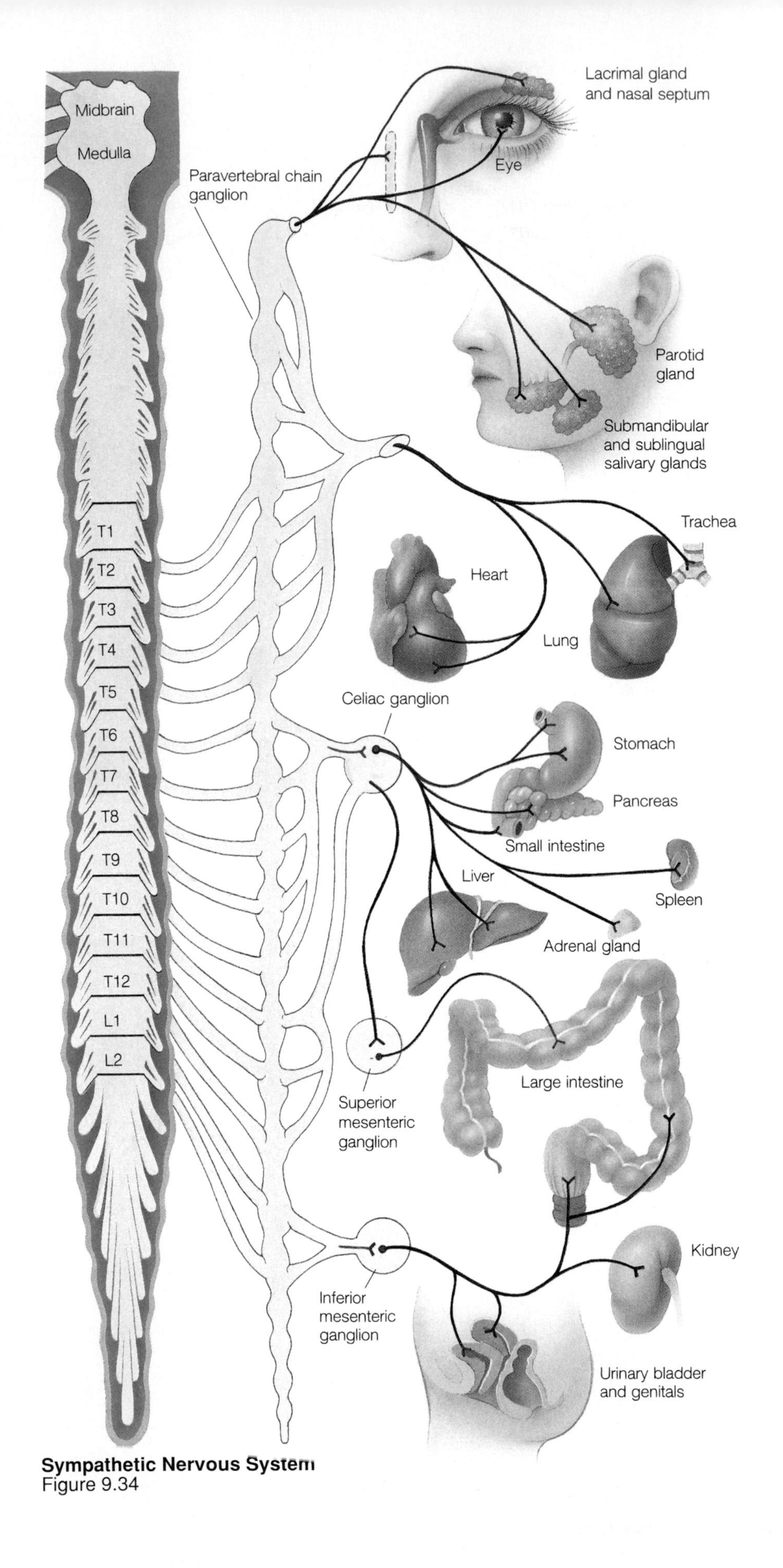
Midbrain
Medulla
Paravertebral chain ganglion
Lacrimal gland and nasal septum
Eye
Parotid gland
Submandibular and sublingual salivary glands
Trachea
Heart
Lung
T1
T2
T3
T4
T5
T6
T7
T8
T9
T10
T11
T12
L1
L2
Celiac ganglion
Stomach
Pancreas
Small intestine
Liver
Spleen
Adrenal gland
Large intestine
Superior mesenteric ganglion
Kidney
Inferior mesenteric ganglion
Urinary bladder and genitals

Sympathetic Nervous System
Figure 9.34

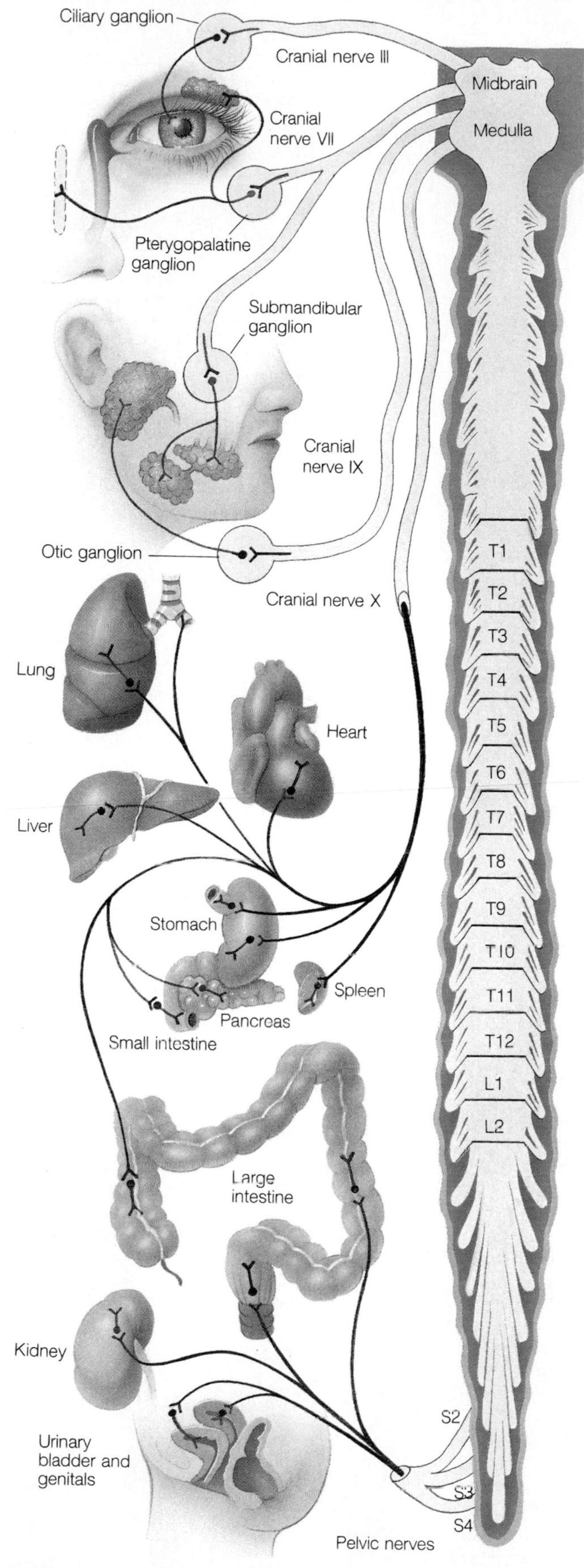
Ciliary ganglion
Cranial nerve III
Midbrain
Cranial nerve VII
Medulla
Pterygopalatine ganglion
Submandibular ganglion
Cranial nerve IX
Otic ganglion
Cranial nerve X
Lung
Heart
Liver
Stomach
Spleen
Pancreas
Small intestine
Large intestine
Kidney
Urinary bladder and genitals
Pelvic nerves
T1
T2
T3
T4
T5
T6
T7
T8
T9
T10
T11
T12
L1
L2
S2
S3
S4

Parasympathetic Nervous System
Figure 9.35

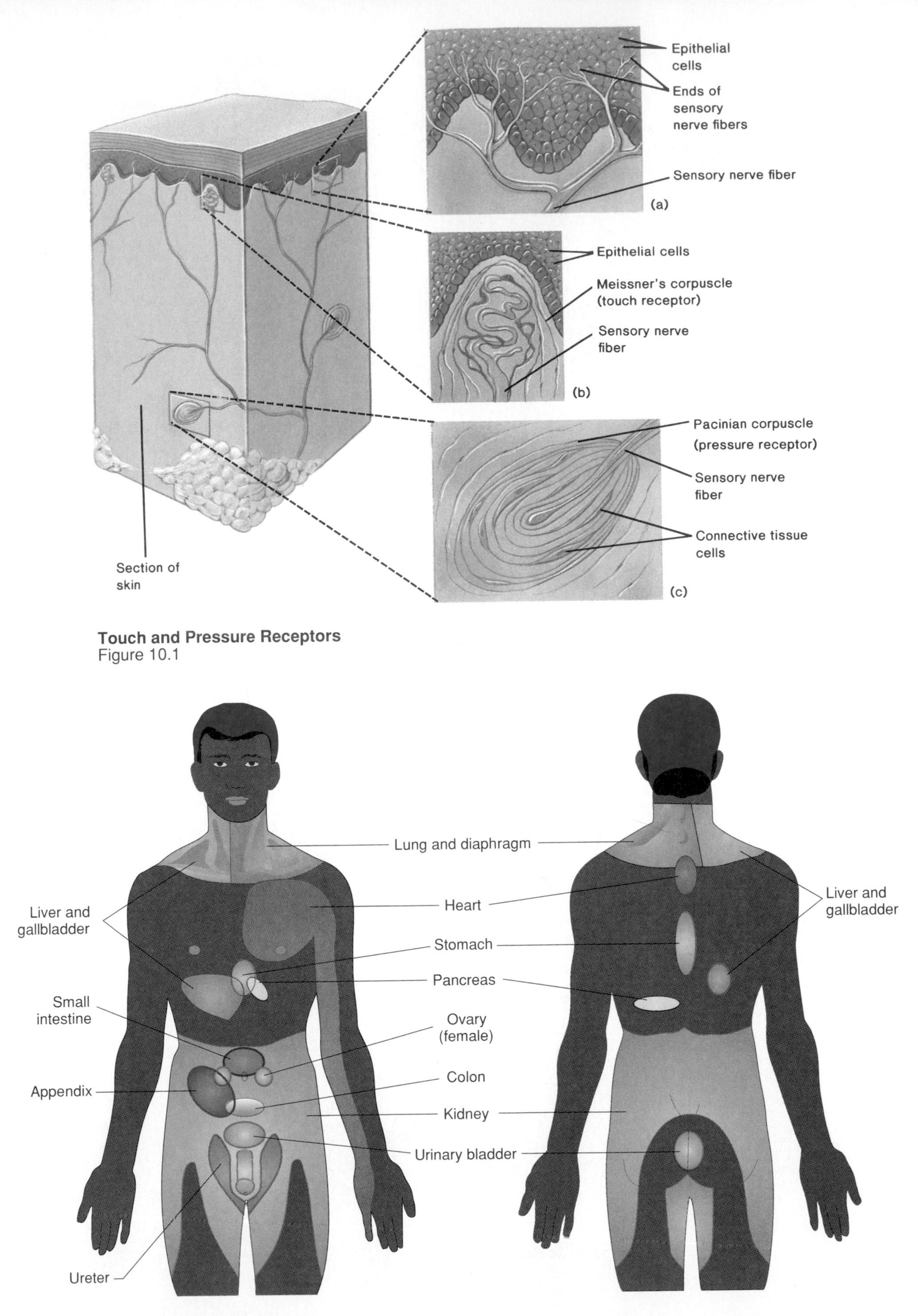
Epithelial cells
Ends of sensory nerve fibers
Sensory nerve fiber
(a)
Epithelial cells
Meissner's corpuscle (touch receptor)
Sensory nerve fiber
(b)
Pacinian corpuscle (pressure receptor)
Sensory nerve fiber
Connective tissue cells
(c)
Section of skin

Touch and Pressure Receptors
Figure 10.1

Lung and diaphragm
Liver and gallbladder
Heart
Stomach
Pancreas
Small intestine
Ovary (female)
Appendix
Colon
Kidney
Urinary bladder
Ureter
Liver and gallbladder

Surface Regions of Visceral Pain
Figure 10.2

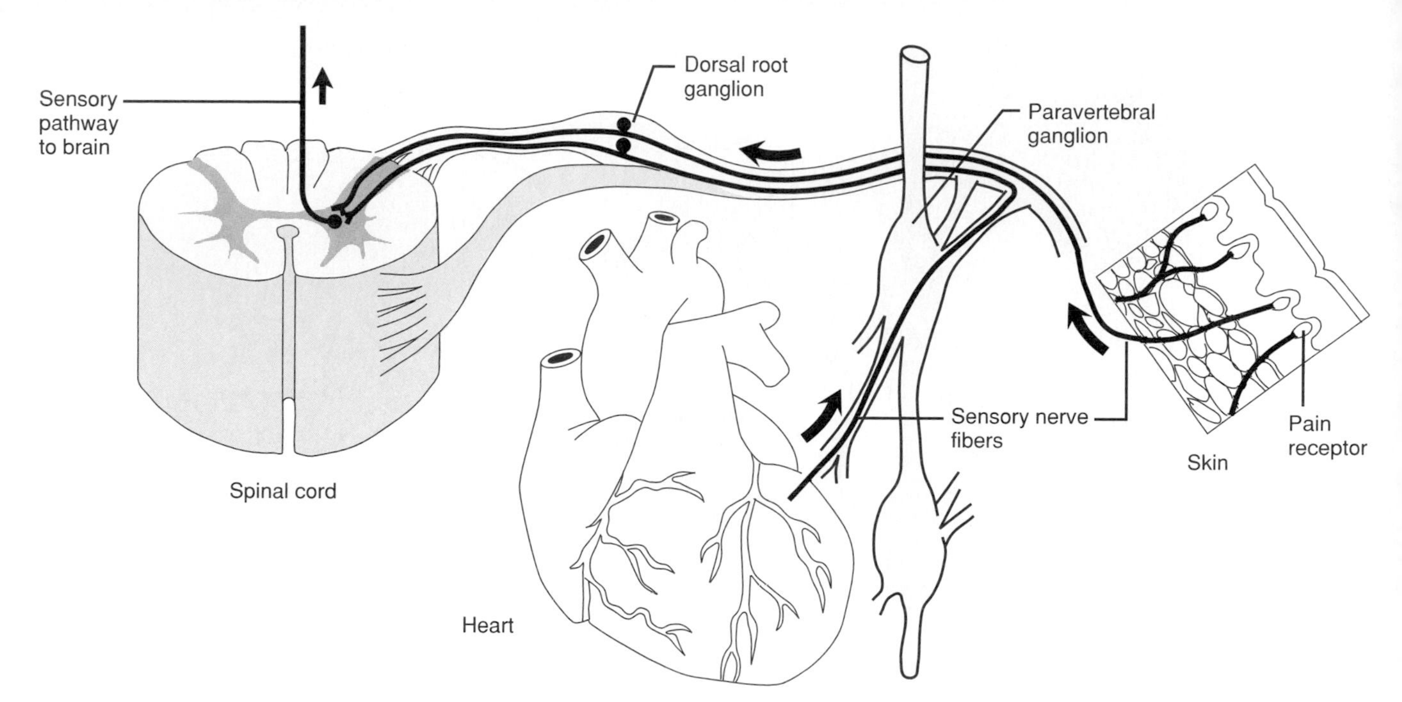
Sensory pathway to brain
Dorsal root ganglion
Paravertebral ganglion
Spinal cord
Heart
Sensory nerve fibers
Skin
Pain receptor

Referred Pain Pathway
Figure 10.3

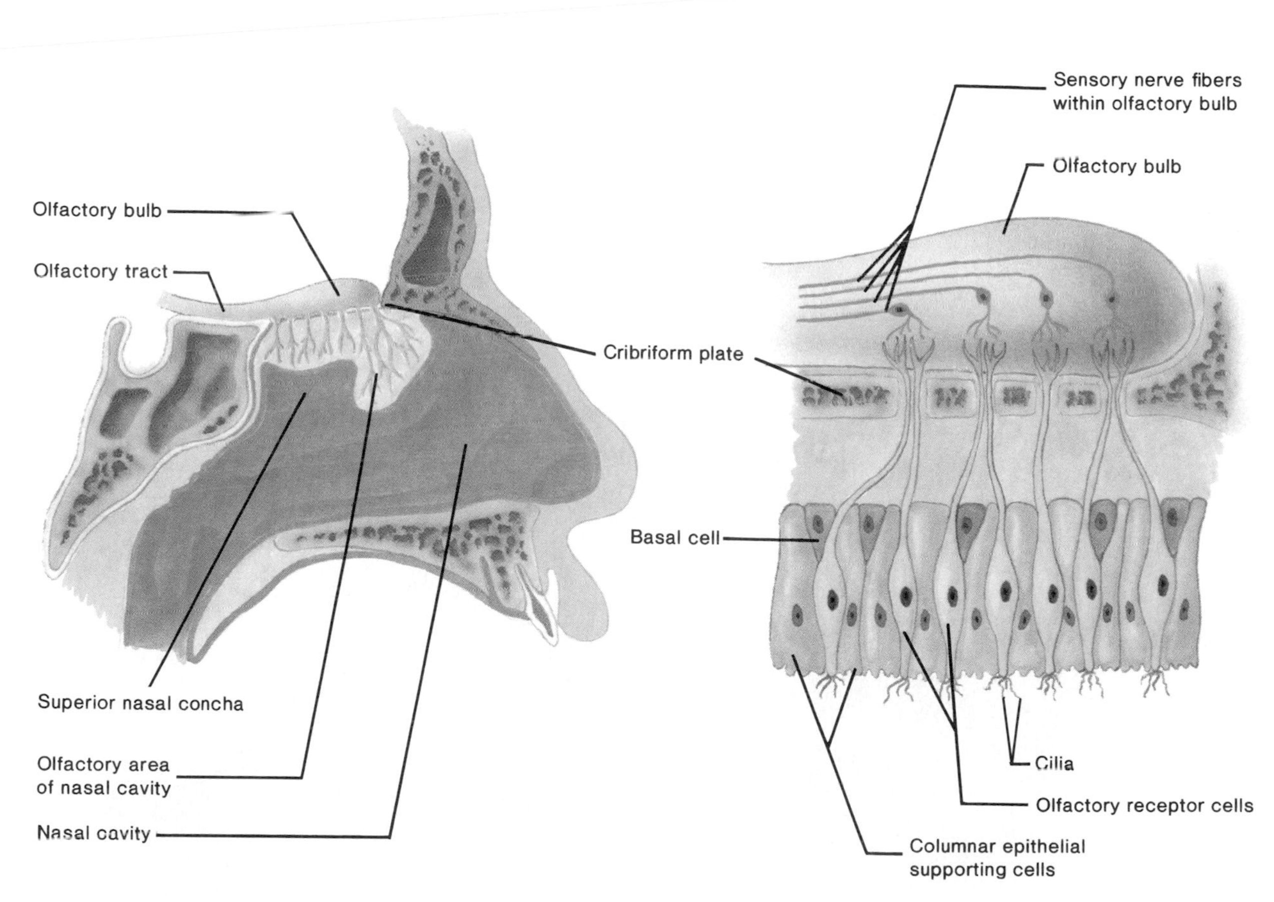
Olfactory bulb
Olfactory tract
Cribriform plate
Superior nasal concha
Olfactory area of nasal cavity
Nasal cavity
Sensory nerve fibers within olfactory bulb
Olfactory bulb
Basal cell
Cilia
Olfactory receptor cells
Columnar epithelial supporting cells

Olfactory Receptors
Figure 10.4

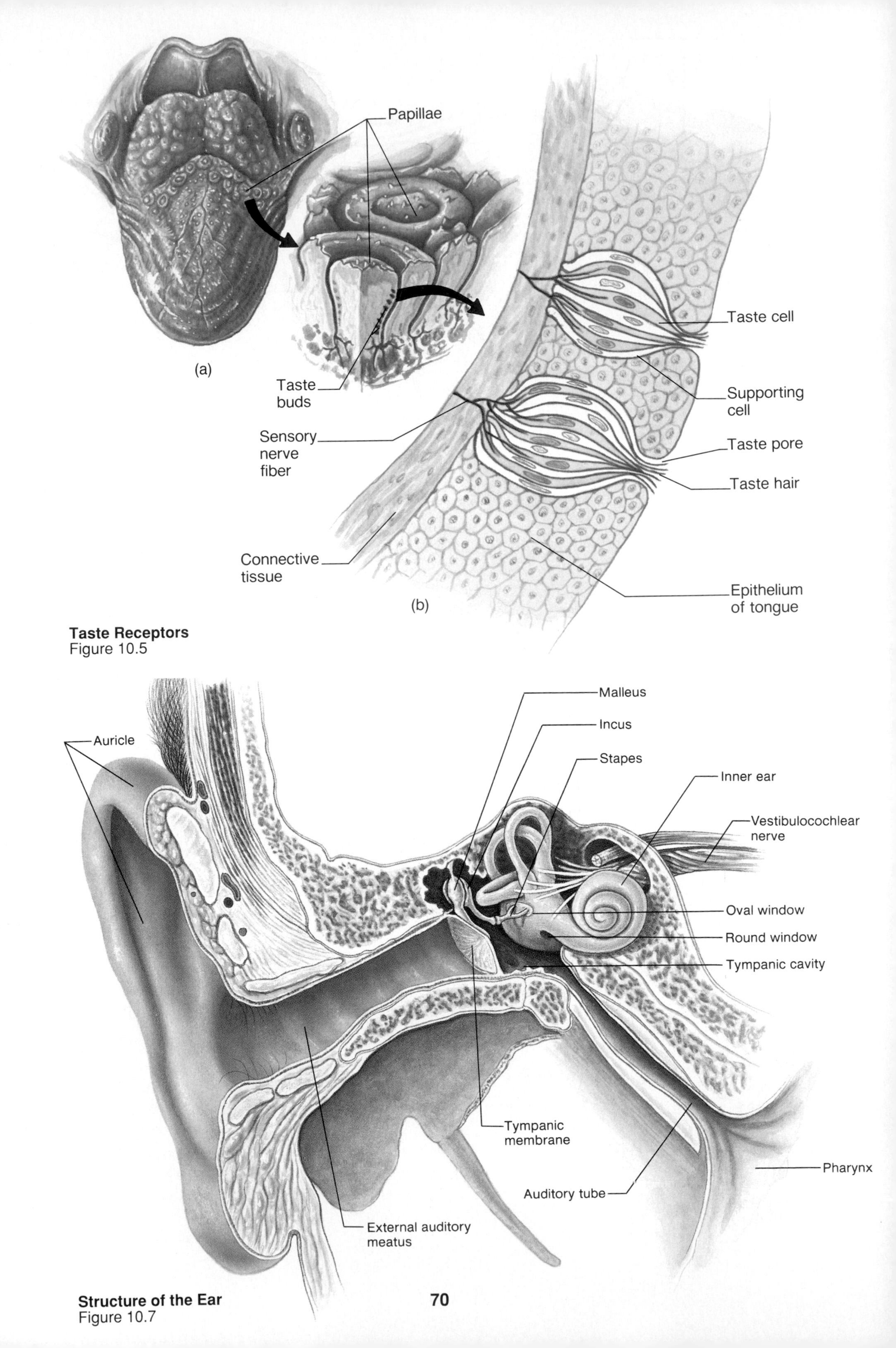
Papillae
(a)
Taste buds
Taste cell
Supporting cell
Sensory nerve fiber
Taste pore
Taste hair
Connective tissue
(b)
Epithelium of tongue

Taste Receptors
Figure 10.5

Malleus
Incus
Auricle
Stapes
Inner ear
Vestibulocochlear nerve
Oval window
Round window
Tympanic cavity
Tympanic membrane
Pharynx
Auditory tube
External auditory meatus

Structure of the Ear
Figure 10.7

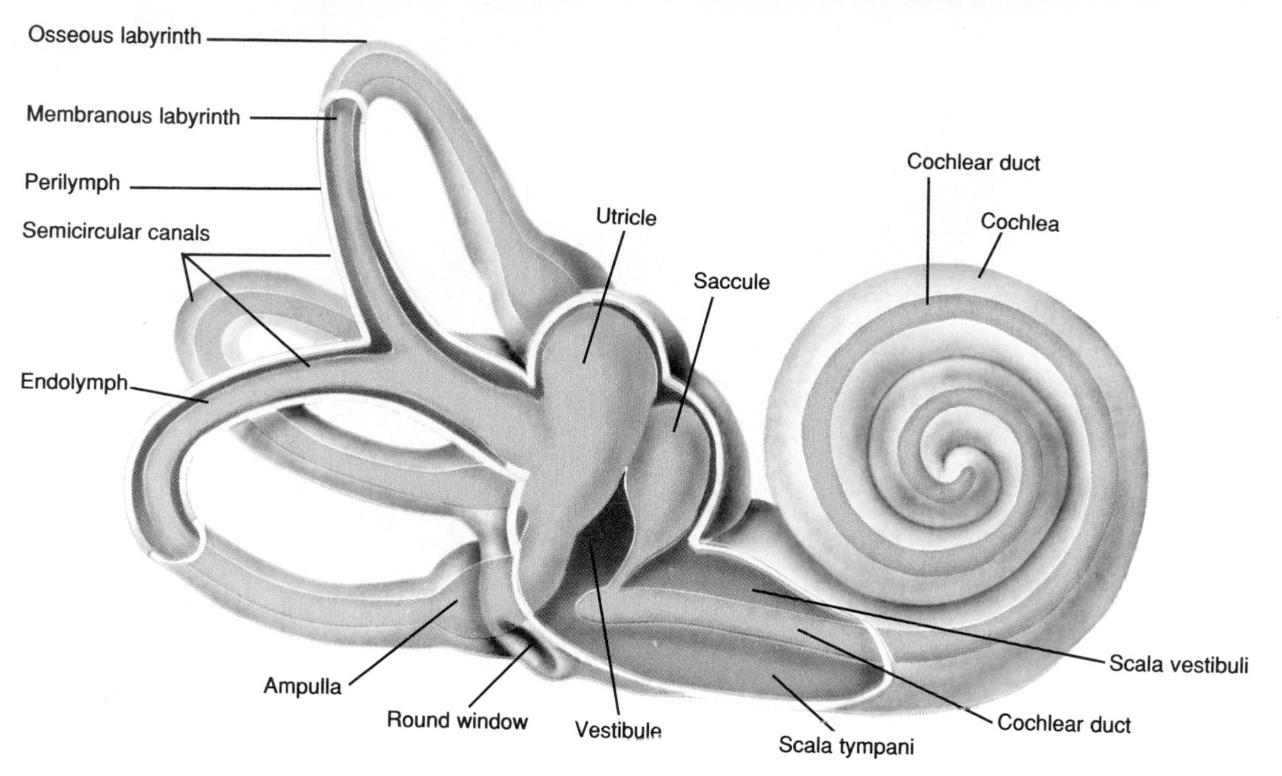
Osseous labyrinth
Membranous labyrinth
Perilymph
Semicircular canals
Endolymph
Utricle
Saccule
Cochlear duct
Cochlea
Ampulla
Round window
Vestibule
Scala tympani
Cochlear duct
Scala vestibuli

Inner Ear Structure
Figure 10.9

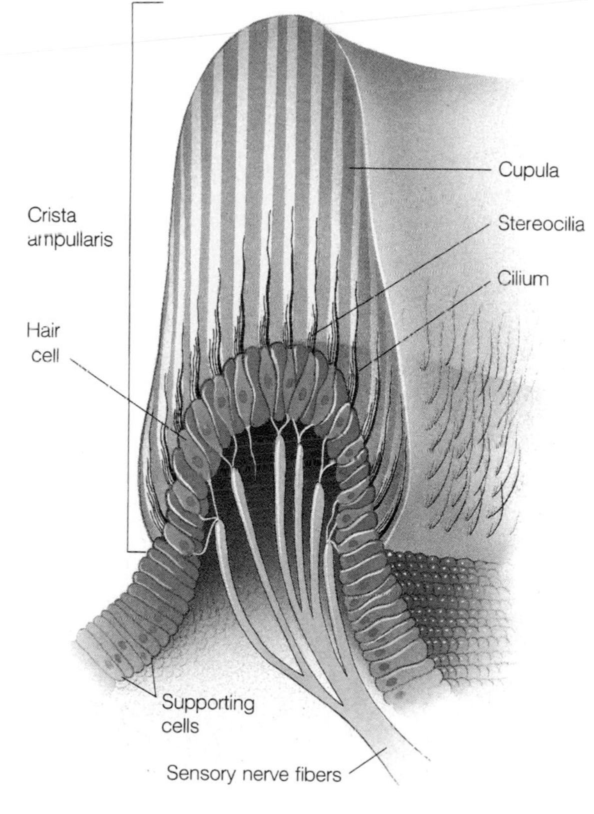
Cupula
Stereocilia
Cilium
Crista ampullaris
Hair cell
Supporting cells
Sensory nerve fibers

Crista Ampullaris
Figure 10.13

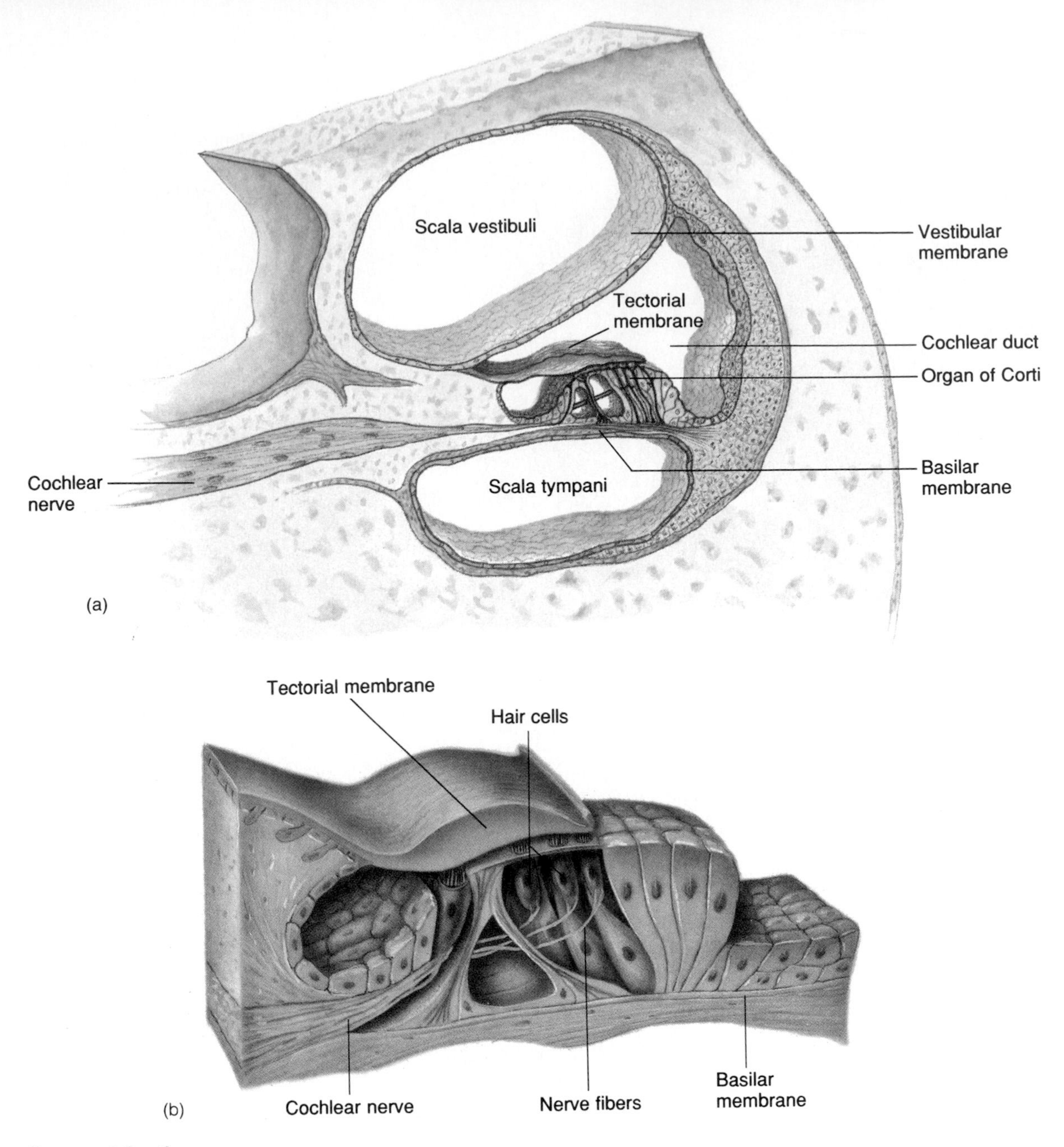
Scala vestibuli
Vestibular membrane
Tectorial membrane
Cochlear duct
Organ of Corti
Cochlear nerve
Basilar membrane
Scala tympani
(a)
Tectorial membrane
Hair cells
Cochlear nerve
Nerve fibers
Basilar membrane
(b)

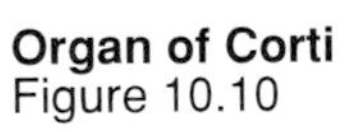

Organ of Corti
Figure 10.10

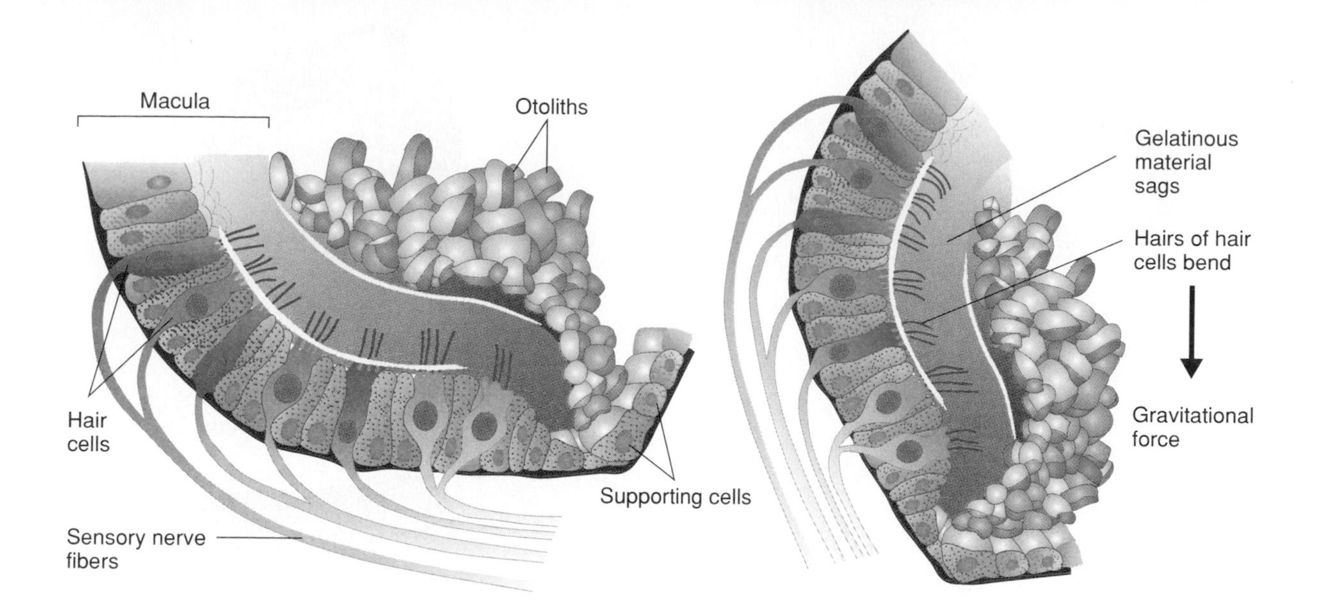
Macula
Otoliths
Hair cells
Sensory nerve fibers
Supporting cells
(a)
Gelatinous material sags
Hairs of hair cells bend
Gravitational force
(b)

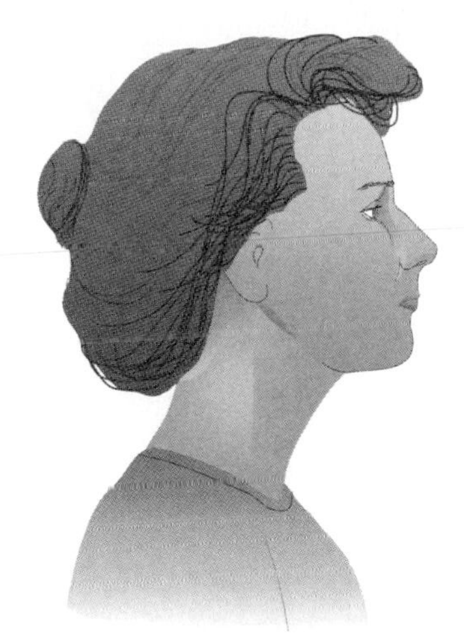

Head upright

Head bent forward

Static Equilibrium
Figure 10.12

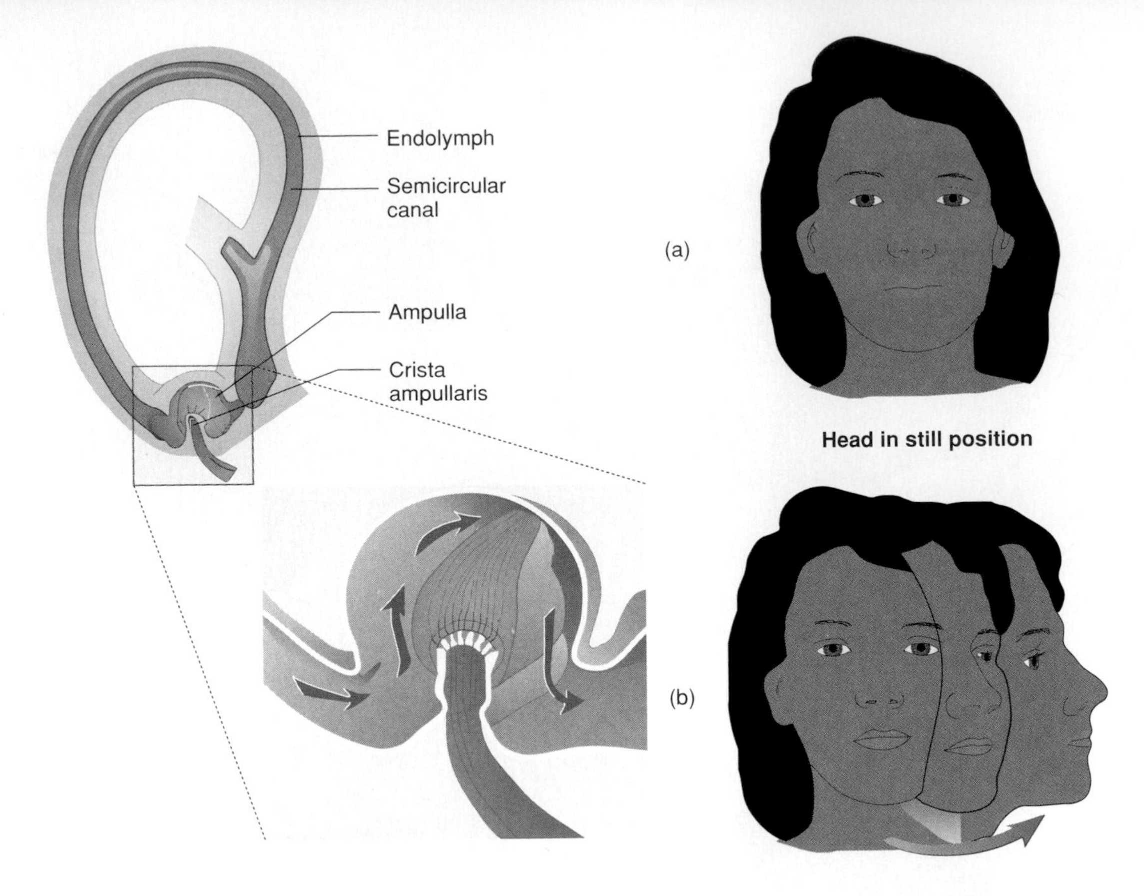
Endolymph
Semicircular canal
Ampulla
Crista ampullaris
(a)
Head in still position
(b)
Head rotating

Dynamic Equilibrium
Figure 10.14

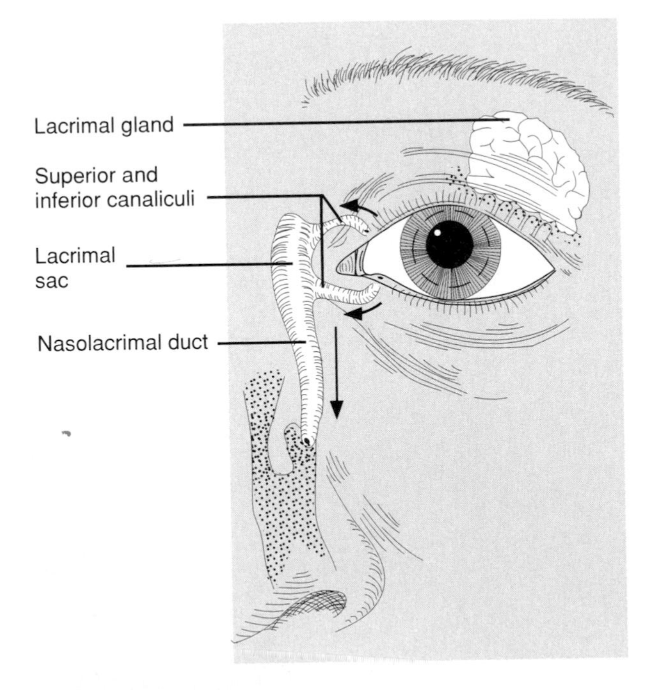
Lacrimal gland
Superior and inferior canaliculi
Lacrimal sac
Nasolacrimal duct

Lacrimal Apparatus
Figure 10.16

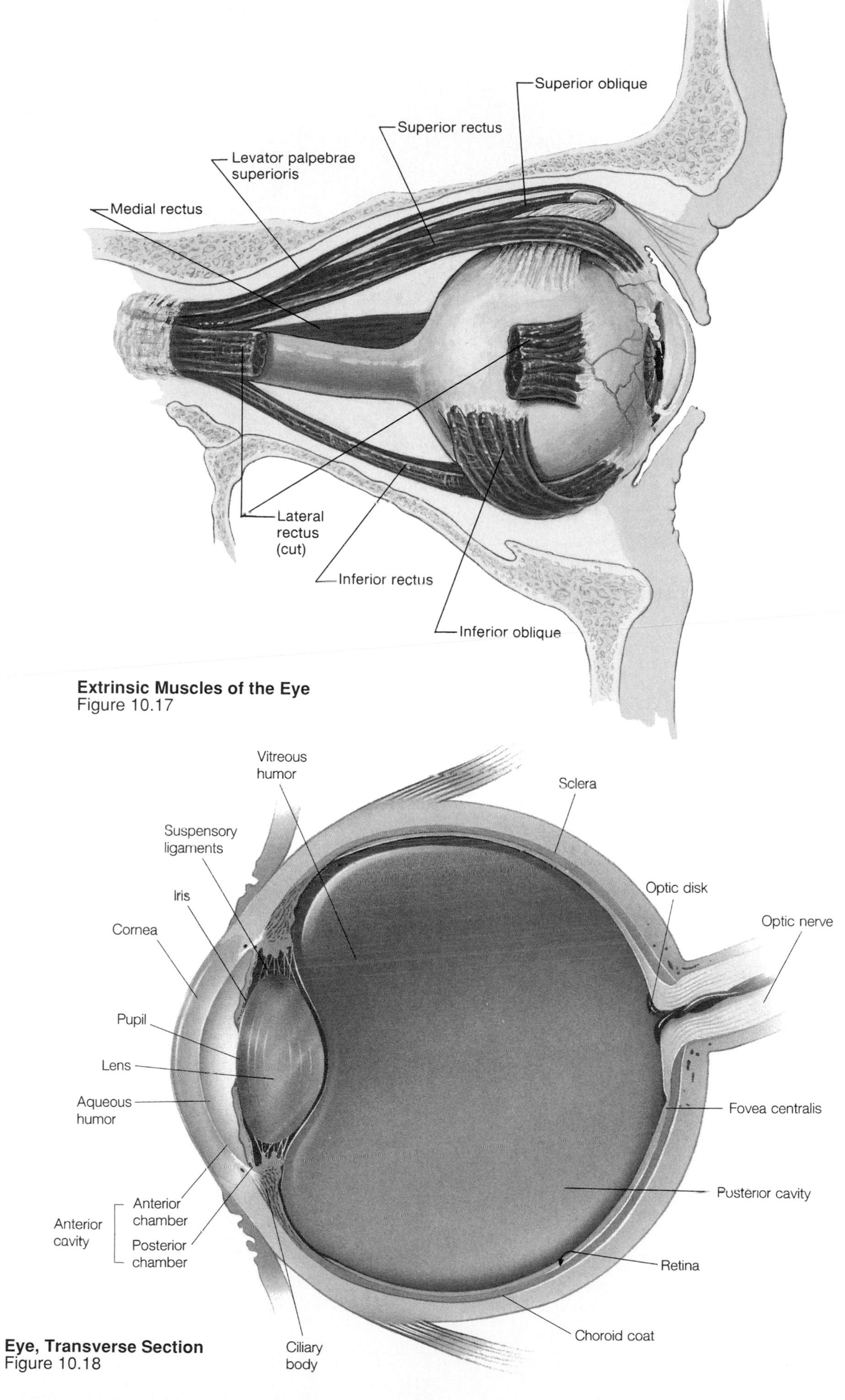
Superior oblique
Superior rectus
Levator palpebrae superioris
Medial rectus
Lateral rectus (cut)
Inferior rectus
Inferior oblique

Extrinsic Muscles of the Eye
Figure 10.17

Vitreous humor
Sclera
Suspensory ligaments
Optic disk
Iris
Optic nerve
Cornea
Pupil
Lens
Aqueous humor
Fovea centralis
Anterior cavity
Anterior chamber
Posterior chamber
Posterior cavity
Retina
Choroid coat
Ciliary body

Eye, Transverse Section
Figure 10.18

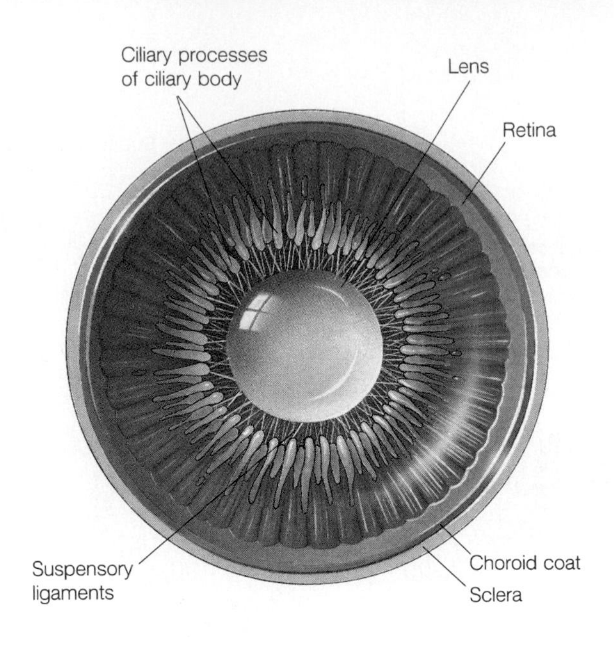
Ciliary processes
of ciliary body
Lens
Retina
Suspensory
ligaments
Choroid coat
Sclera

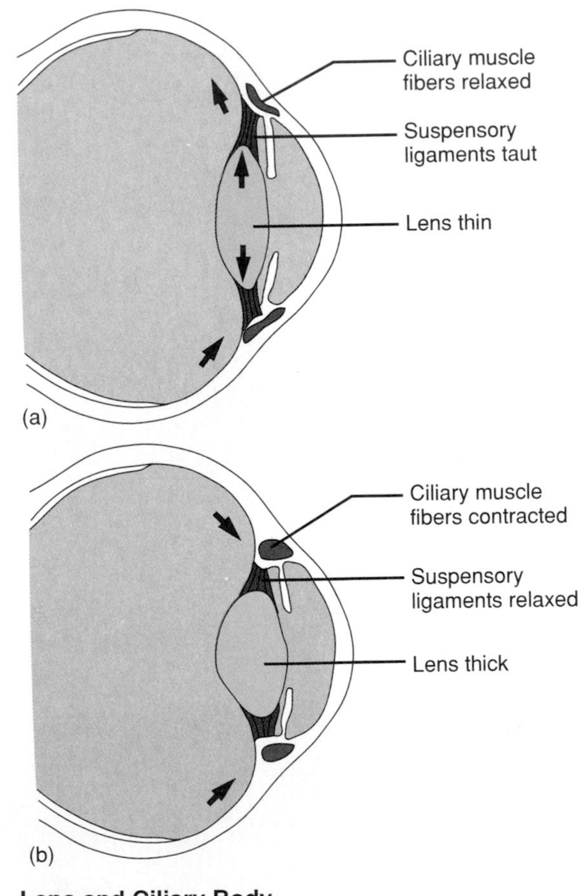
Ciliary muscle
fibers relaxed
Suspensory
ligaments taut
Lens thin
(a)
Ciliary muscle
fibers contracted
Suspensory
ligaments relaxed
Lens thick
(b)

Lens and Ciliary Body
Figures 10.19 and 10.20

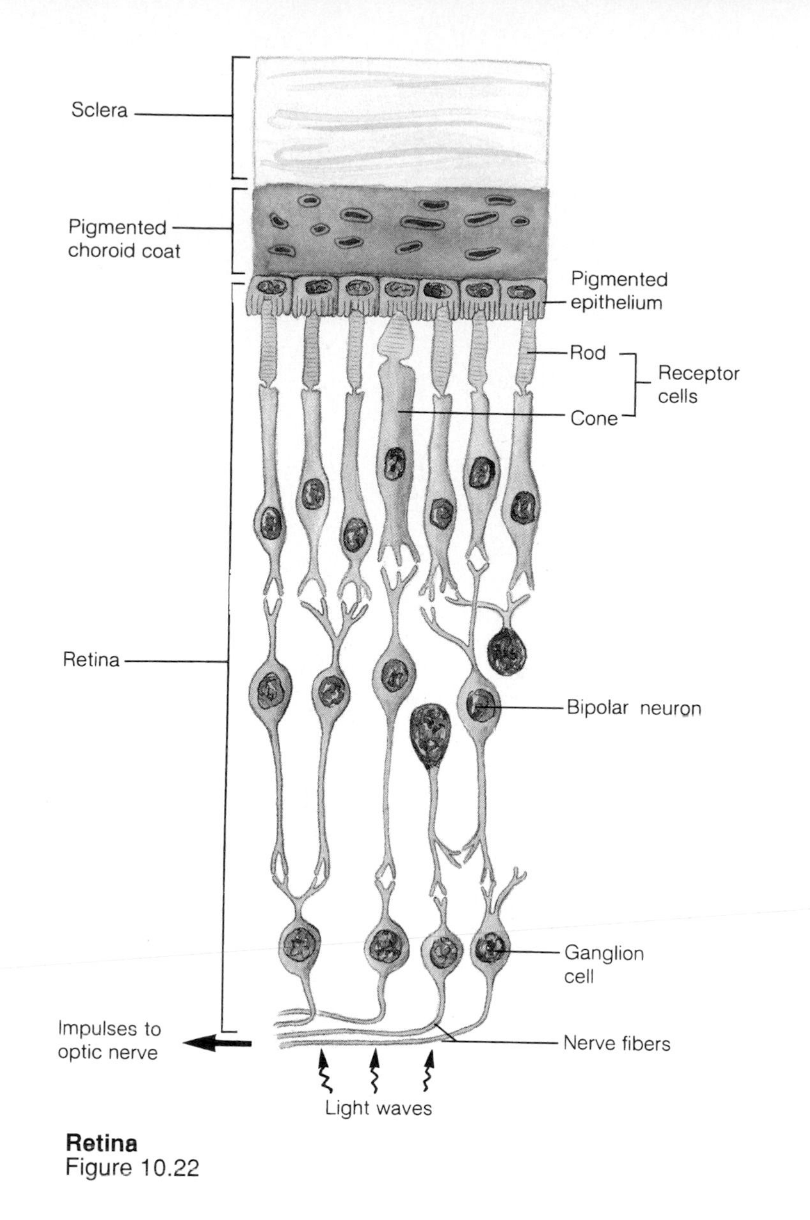
Sclera
Pigmented
choroid coat
Pigmented
epithelium
Rod
Receptor
cells
Cone
Retina
Bipolar neuron
Ganglion
cell
Impulses to
optic nerve
Nerve fibers
Light waves

Retina
Figure 10.22

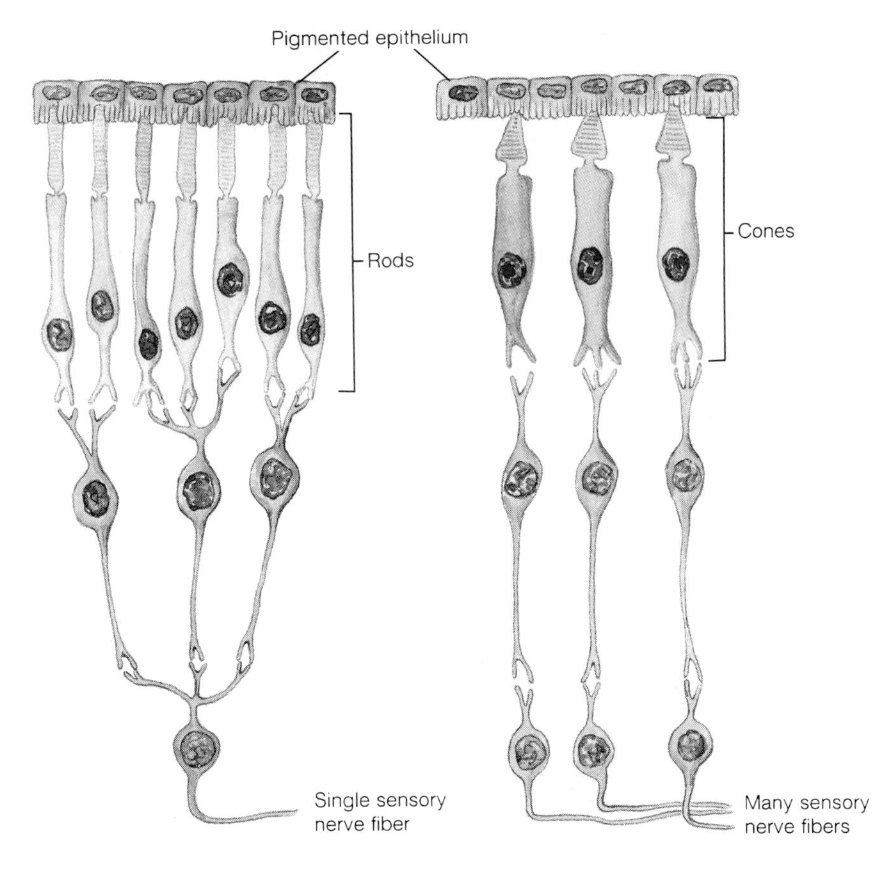
Pigmented epithelium
Rods
Cones
Single sensory
nerve fiber
Many sensory
nerve fibers

Rods and Cones
Figure 10.26 (a–b)

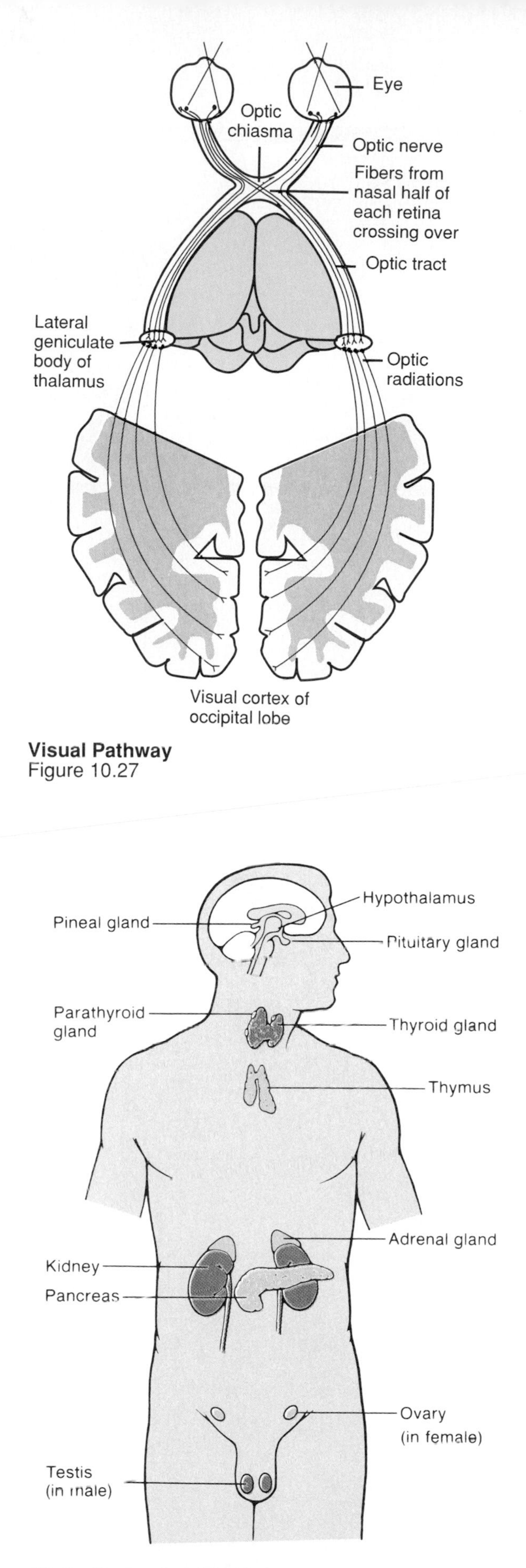
Eye
Optic chiasma
Optic nerve
Fibers from nasal half of each retina crossing over
Optic tract
Lateral geniculate body of thalamus
Optic radiations
Visual cortex of occipital lobe
Hypothalamus
Pineal gland
Pituitary gland
Parathyroid gland
Thyroid gland
Thymus
Adrenal gland
Kidney
Pancreas
Ovary (in female)
Testis (in male)

Visual Pathway
Figure 10.27

Major Endocrine Glands
Figure 11.1

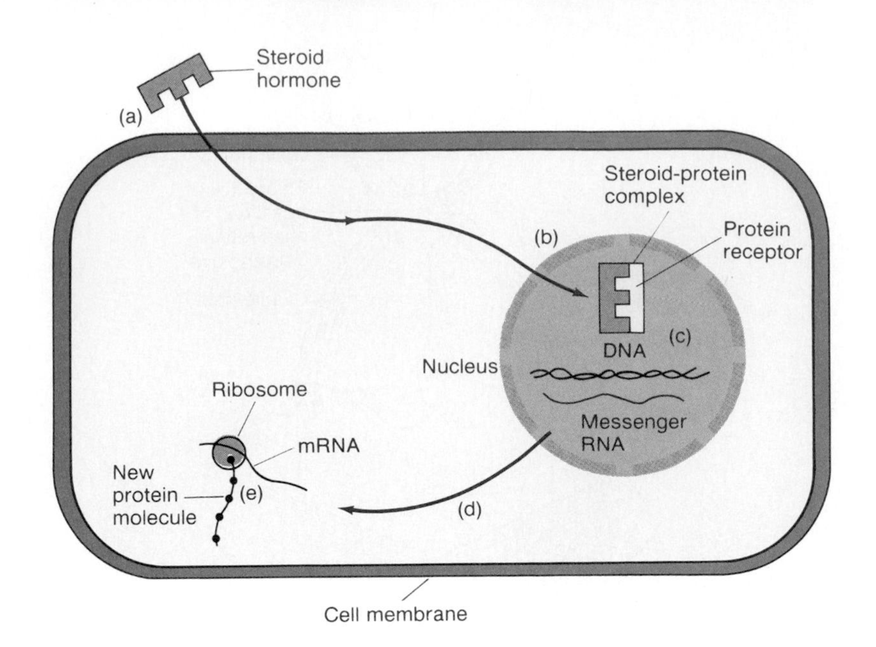
Steroid
hormone
(a)
Steroid-protein
complex
Protein
receptor
(b)
(c)
DNA
Nucleus
Ribosome
Messenger
RNA
mRNA
New
protein
molecule
(e)
(d)
Cell membrane

Steroid Hormone Action
Figure 11.2

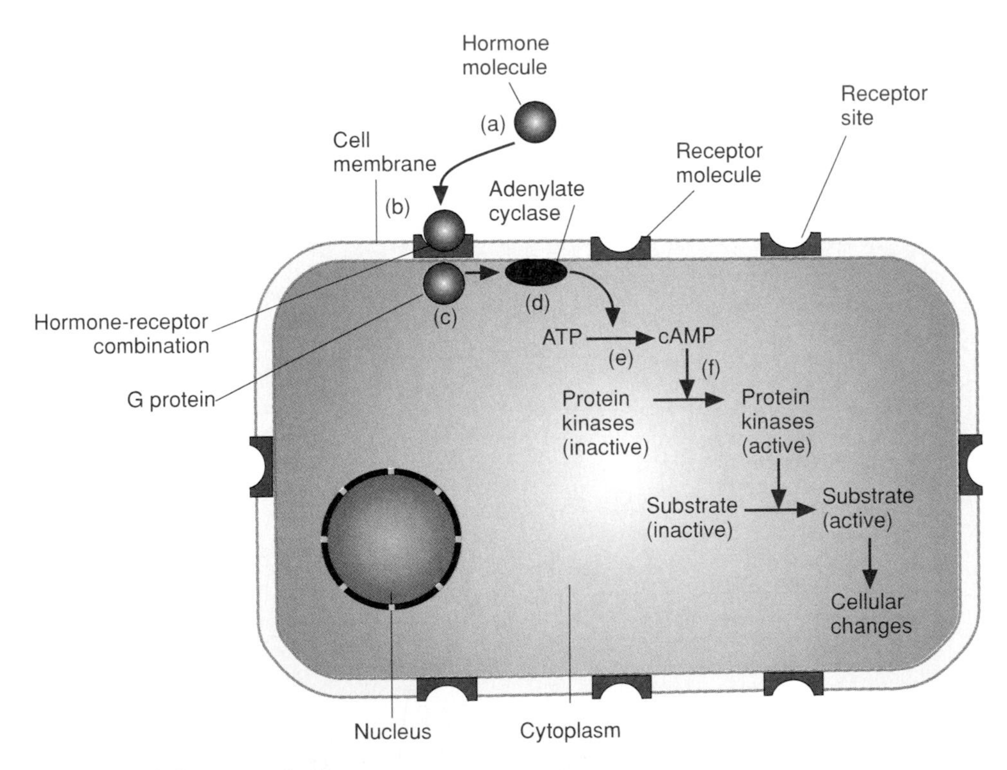
Hormone
molecule
(a)
Receptor
site
Cell
membrane
Receptor
molecule
Adenylate
cyclase
(b)
(d)
Hormone-receptor
combination
(c)
ATP
cAMP
(e)
(f)
G protein
Protein
kinases
(inactive)
Protein
kinases
(active)
Substrate
(inactive)
Substrate
(active)
Cellular
changes
Nucleus
Cytoplasm

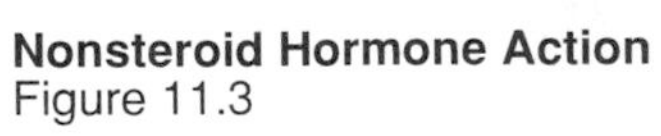
Nonsteroid Hormone Action
Figure 11.3

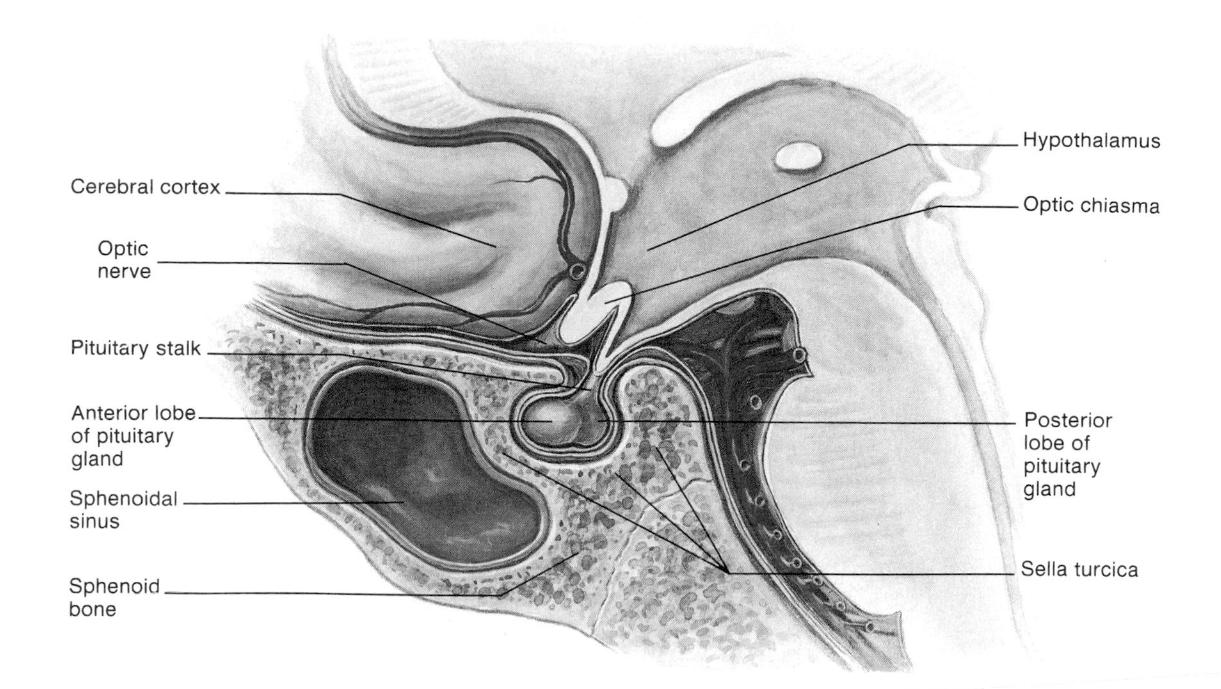
Cerebral cortex
Optic nerve
Pituitary stalk
Anterior lobe of pituitary gland
Sphenoidal sinus
Sphenoid bone
Hypothalamus
Optic chiasma
Posterior lobe of pituitary gland
Sella turcica

Pituitary Gland
Figures 11.6 and 11.7

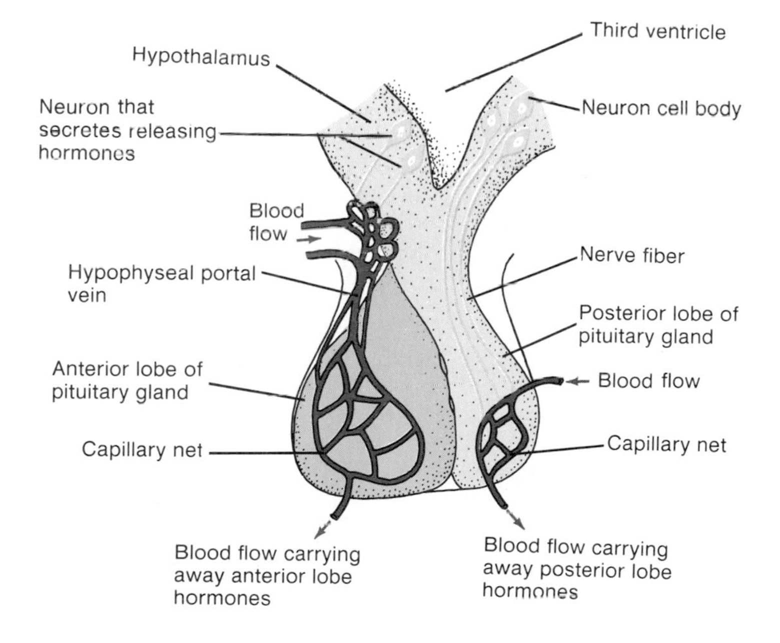
Third ventricle
Hypothalamus
Neuron that secretes releasing hormones
Neuron cell body
Blood flow
Hypophyseal portal vein
Nerve fiber
Posterior lobe of pituitary gland
Anterior lobe of pituitary gland
Blood flow
Capillary net
Capillary net
Blood flow carrying away anterior lobe hormones
Blood flow carrying away posterior lobe hormones

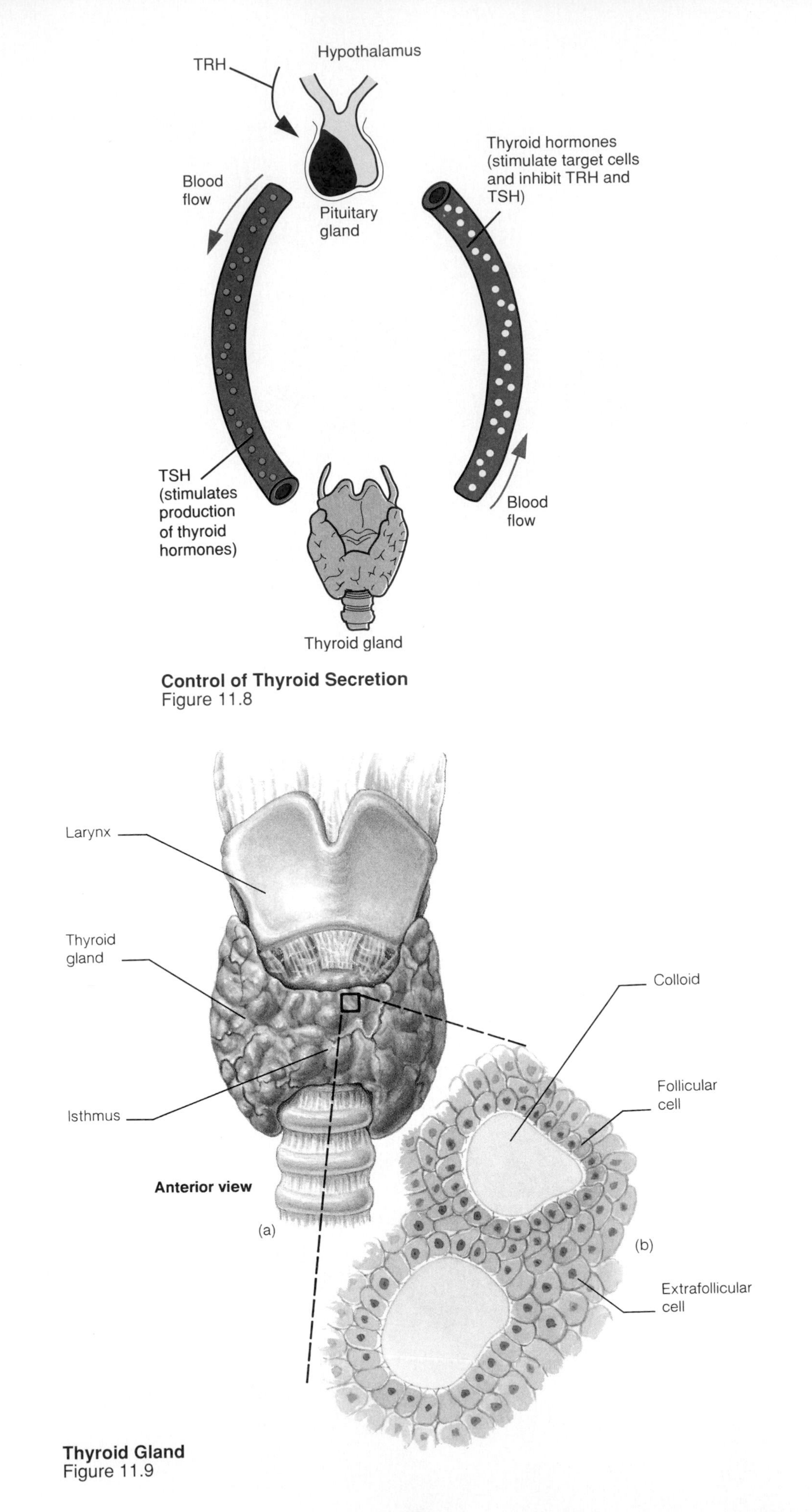

Hypothalamus
TRH
Blood flow
Pituitary gland
Thyroid hormones (stimulate target cells and inhibit TRH and TSH)
TSH (stimulates production of thyroid hormones)
Blood flow
Thyroid gland

Control of Thyroid Secretion
Figure 11.8

Larynx
Thyroid gland
Isthmus
Anterior view
(a)
Colloid
Follicular cell
(b)
Extrafollicular cell

Thyroid Gland
Figure 11.9

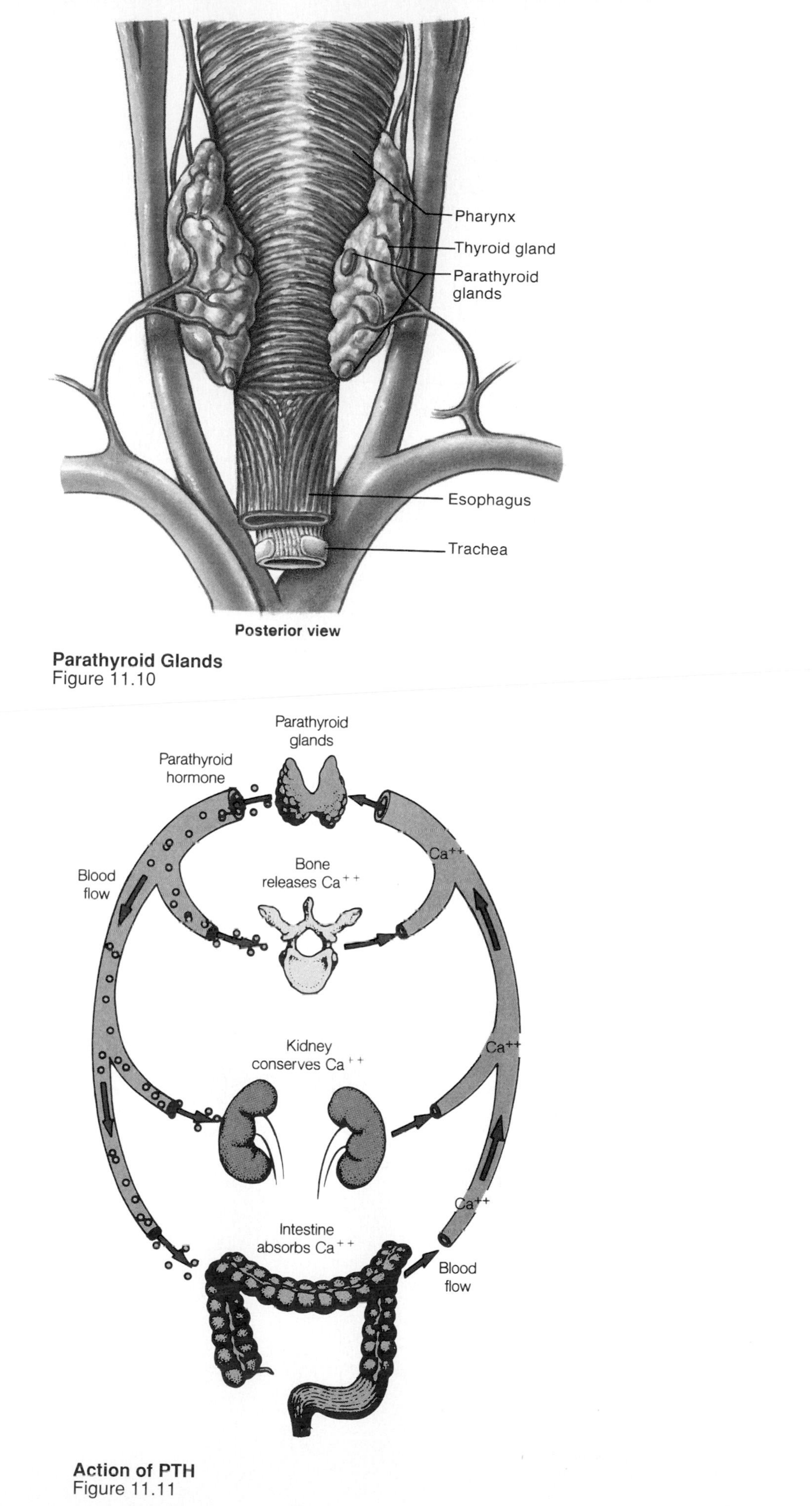
Pharynx
Thyroid gland
Parathyroid glands
Esophagus
Trachea
Posterior view
Parathyroid glands
Parathyroid hormone
Blood flow
Bone releases Ca^{++}
Ca^{++}
Kidney conserves Ca^{++}
Ca^{++}
Ca^{++}
Intestine absorbs Ca^{++}
Blood flow

Parathyroid Glands
Figure 11.10

Action of PTH
Figure 11.11

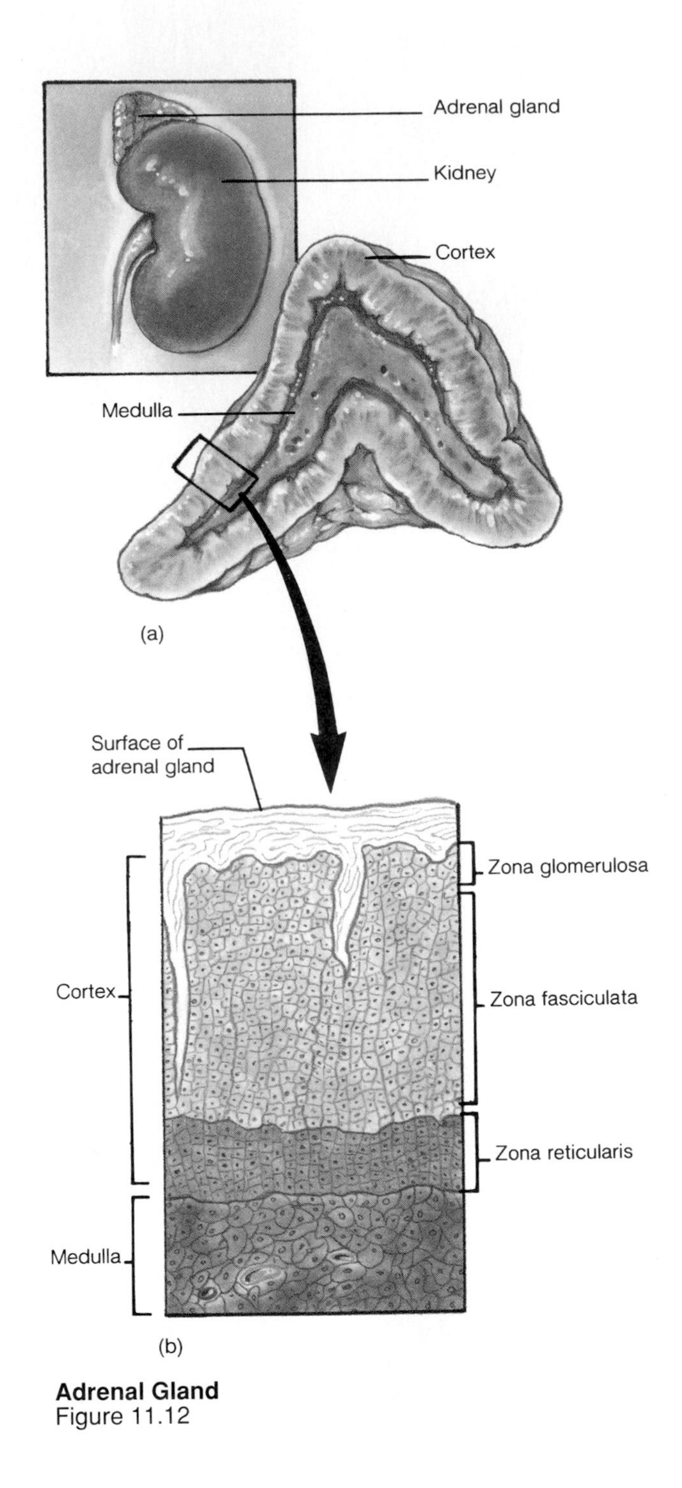
Adrenal gland
Kidney
Cortex
Medulla
(a)
Surface of adrenal gland
Zona glomerulosa
Cortex
Zona fasciculata
Zona reticularis
Medulla
(b)

Adrenal Gland
Figure 11.12

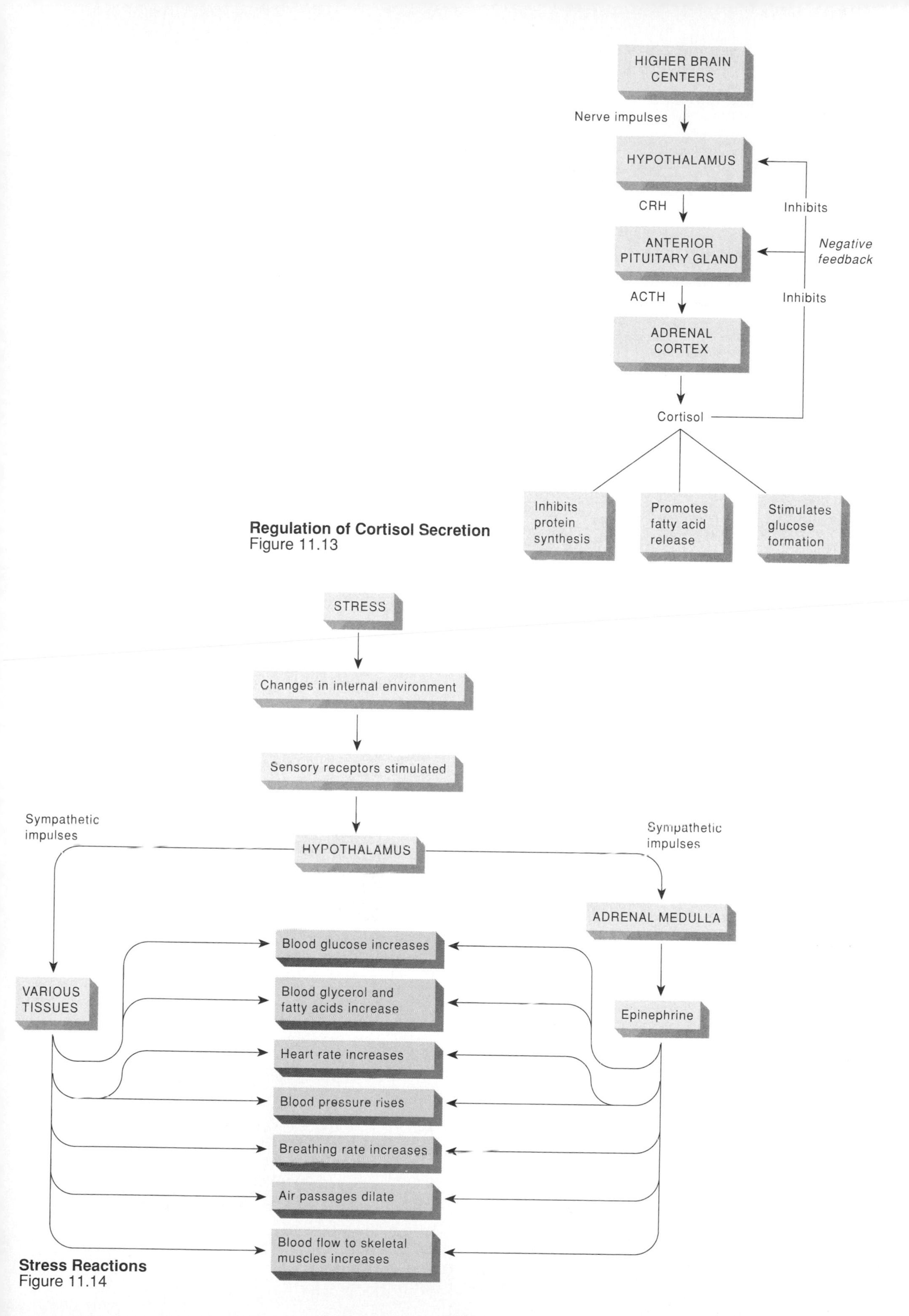
HIGHER BRAIN CENTERS
Nerve impulses
HYPOTHALAMUS
CRH
Inhibits
ANTERIOR PITUITARY GLAND
Negative feedback
ACTH
Inhibits
ADRENAL CORTEX
Cortisol
Inhibits protein synthesis
Promotes fatty acid release
Stimulates glucose formation

Regulation of Cortisol Secretion
Figure 11.13

STRESS
Changes in internal environment
Sensory receptors stimulated
Sympathetic impulses
HYPOTHALAMUS
Sympathetic impulses
ADRENAL MEDULLA
VARIOUS TISSUES
Epinephrine
Blood glucose increases
Blood glycerol and fatty acids increase
Heart rate increases
Blood pressure rises
Breathing rate increases
Air passages dilate
Blood flow to skeletal muscles increases

Stress Reactions
Figure 11.14

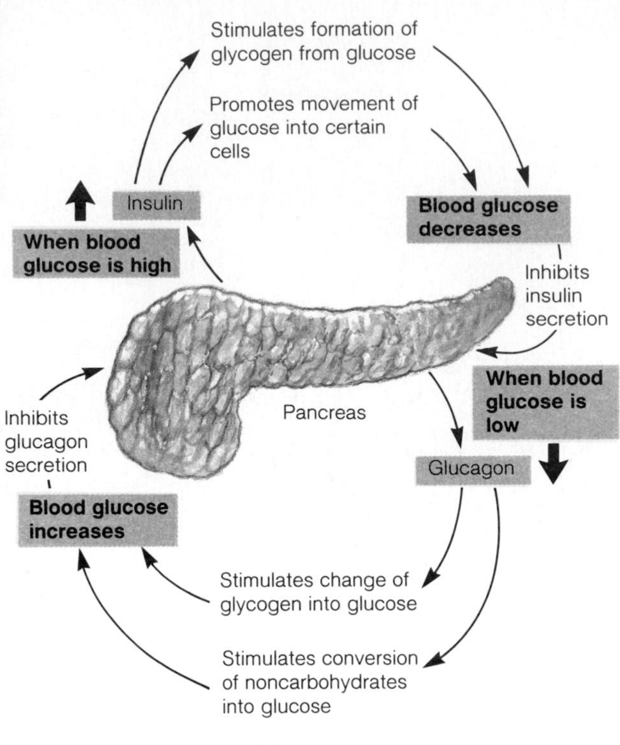
Stimulates formation of glycogen from glucose
Promotes movement of glucose into certain cells
Insulin
When blood glucose is high
Blood glucose decreases
Inhibits insulin secretion
When blood glucose is low
Pancreas
Inhibits glucagon secretion
Glucagon
Blood glucose increases
Stimulates change of glycogen into glucose
Stimulates conversion of noncarbohydrates into glucose

Regulation of Blood Glucose
Figure 11.18

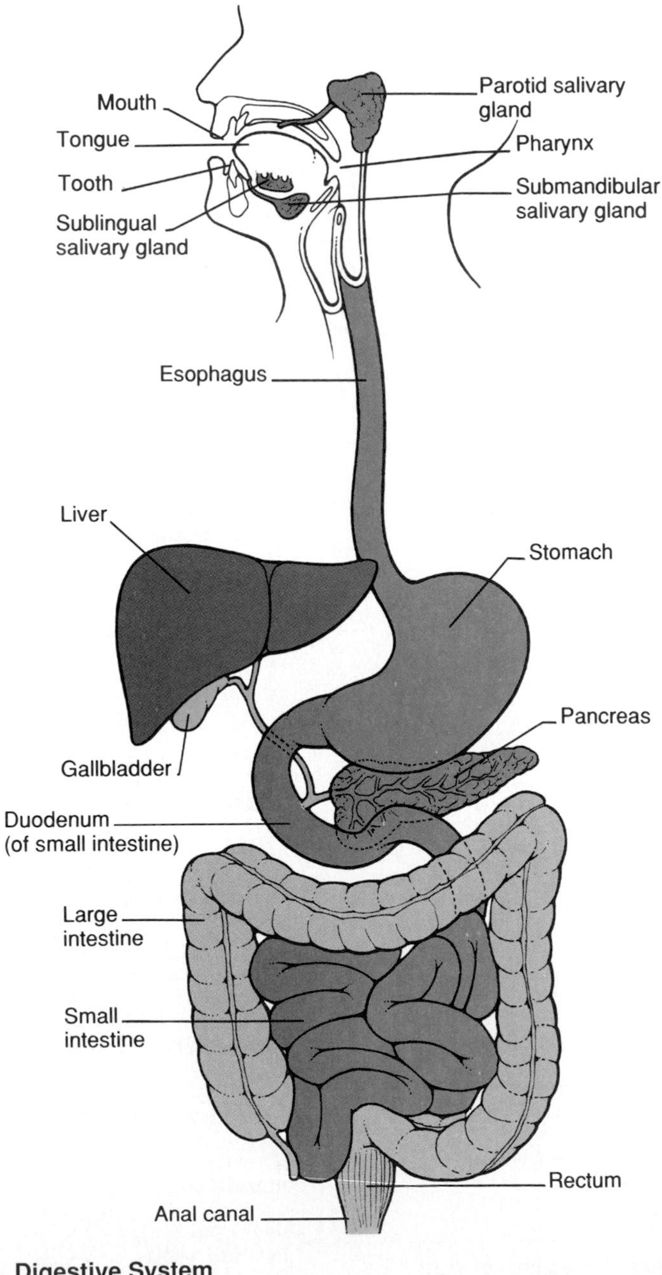
Mouth
Tongue
Tooth
Sublingual salivary gland
Parotid salivary gland
Pharynx
Submandibular salivary gland
Esophagus
Liver
Stomach
Pancreas
Gallbladder
Duodenum (of small intestine)
Large intestine
Small intestine
Rectum
Anal canal

Digestive System
Figure 12.1

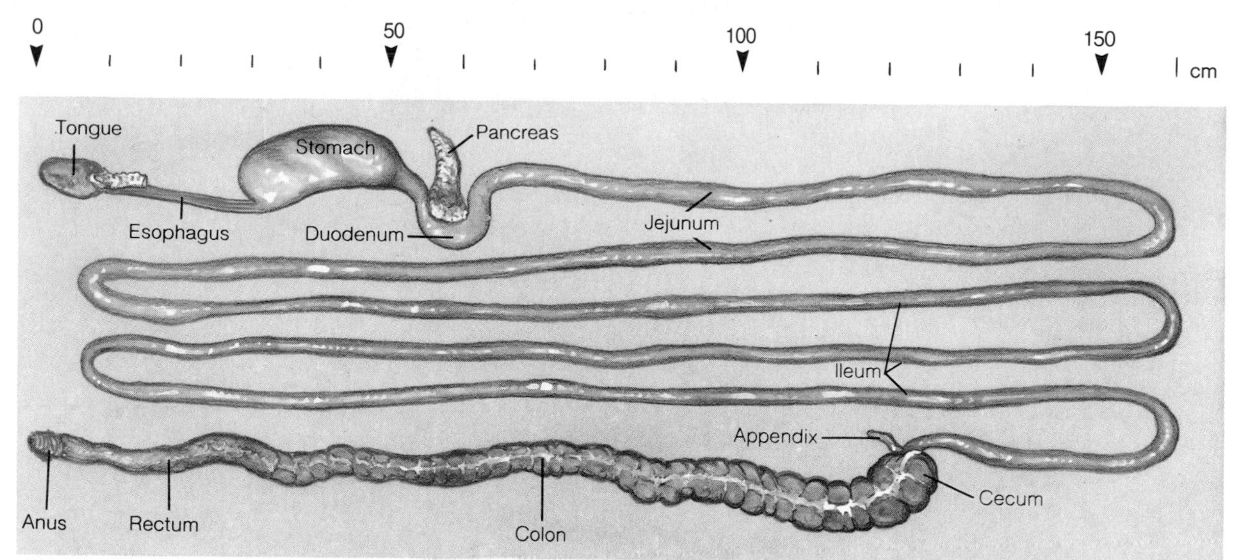
0
50
100
150
cm
Tongue
Stomach
Pancreas
Esophagus
Duodenum
Jejunum
Ileum
Appendix
Cecum
Anus
Rectum
Colon

Alimentary Canal
Figure 12.2

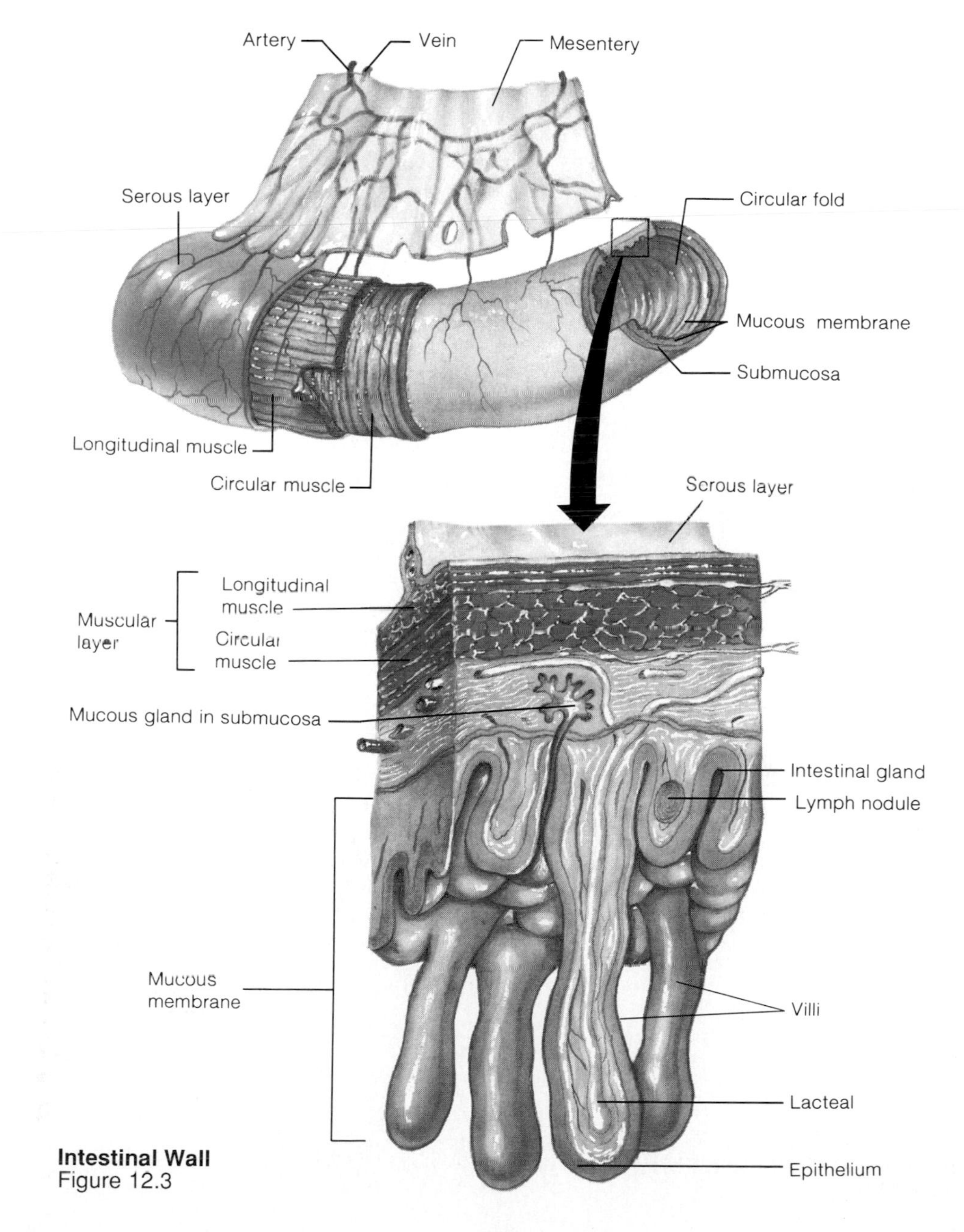
Artery
Vein
Mesentery
Serous layer
Circular fold
Mucous membrane
Submucosa
Longitudinal muscle
Circular muscle
Serous layer
Muscular layer
Longitudinal muscle
Circular muscle
Mucous gland in submucosa
Intestinal gland
Lymph nodule
Mucous membrane
Villi
Lacteal
Epithelium

Intestinal Wall
Figure 12.3

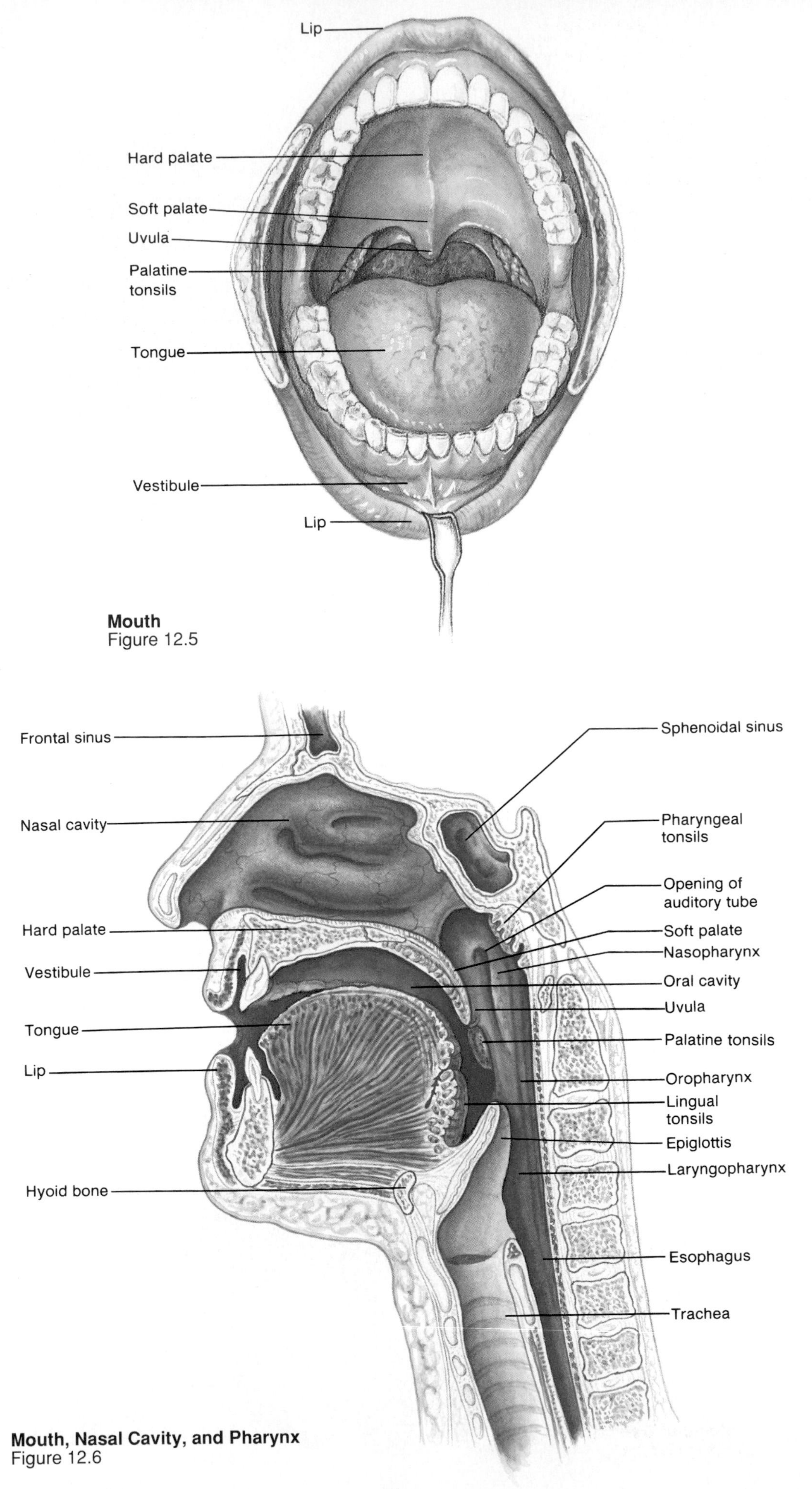

Lip
Hard palate
Soft palate
Uvula
Palatine tonsils
Tongue
Vestibule
Lip

Mouth
Figure 12.5

Frontal sinus
Nasal cavity
Hard palate
Vestibule
Tongue
Lip
Hyoid bone
Sphenoidal sinus
Pharyngeal tonsils
Opening of auditory tube
Soft palate
Nasopharynx
Oral cavity
Uvula
Palatine tonsils
Oropharynx
Lingual tonsils
Epiglottis
Laryngopharynx
Esophagus
Trachea

Mouth, Nasal Cavity, and Pharynx
Figure 12.6

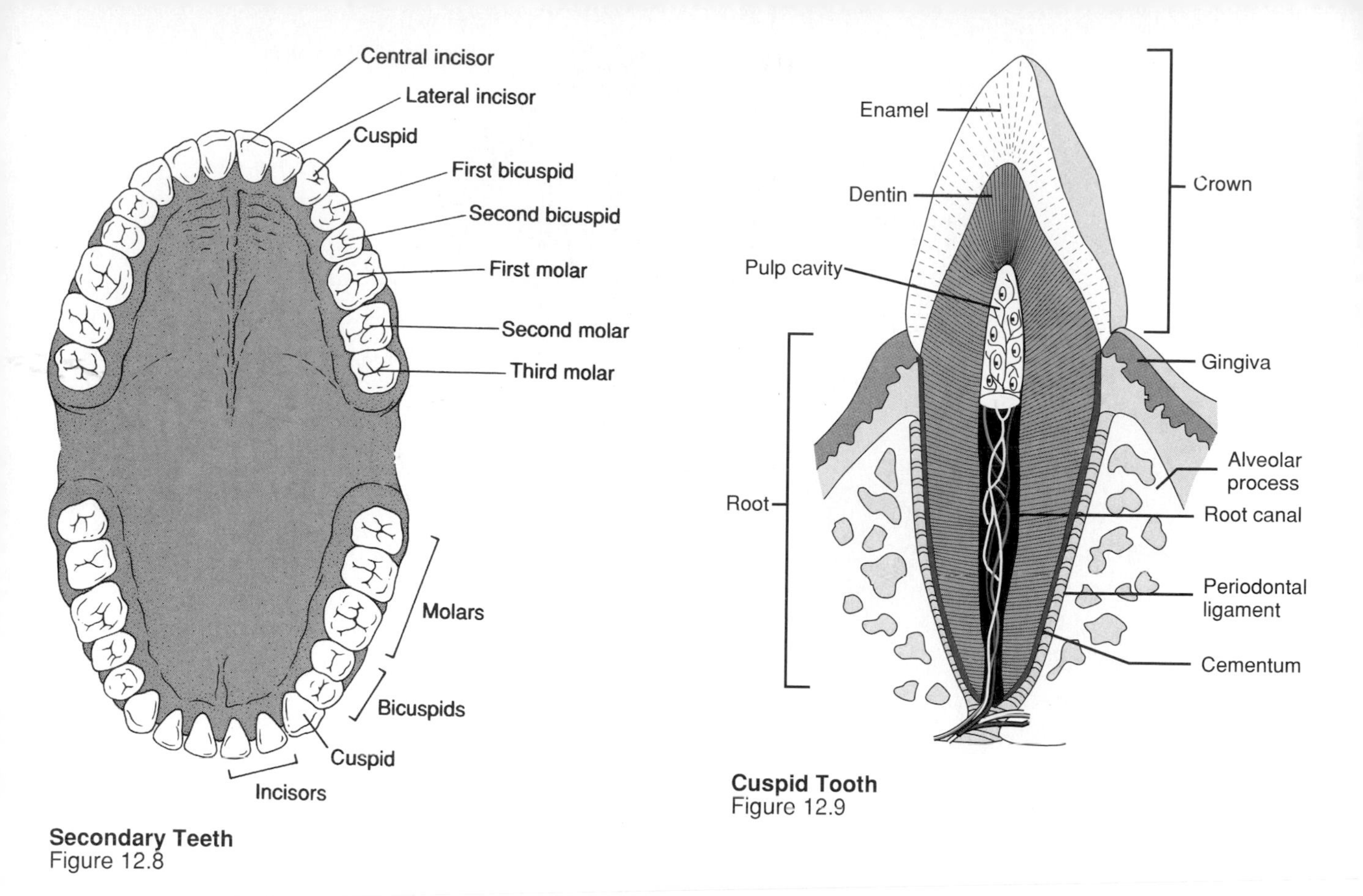
Central incisor
Lateral incisor
Cuspid
First bicuspid
Second bicuspid
First molar
Second molar
Third molar
Molars
Bicuspids
Cuspid
Incisors
Enamel
Dentin
Crown
Pulp cavity
Gingiva
Root
Alveolar process
Root canal
Periodontal ligament
Cementum

Secondary Teeth
Figure 12.8

Cuspid Tooth
Figure 12.9

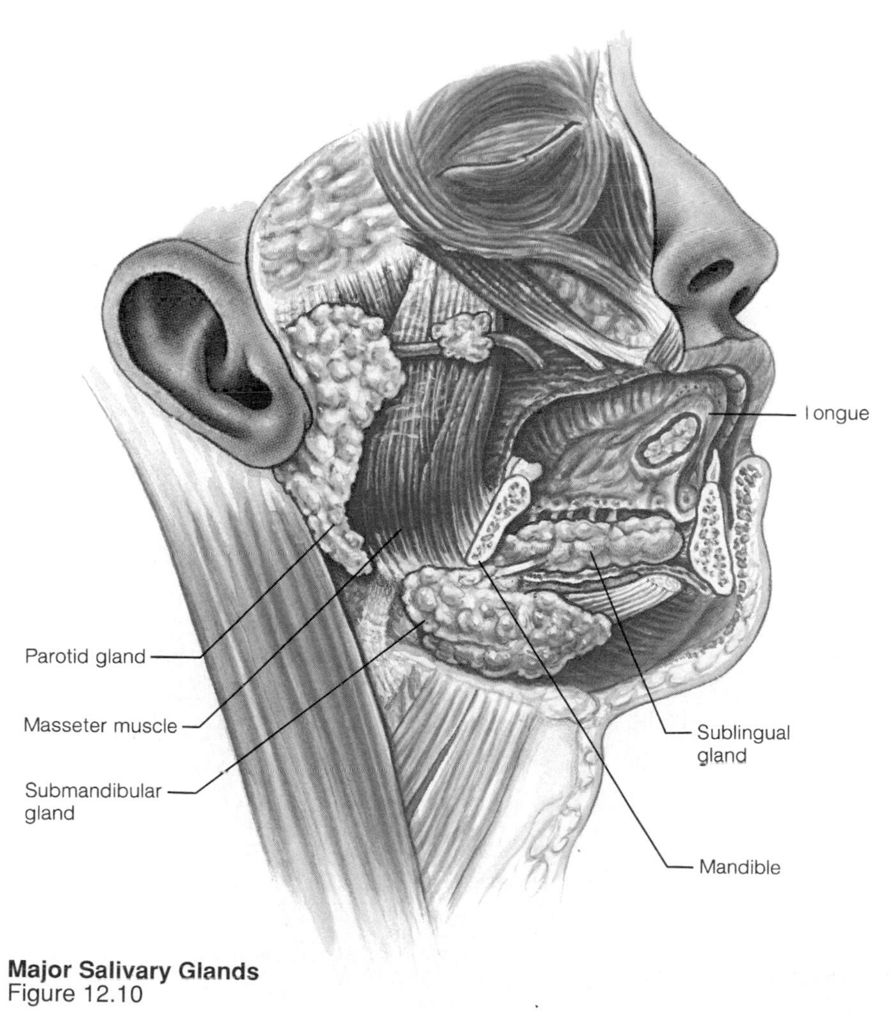
Tongue
Parotid gland
Masseter muscle
Submandibular gland
Sublingual gland
Mandible

Major Salivary Glands
Figure 12.10

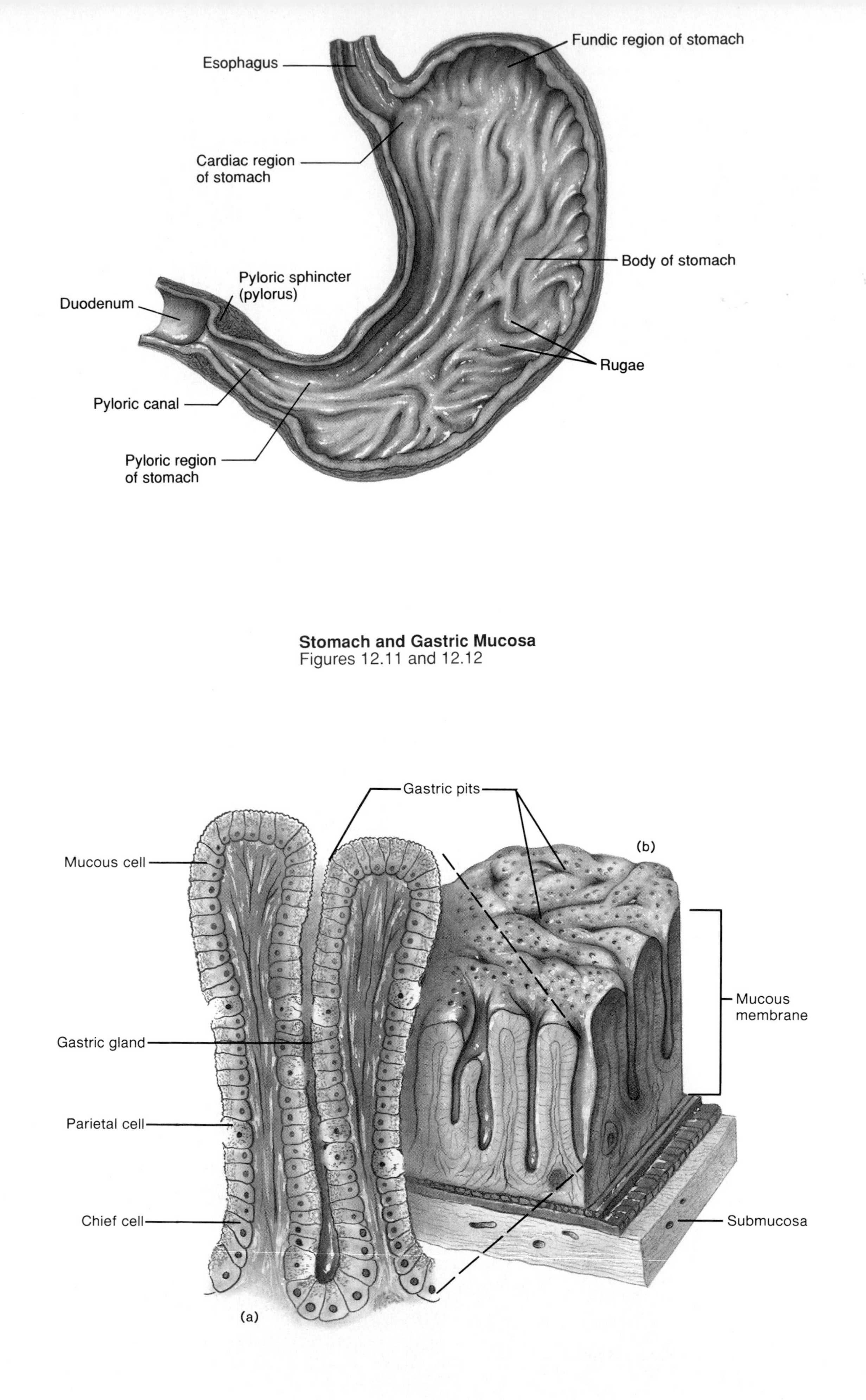
Esophagus
Fundic region of stomach
Cardiac region of stomach
Body of stomach
Pyloric sphincter (pylorus)
Duodenum
Rugae
Pyloric canal
Pyloric region of stomach
Gastric pits
(b)
Mucous cell
Mucous membrane
Gastric gland
Parietal cell
Chief cell
Submucosa
(a)

Stomach and Gastric Mucosa
Figures 12.11 and 12.12

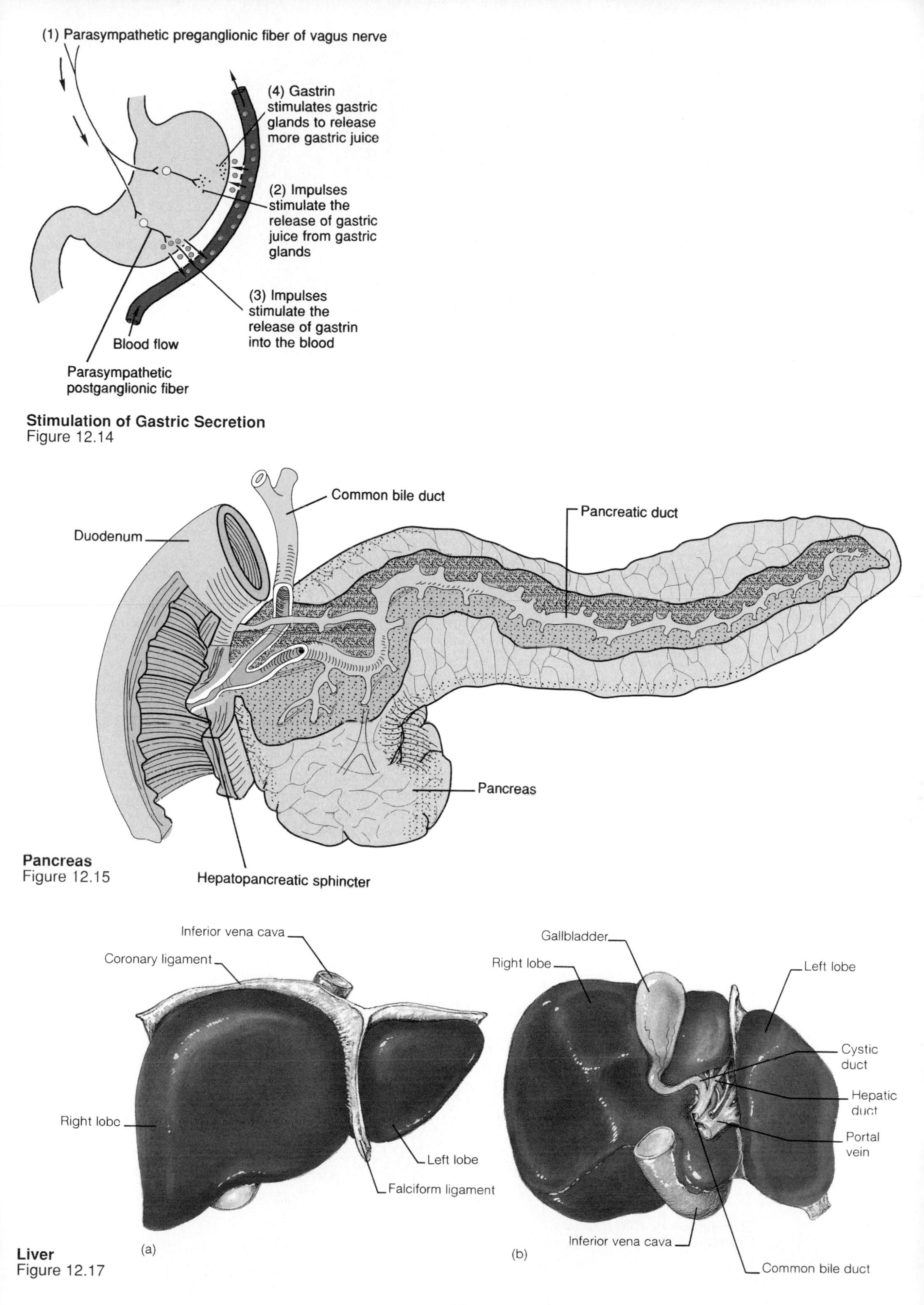

(1) Parasympathetic preganglionic fiber of vagus nerve
(4) Gastrin stimulates gastric glands to release more gastric juice
(2) Impulses stimulate the release of gastric juice from gastric glands
(3) Impulses stimulate the release of gastrin into the blood
Blood flow
Parasympathetic postganglionic fiber

Stimulation of Gastric Secretion
Figure 12.14

Common bile duct
Pancreatic duct
Duodenum
Pancreas
Hepatopancreatic sphincter

Pancreas
Figure 12.15

Inferior vena cava
Coronary ligament
Right lobe
Left lobe
Falciform ligament
(a)
Gallbladder
Right lobe
Left lobe
Cystic duct
Hepatic duct
Portal vein
Inferior vena cava
Common bile duct
(b)

Liver
Figure 12.17

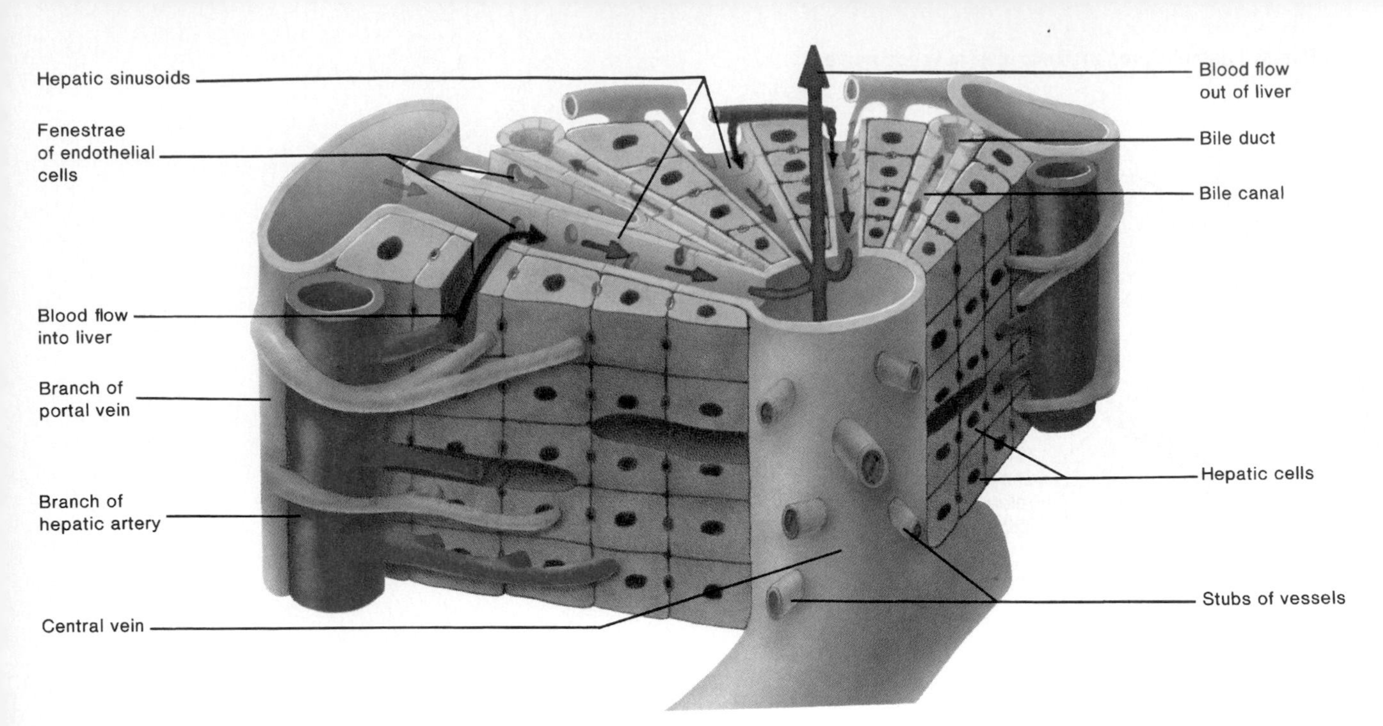

Hepatic sinusoids
Fenestrae of endothelial cells
Blood flow into liver
Branch of portal vein
Branch of hepatic artery
Central vein
Blood flow out of liver
Bile duct
Bile canal
Hepatic cells
Stubs of vessels

Hepatic Lobule
Figure 12.18

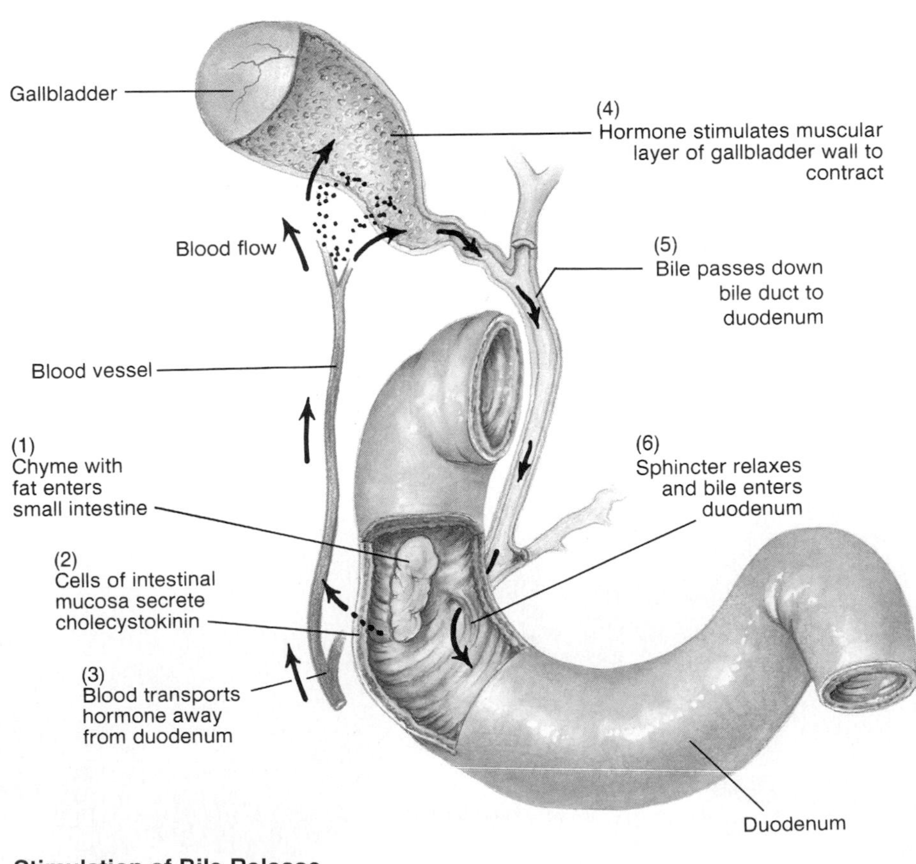

Gallbladder
(4) Hormone stimulates muscular layer of gallbladder wall to contract
Blood flow
(5) Bile passes down bile duct to duodenum
Blood vessel
(1) Chyme with fat enters small intestine
(6) Sphincter relaxes and bile enters duodenum
(2) Cells of intestinal mucosa secrete cholecystokinin
(3) Blood transports hormone away from duodenum
Duodenum

Stimulation of Bile Release
Figure 12.21

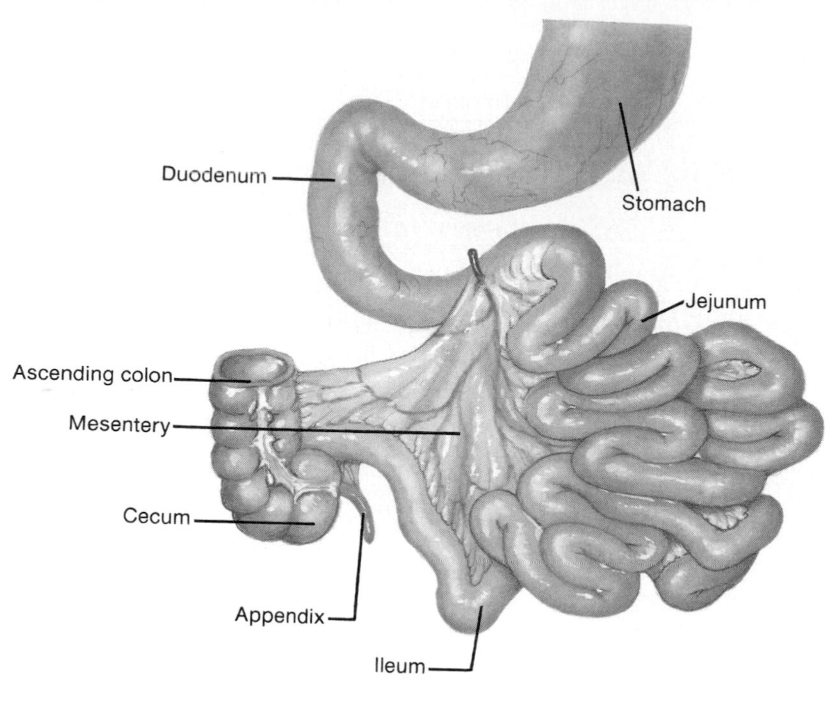

Duodenum
Stomach
Jejunum
Ascending colon
Mesentery
Cecum
Appendix
Ileum

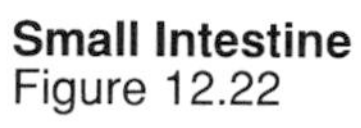

Small Intestine
Figure 12.22

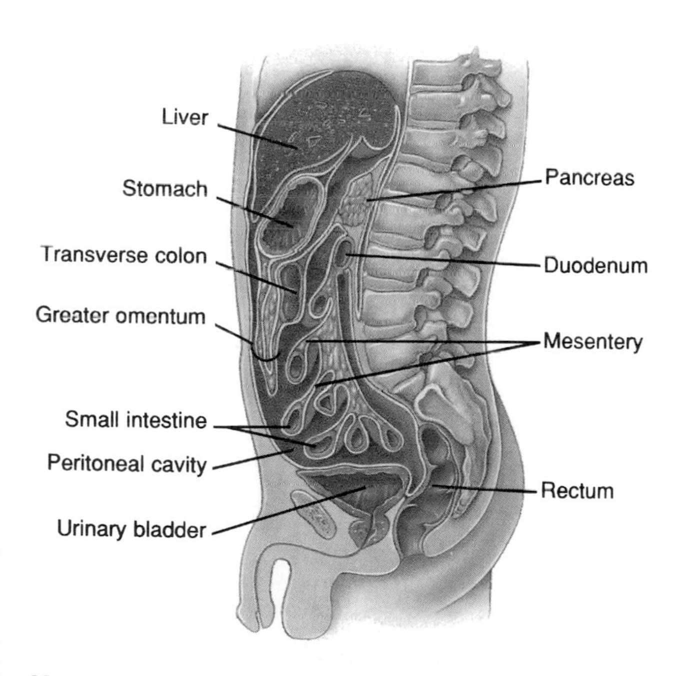

Liver
Stomach
Transverse colon
Greater omentum
Small intestine
Peritoneal cavity
Urinary bladder
Pancreas
Duodenum
Mesentery
Rectum

Mesentery
Figure 12.24

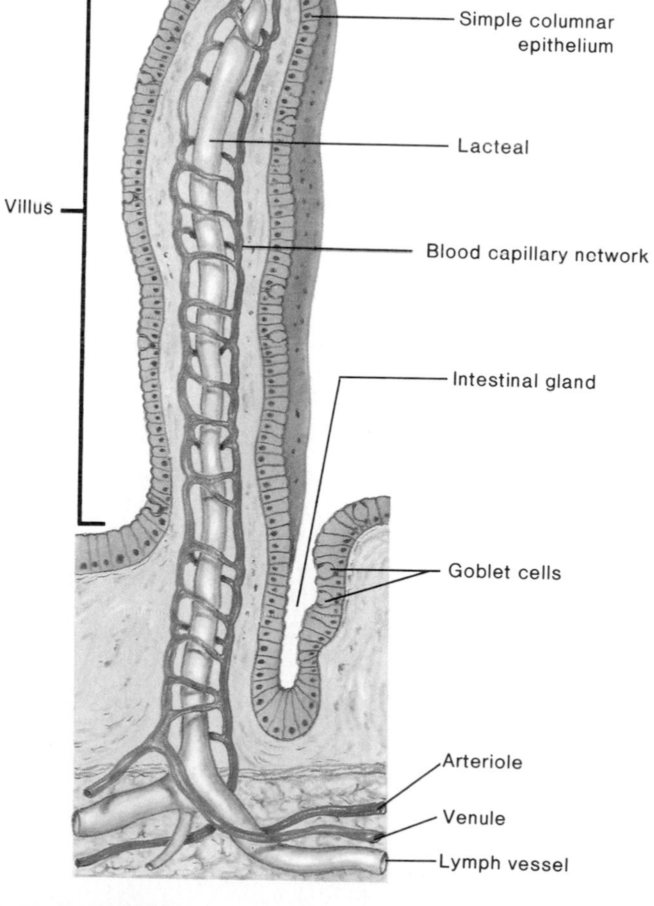

Villus
Simple columnar epithelium
Lacteal
Blood capillary network
Intestinal gland
Goblet cells
Arteriole
Venule
Lymph vessel

Intestinal Villus
Figure 12.25

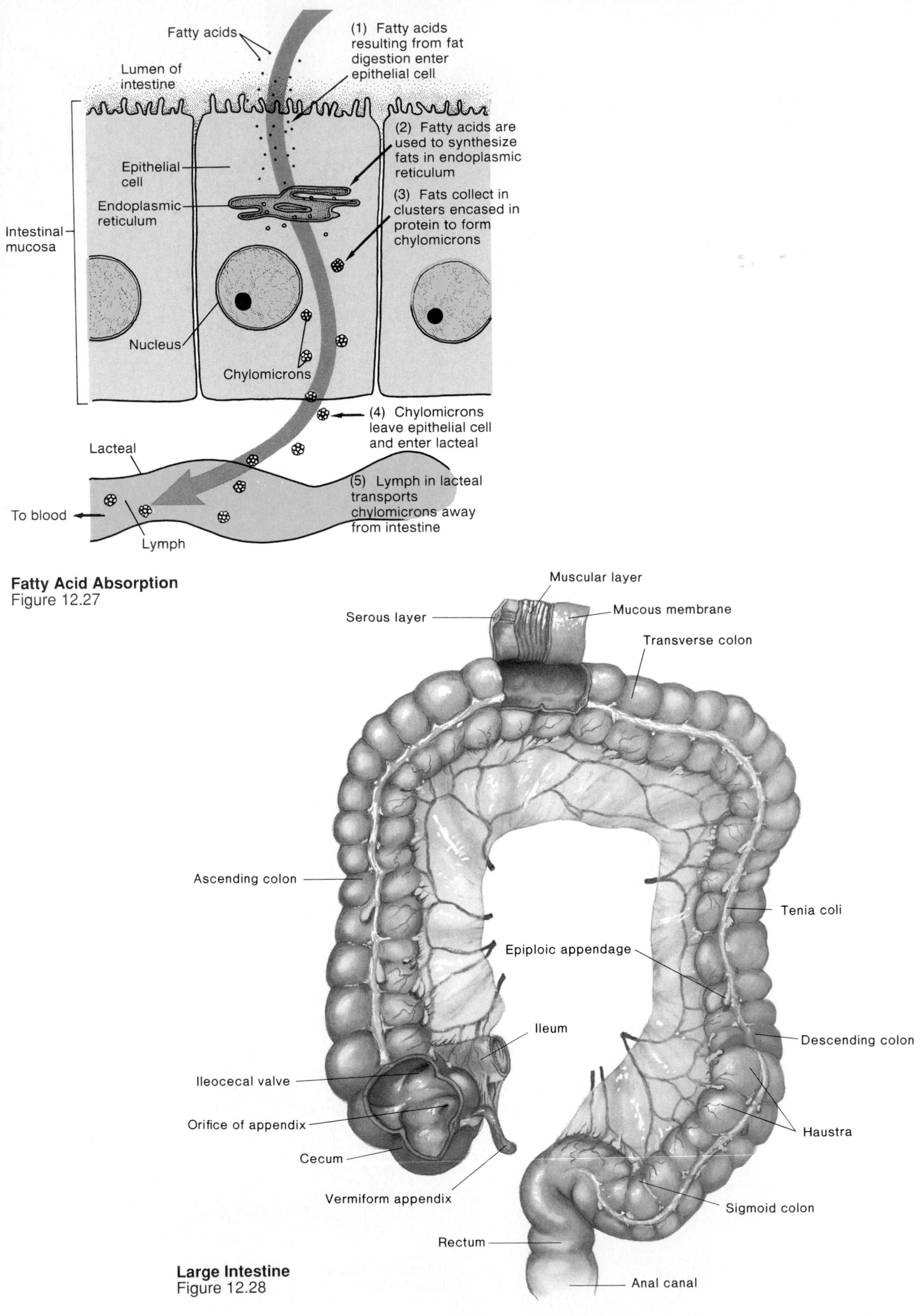
Fatty acids
Lumen of intestine
(1) Fatty acids resulting from fat digestion enter epithelial cell
(2) Fatty acids are used to synthesize fats in endoplasmic reticulum
Epithelial cell
Endoplasmic reticulum
(3) Fats collect in clusters encased in protein to form chylomicrons
Intestinal mucosa
Nucleus
Chylomicrons
(4) Chylomicrons leave epithelial cell and enter lacteal
Lacteal
(5) Lymph in lacteal transports chylomicrons away from intestine
To blood
Lymph
Muscular layer
Serous layer
Mucous membrane
Transverse colon
Ascending colon
Tenia coli
Epiploic appendage
Ileum
Descending colon
Ileocecal valve
Orifice of appendix
Haustra
Cecum
Vermiform appendix
Sigmoid colon
Rectum
Anal canal

Fatty Acid Absorption
Figure 12.27

Large Intestine
Figure 12.28

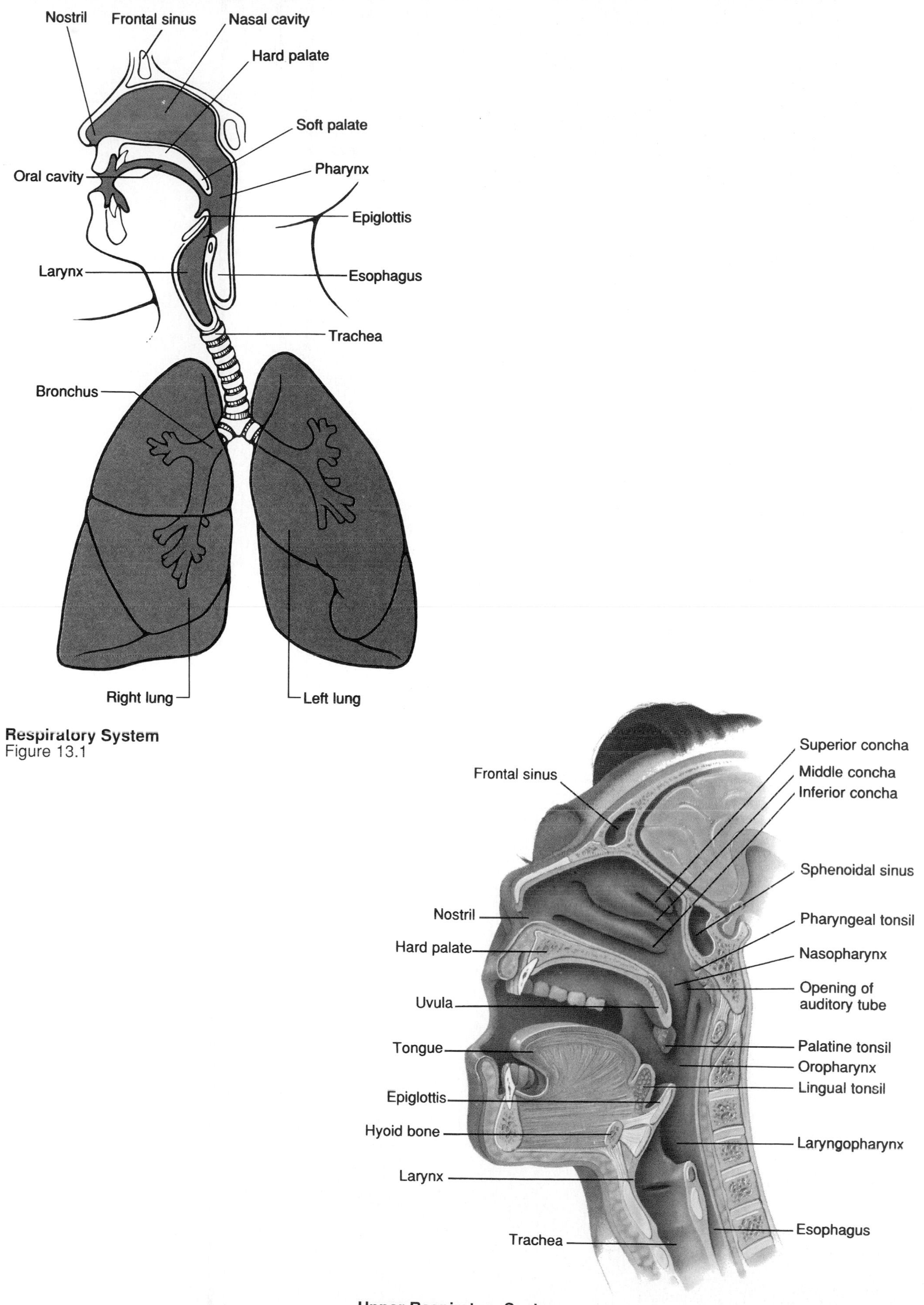
Nostril
Frontal sinus
Nasal cavity
Hard palate
Soft palate
Oral cavity
Pharynx
Epiglottis
Larynx
Esophagus
Trachea
Bronchus
Right lung
Left lung

Respiratory System
Figure 13.1

Superior concha
Middle concha
Inferior concha
Frontal sinus
Sphenoidal sinus
Nostril
Pharyngeal tonsil
Hard palate
Nasopharynx
Opening of
auditory tube
Uvula
Tongue
Palatine tonsil
Oropharynx
Lingual tonsil
Epiglottis
Hyoid bone
Laryngopharynx
Larynx
Esophagus
Trachea

Upper Respiratory System
Figure 13.2

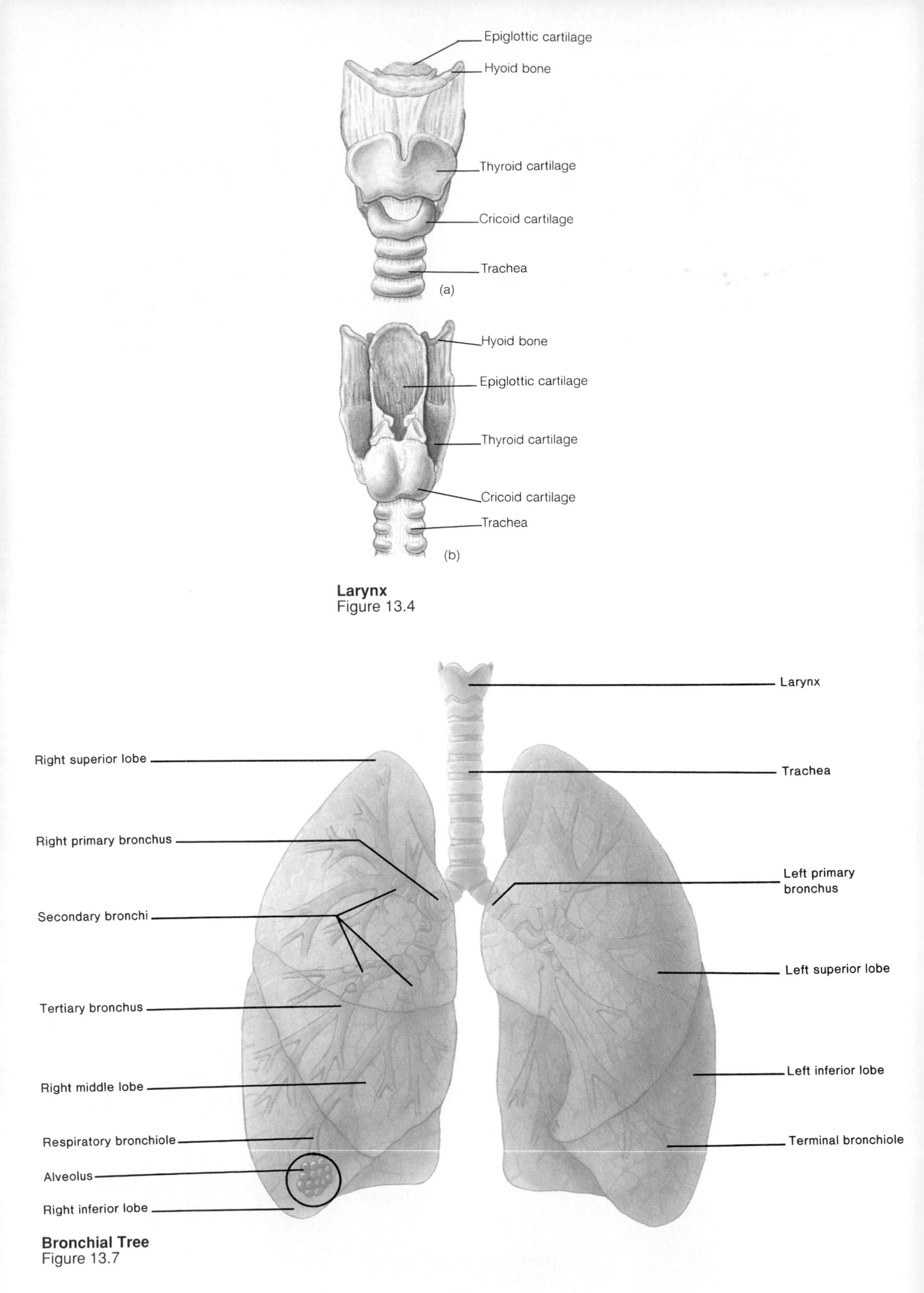
Epiglottic cartilage
Hyoid bone
Thyroid cartilage
Cricoid cartilage
Trachea
(a)
Hyoid bone
Epiglottic cartilage
Thyroid cartilage
Cricoid cartilage
Trachea
(b)

Larynx
Figure 13.4

Larynx
Right superior lobe
Trachea
Right primary bronchus
Left primary bronchus
Secondary bronchi
Left superior lobe
Tertiary bronchus
Right middle lobe
Left inferior lobe
Respiratory bronchiole
Terminal bronchiole
Alveolus
Right inferior lobe

Bronchial Tree
Figure 13.7

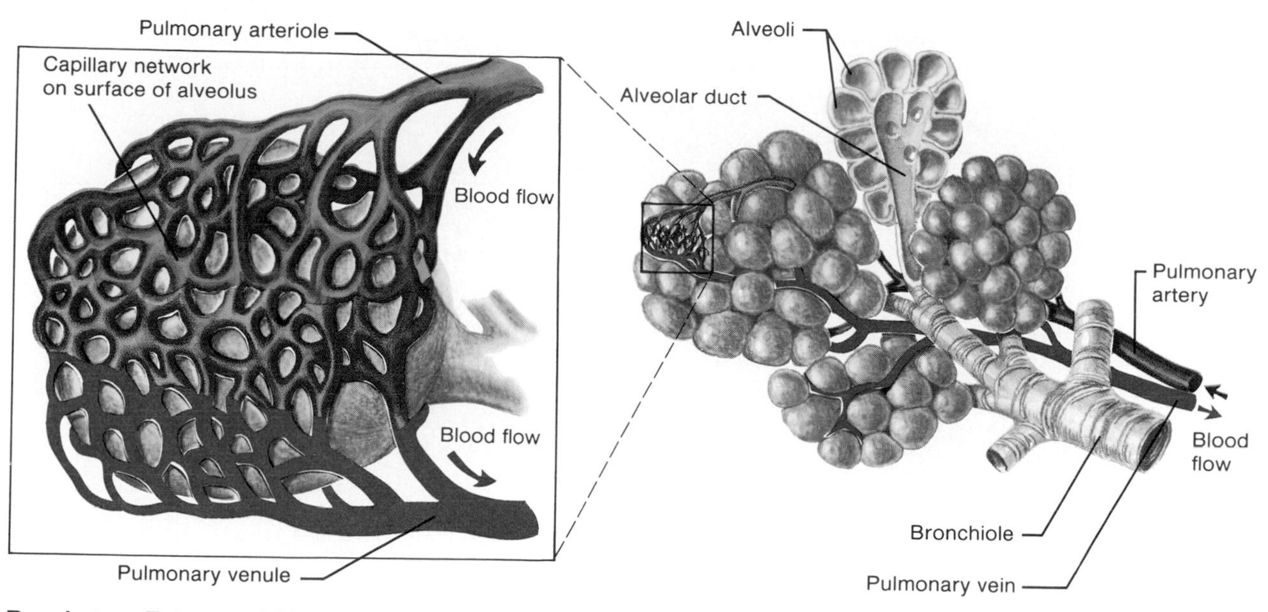
Pulmonary arteriole
Capillary network on surface of alveolus
Blood flow
Blood flow
Pulmonary venule
Alveoli
Alveolar duct
Pulmonary artery
Blood flow
Bronchiole
Pulmonary vein

Respiratory Tubes and Alveoli
Figure 13.8

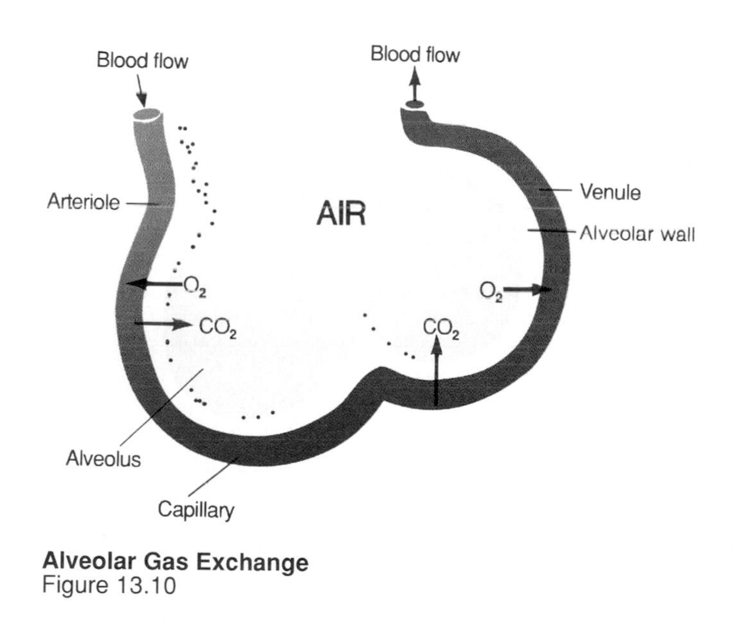
Blood flow
Blood flow
Arteriole
AIR
Venule
Alvcolar wall
O_2
CO_2
O_2
CO_2
Alveolus
Capillary

Alveolar Gas Exchange
Figure 13.10

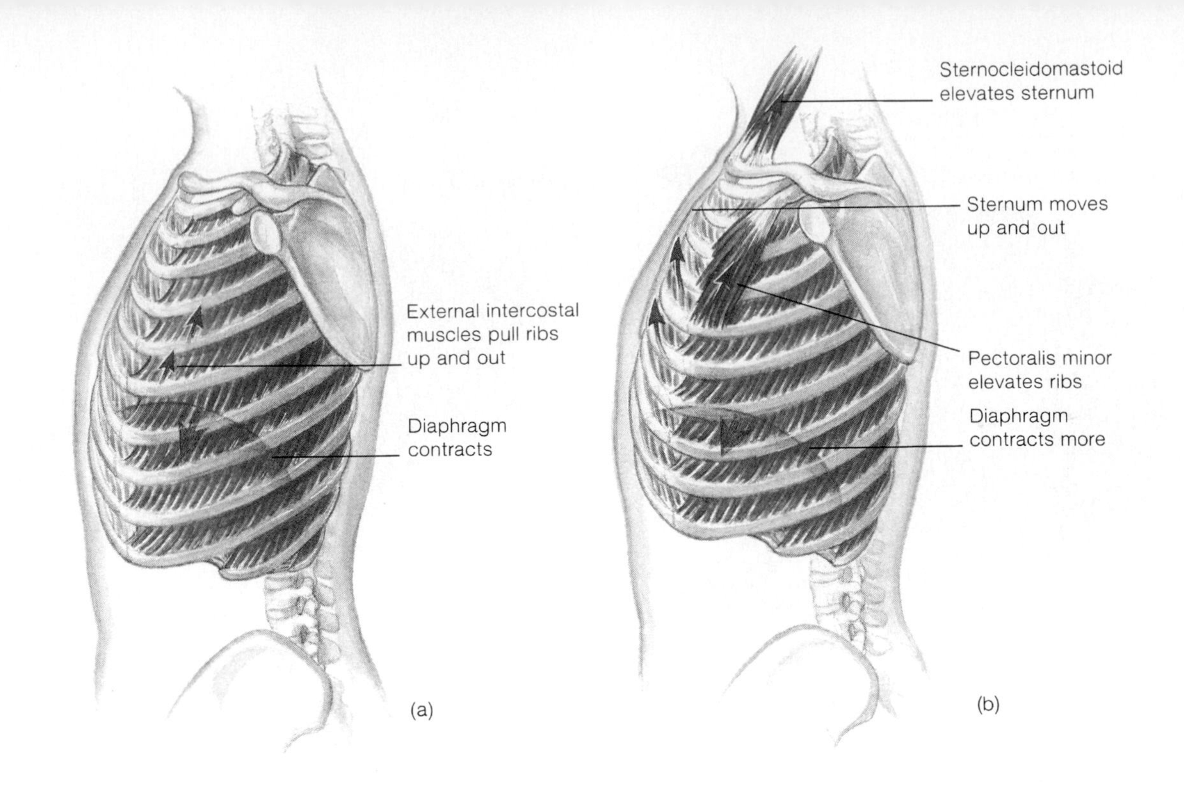
External intercostal
muscles pull ribs
up and out
Diaphragm
contracts
(a)
Sternocleidomastoid
elevates sternum
Sternum moves
up and out
Pectoralis minor
elevates ribs
Diaphragm
contracts more
(b)

Inspiration and Expiration
Figures 13.14 (a–b) and 13.15 (a–b)

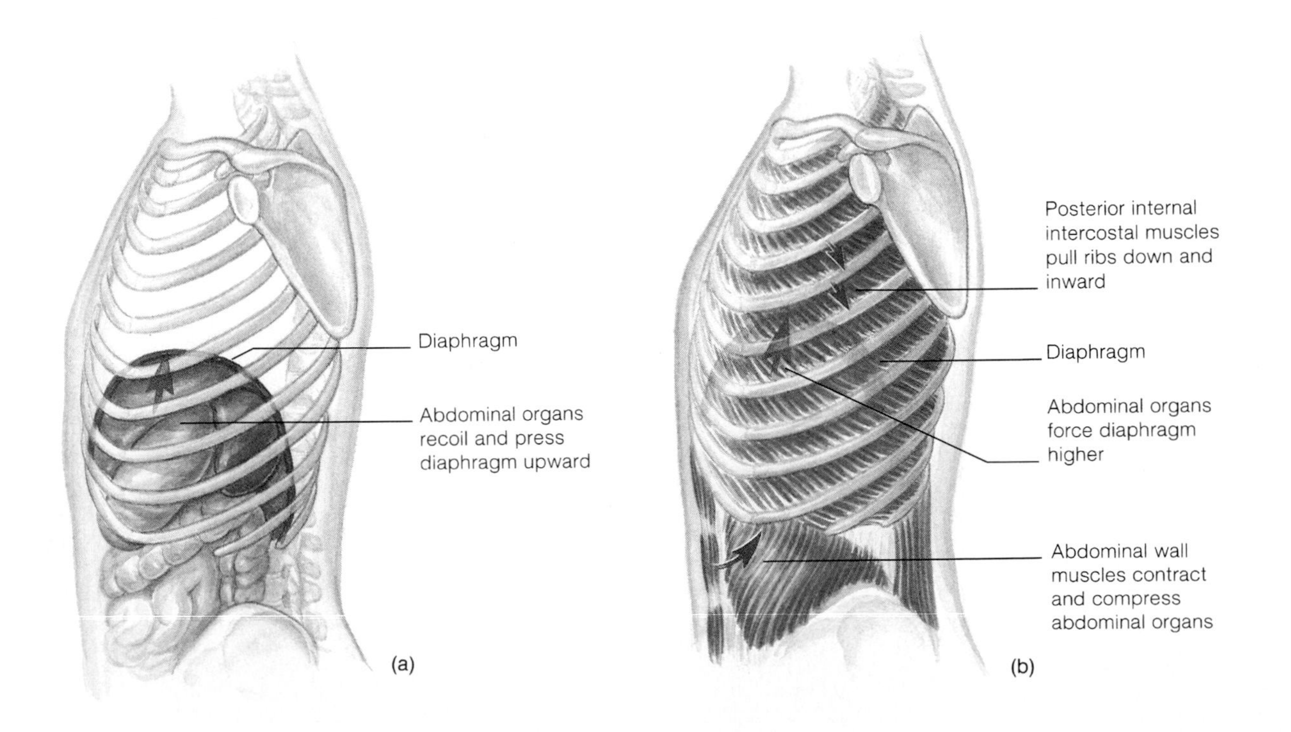
Diaphragm
Abdominal organs
recoil and press
diaphragm upward
(a)
Posterior internal
intercostal muscles
pull ribs down and
inward
Diaphragm
Abdominal organs
force diaphragm
higher
Abdominal wall
muscles contract
and compress
abdominal organs
(b)

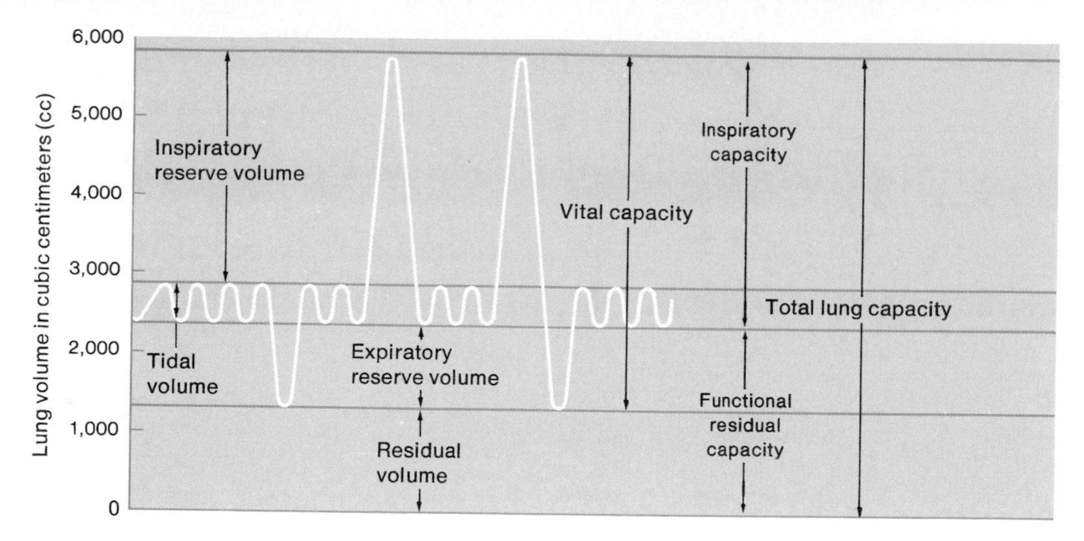
6,000
5,000
4,000
3,000
2,000
1,000
0
Lung volume in cubic centimeters (cc)
Inspiratory reserve volume
Tidal volume
Expiratory reserve volume
Residual volume
Vital capacity
Inspiratory capacity
Total lung capacity
Functional residual capacity

Respiratory Air Volumes and Capacities
Figure 13.16

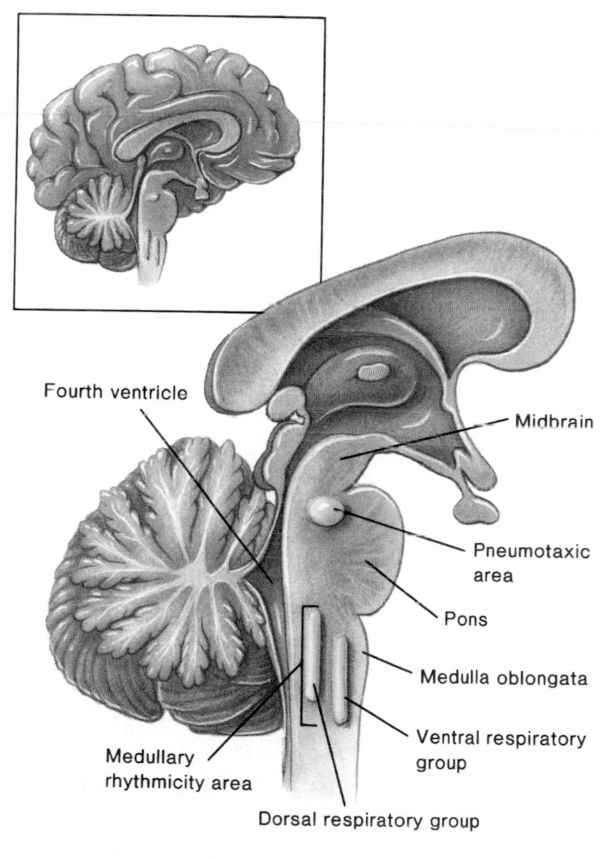
Fourth ventricle
Midbrain
Pneumotaxic area
Pons
Medulla oblongata
Ventral respiratory group
Medullary rhythmicity area
Dorsal respiratory group

Respiratory Center
Figure 13.17

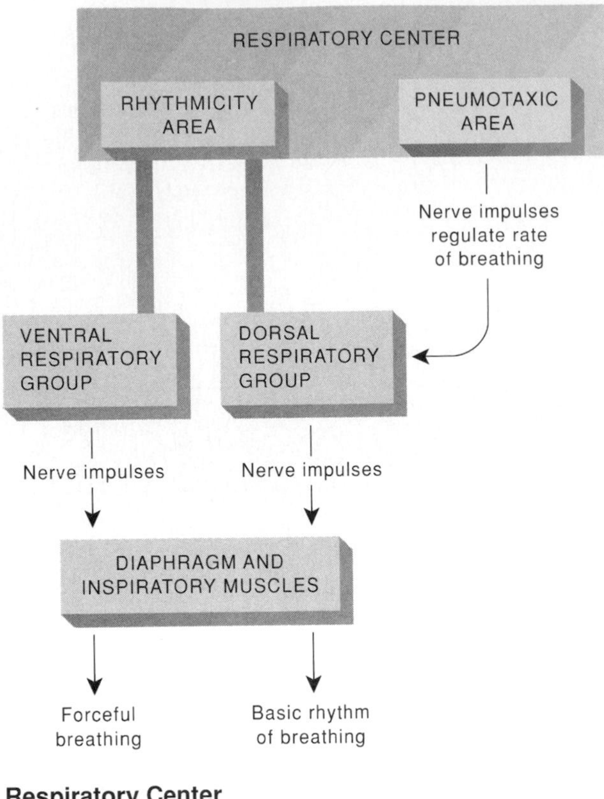
RESPIRATORY CENTER
RHYTHMICITY AREA
PNEUMOTAXIC AREA
Nerve impulses regulate rate of breathing
VENTRAL RESPIRATORY GROUP
DORSAL RESPIRATORY GROUP
Nerve impulses
Nerve impulses
DIAPHRAGM AND INSPIRATORY MUSCLES
Forceful breathing
Basic rhythm of breathing

Respiratory Center
Figure 13.18

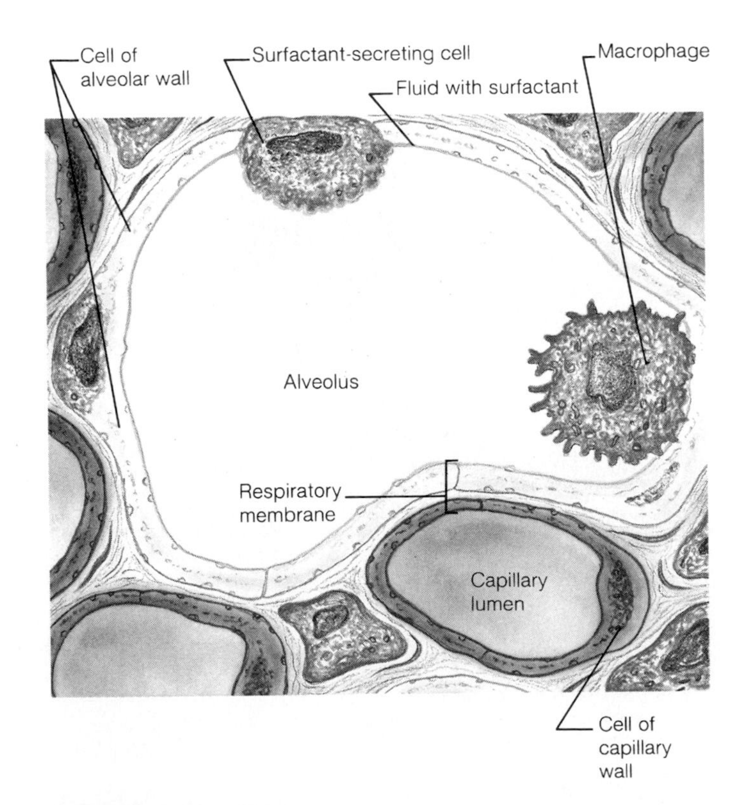
Cell of alveolar wall
Surfactant-secreting cell
Fluid with surfactant
Macrophage
Alveolus
Respiratory membrane
Capillary lumen
Cell of capillary wall

Respiratory Membrane
Figure 13.20

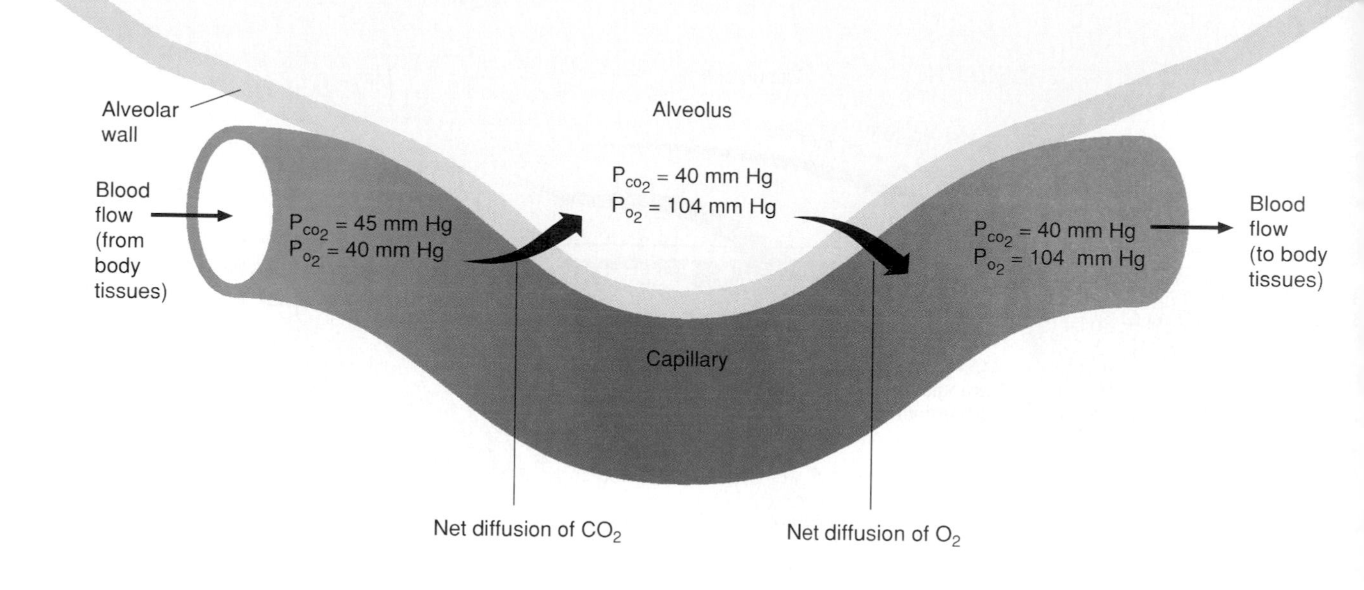

Alveolar wall
Alveolus
Blood flow (from body tissues)
P_{CO_2} = 45 mm Hg
P_{O_2} = 40 mm Hg
P_{CO_2} = 40 mm Hg
P_{O_2} = 104 mm Hg
P_{CO_2} = 40 mm Hg
P_{O_2} = 104 mm Hg
Blood flow (to body tissues)
Capillary
Net diffusion of CO_2
Net diffusion of O_2

Oxygen Transport
Figures 13.21 and 13.22

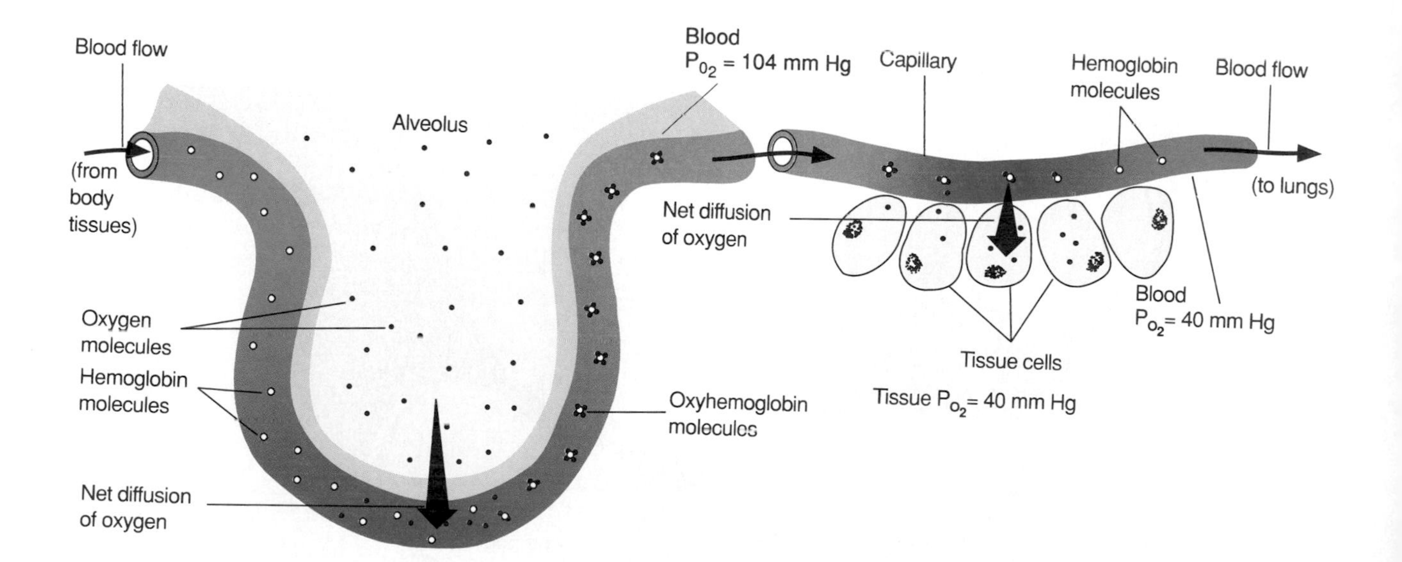

Blood flow
(from body tissues)
Alveolus
Oxygen molecules
Hemoglobin molecules
Net diffusion of oxygen
Oxyhemoglobin molecules
Blood
P_{O_2} = 104 mm Hg
Capillary
Hemoglobin molecules
Blood flow
(to lungs)
Net diffusion of oxygen
Blood
P_{O_2}= 40 mm Hg
Tissue cells
Tissue P_{O_2}= 40 mm Hg

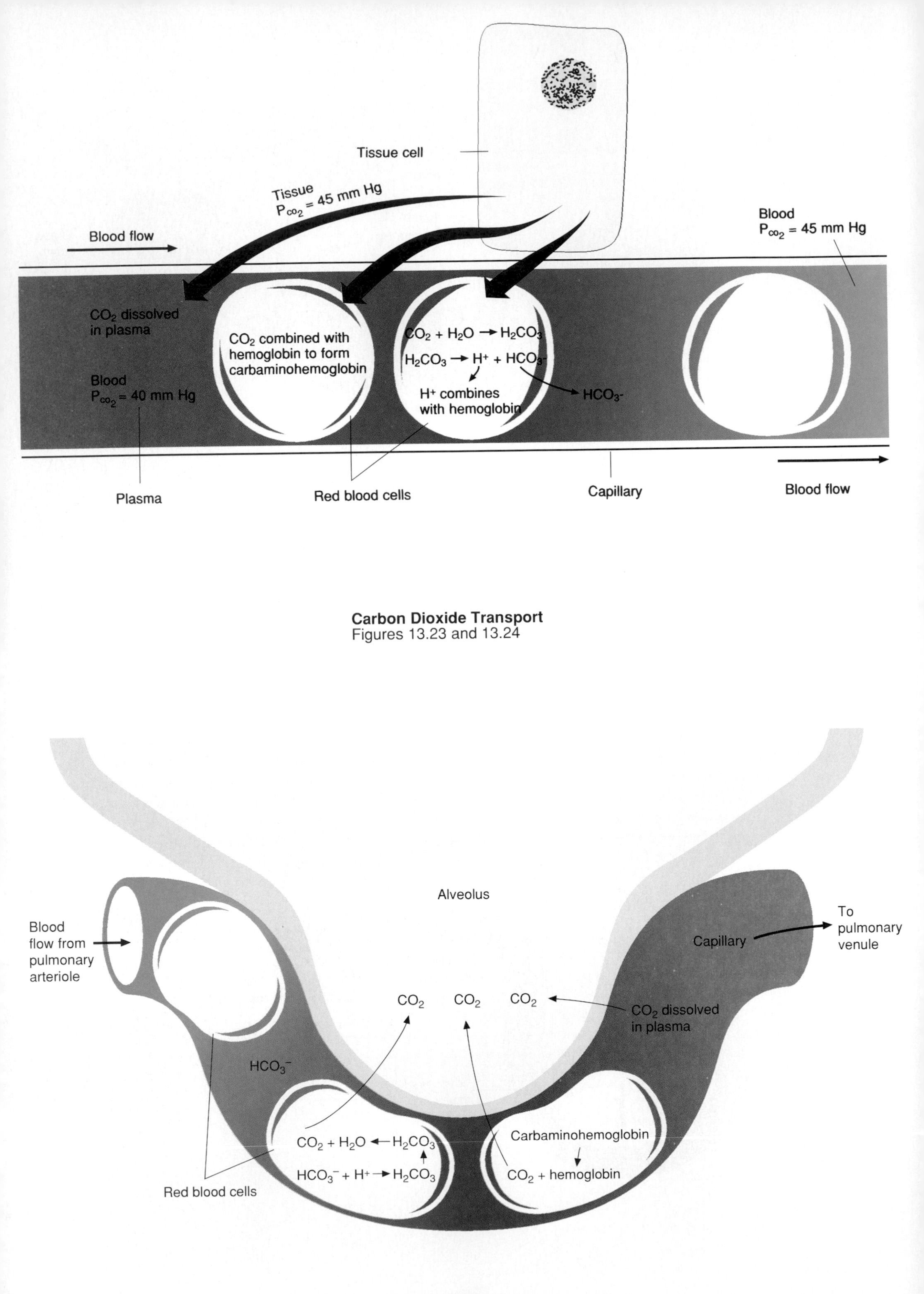
Tissue cell
Tissue
P_{CO_2} = 45 mm Hg
Blood flow
Blood
P_{CO_2} = 45 mm Hg
CO_2 dissolved in plasma
CO_2 combined with hemoglobin to form carbaminohemoglobin
$CO_2 + H_2O \rightarrow H_2CO_3$
$H_2CO_3 \rightarrow H^+ + HCO_3^-$
H^+ combines with hemoglobin
HCO_3^-
Blood
P_{CO_2} = 40 mm Hg
Plasma
Red blood cells
Capillary
Blood flow
Alveolus
Blood flow from pulmonary arteriole
Capillary
To pulmonary venule
CO_2
CO_2
CO_2
CO_2 dissolved in plasma
HCO_3^-
$CO_2 + H_2O \leftarrow H_2CO_3$
$HCO_3^- + H^+ \rightarrow H_2CO_3$
Carbaminohemoglobin
CO_2 + hemoglobin
Red blood cells

Carbon Dioxide Transport
Figures 13.23 and 13.24

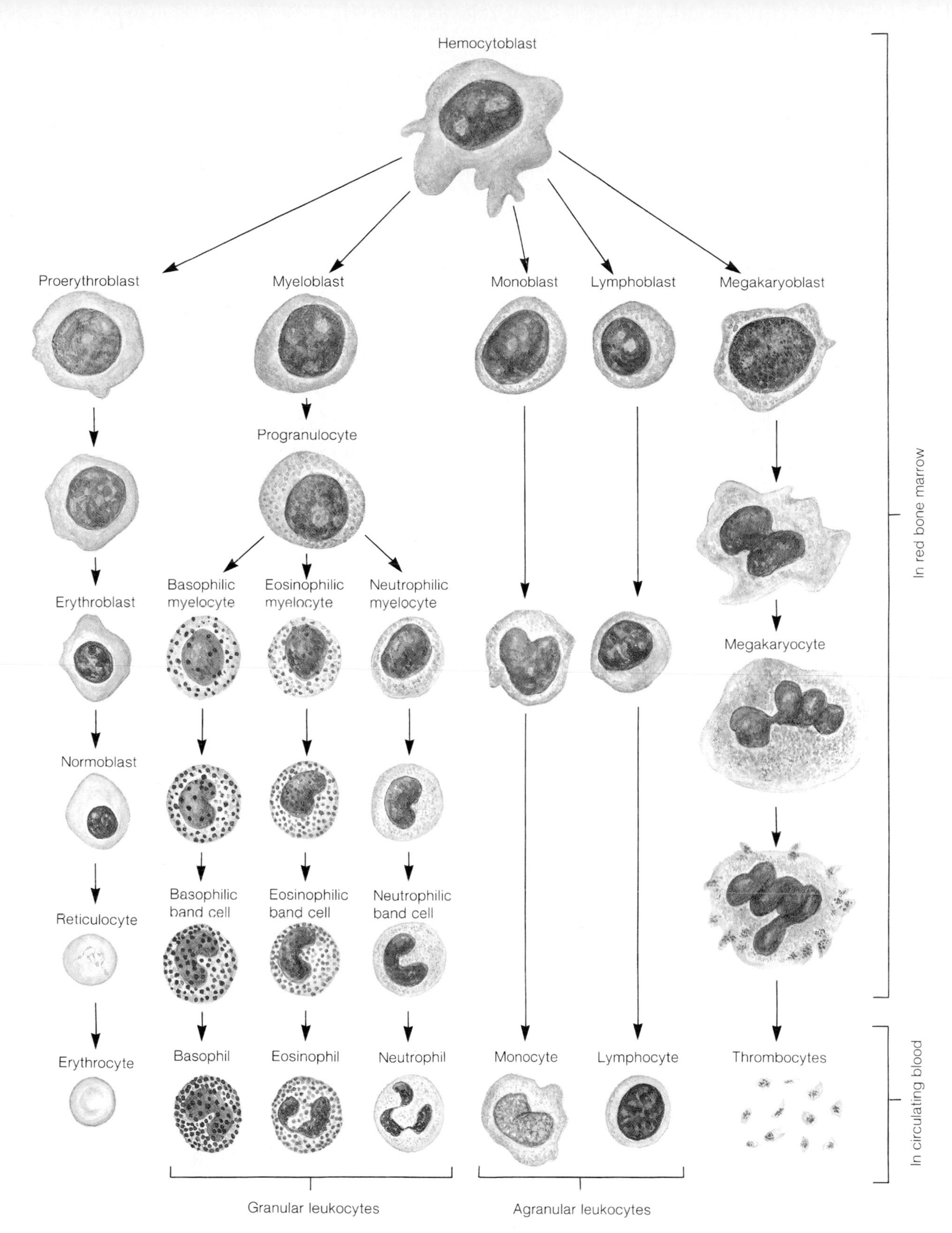
Hemocytoblast
Proerythroblast
Myeloblast
Monoblast
Lymphoblast
Megakaryoblast
Progranulocyte
Erythroblast
Basophilic myelocyte
Eosinophilic myelocyte
Neutrophilic myelocyte
Megakaryocyte
Normoblast
Reticulocyte
Basophilic band cell
Eosinophilic band cell
Neutrophilic band cell
Erythrocyte
Basophil
Eosinophil
Neutrophil
Monocyte
Lymphocyte
Thrombocytes
Granular leukocytes
Agranular leukocytes
In red bone marrow
In circulating blood

Blood Cell Development
Figure 14.4

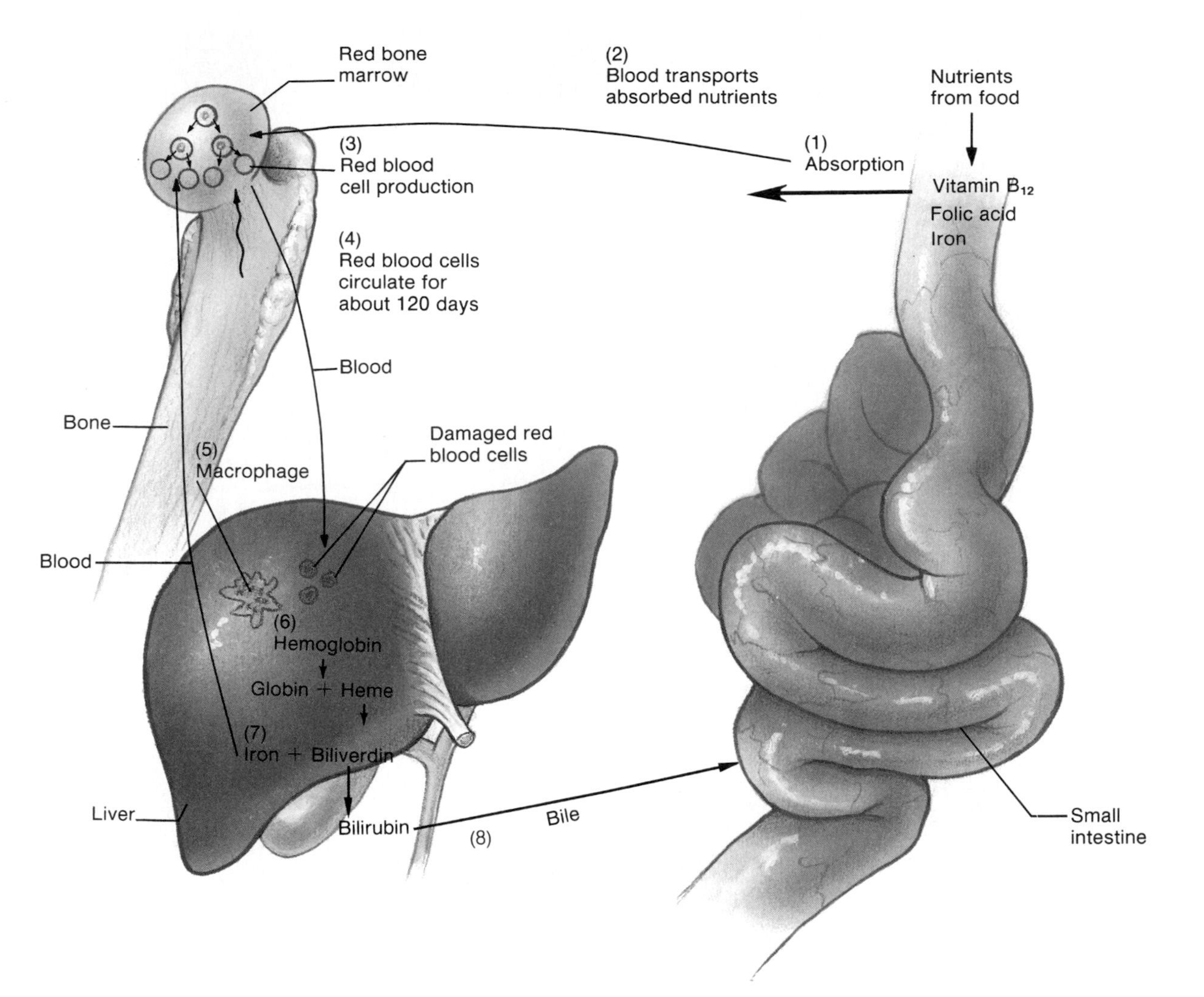
Red bone marrow
(2)
Blood transports absorbed nutrients
Nutrients from food
(1)
Absorption
(3)
Red blood cell production
Vitamin B12
Folic acid
Iron
(4)
Red blood cells circulate for about 120 days
Blood
Bone
(5)
Macrophage
Damaged red blood cells
Blood
(6)
Hemoglobin
Globin + Heme
(7)
Iron + Biliverdin
Liver
Bilirubin
(8)
Bile
Small intestine

Red Blood Cell Life Cycle
Figure 14.5

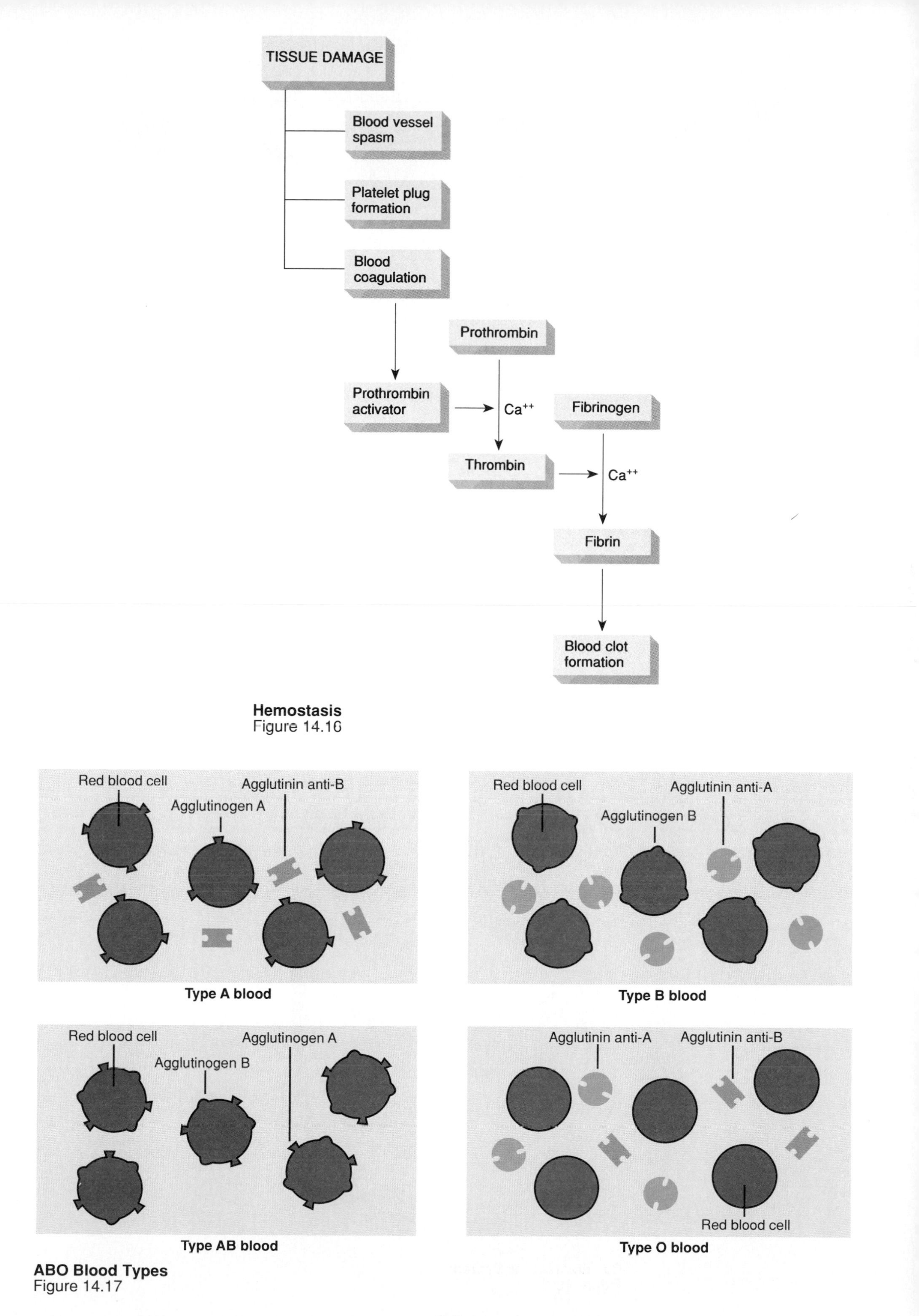

TISSUE DAMAGE
Blood vessel spasm
Platelet plug formation
Blood coagulation
Prothrombin
Prothrombin activator
Ca++
Fibrinogen
Thrombin
Ca++
Fibrin
Blood clot formation

Hemostasis
Figure 14.16

Red blood cell
Agglutinin anti-B
Agglutinogen A
Type A blood
Red blood cell
Agglutinin anti-A
Agglutinogen B
Type B blood
Red blood cell
Agglutinogen A
Agglutinogen B
Type AB blood
Agglutinin anti-A
Agglutinin anti-B
Red blood cell
Type O blood

ABO Blood Types
Figure 14.17

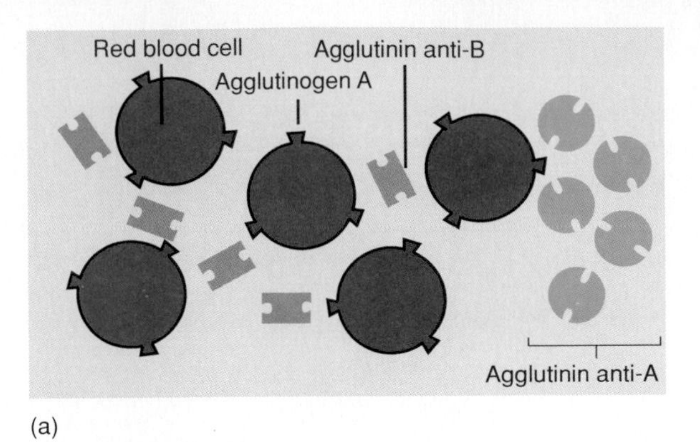
Red blood cell
Agglutinin anti-B
Agglutinogen A
Agglutinin anti-A

(a)

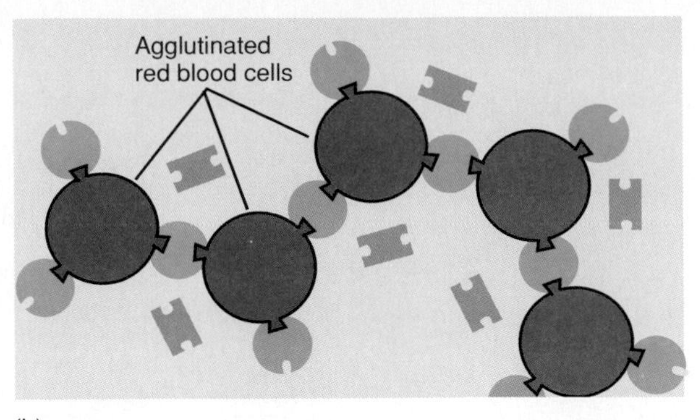
Agglutinated
red blood cells

(b)

ABO Blood Reactions
Figure 14.18

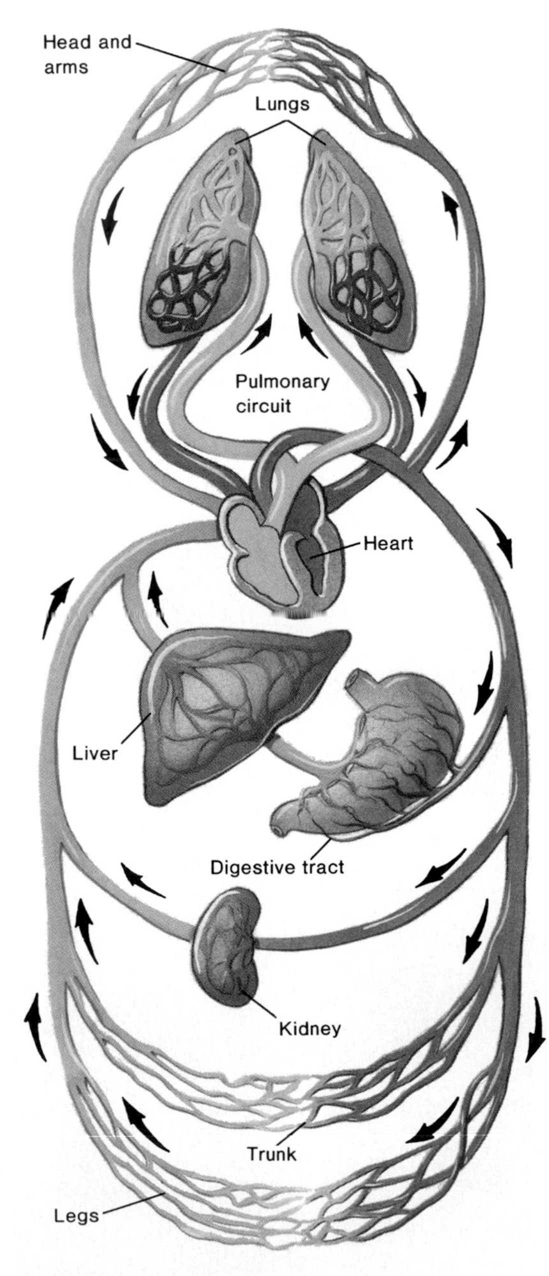
Head and
arms
Lungs
Pulmonary
circuit
Heart
Liver
Digestive tract
Kidney
Trunk
Legs

Cardiovascular System
Figure 15.1

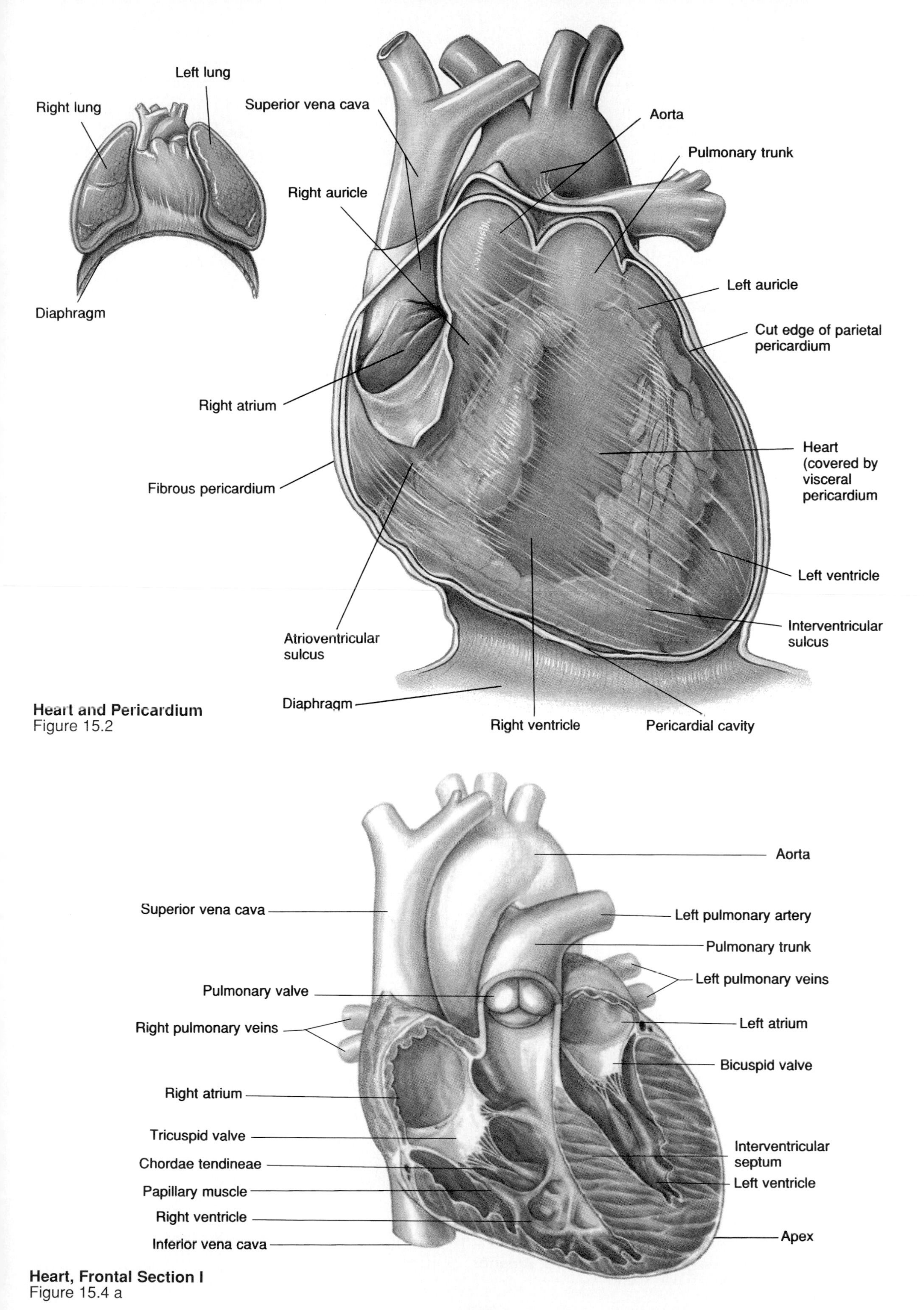
Left lung
Right lung
Diaphragm
Superior vena cava
Aorta
Pulmonary trunk
Right auricle
Left auricle
Cut edge of parietal pericardium
Right atrium
Heart (covered by visceral pericardium
Fibrous pericardium
Left ventricle
Interventricular sulcus
Atrioventricular sulcus
Diaphragm
Right ventricle
Pericardial cavity
Aorta
Superior vena cava
Left pulmonary artery
Pulmonary trunk
Left pulmonary veins
Pulmonary valve
Right pulmonary veins
Left atrium
Bicuspid valve
Right atrium
Tricuspid valve
Interventricular septum
Chordae tendineae
Left ventricle
Papillary muscle
Right ventricle
Apex
Inferior vena cava

Heart and Pericardium
Figure 15.2

Heart, Frontal Section I
Figure 15.4 a

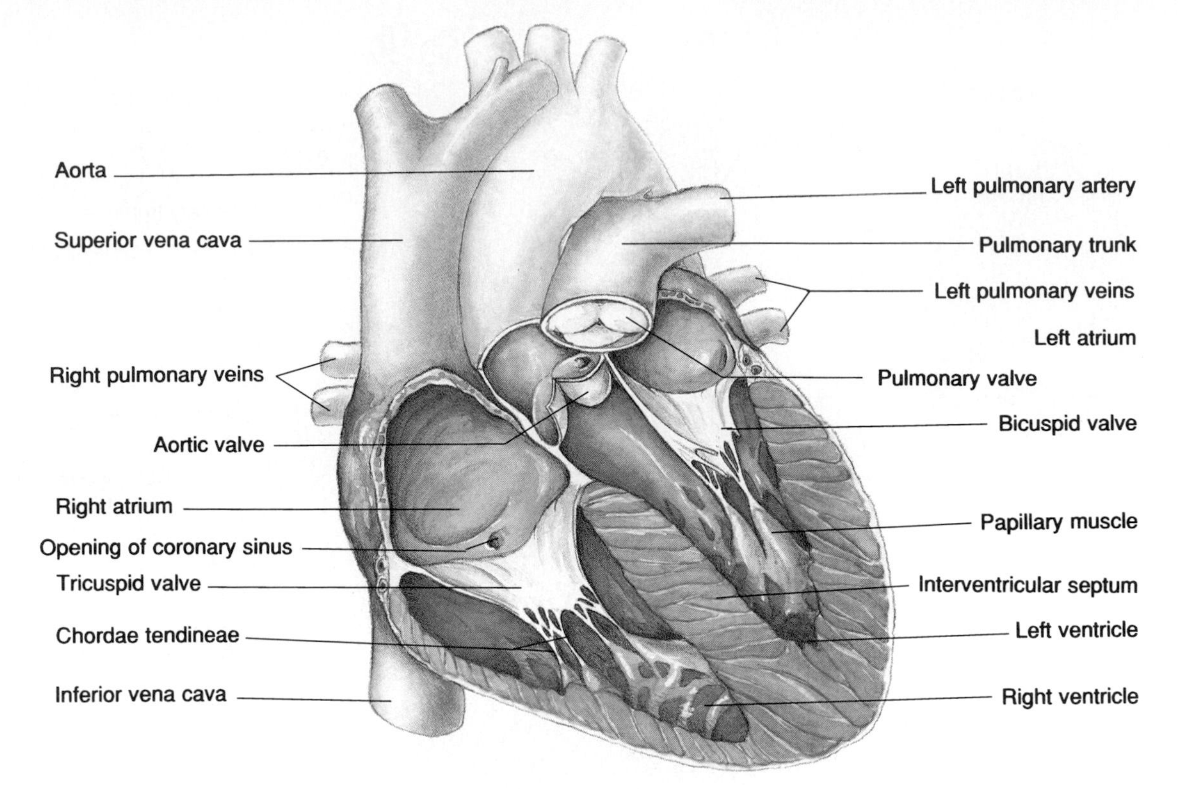
Aorta
Superior vena cava
Right pulmonary veins
Aortic valve
Right atrium
Opening of coronary sinus
Tricuspid valve
Chordae tendineae
Inferior vena cava
Left pulmonary artery
Pulmonary trunk
Left pulmonary veins
Left atrium
Pulmonary valve
Bicuspid valve
Papillary muscle
Interventricular septum
Left ventricle
Right ventricle

Heart, Frontal Section II
Figure 15.4 b

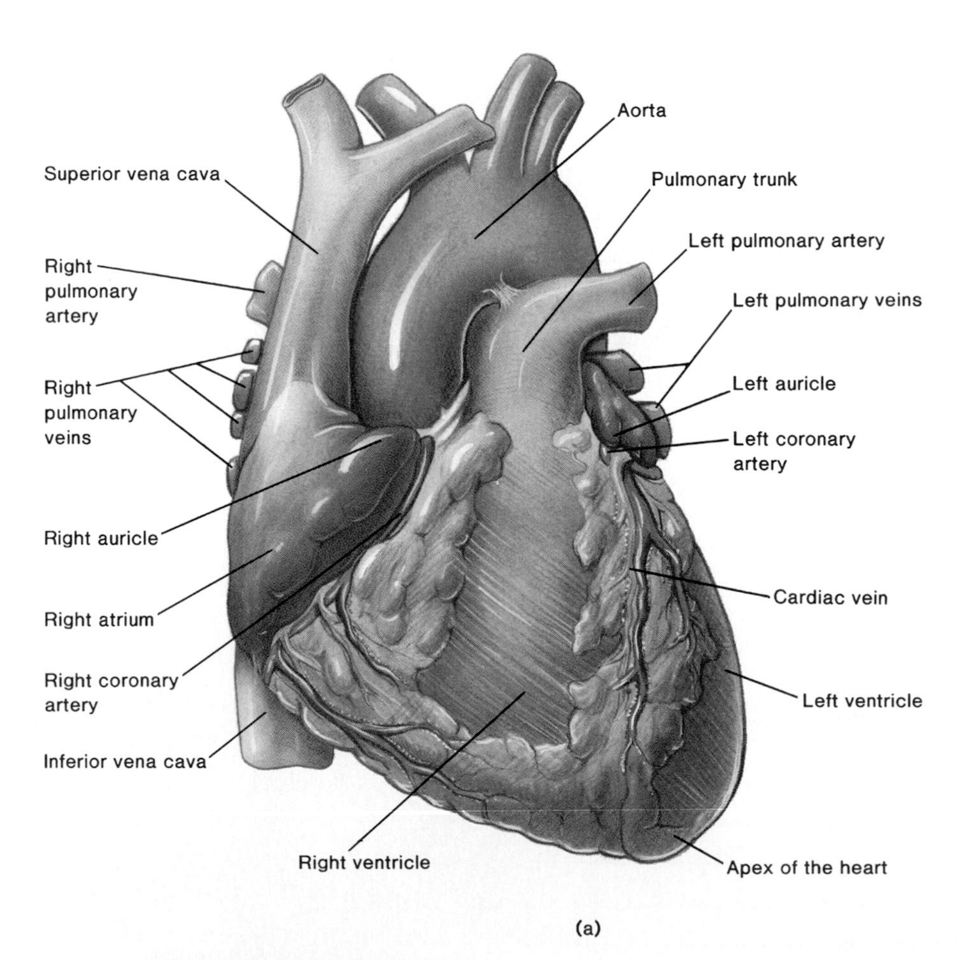
Aorta
Superior vena cava
Pulmonary trunk
Left pulmonary artery
Right pulmonary artery
Left pulmonary veins
Right pulmonary veins
Left auricle
Left coronary artery
Right auricle
Cardiac vein
Right atrium
Right coronary artery
Left ventricle
Inferior vena cava
Right ventricle
Apex of the heart
(a)

Heart and Coronary Vessels, Anterior
Figure 15.9 a

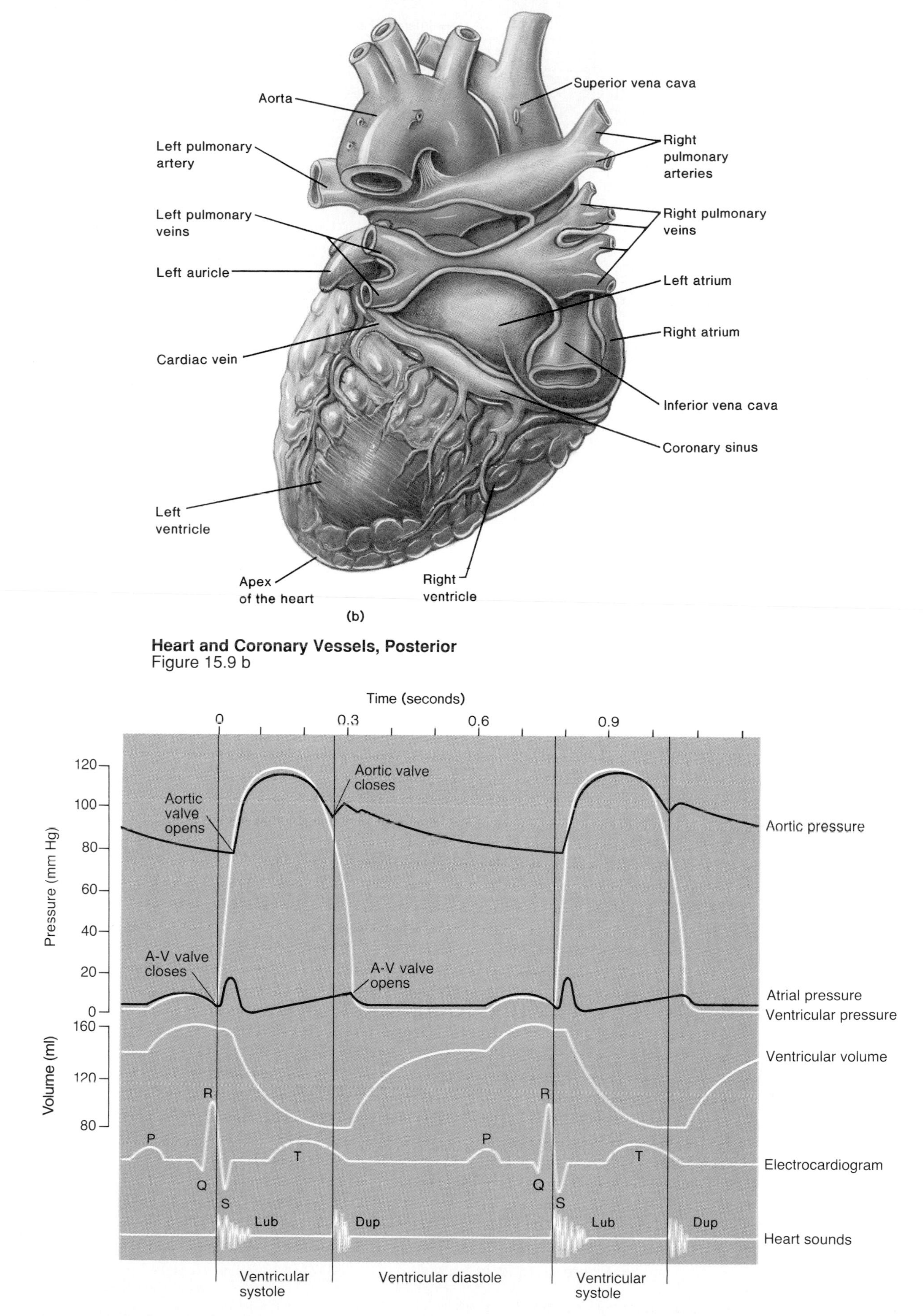

Aorta
Superior vena cava
Left pulmonary artery
Right pulmonary arteries
Left pulmonary veins
Right pulmonary veins
Left auricle
Left atrium
Right atrium
Cardiac vein
Inferior vena cava
Coronary sinus
Left ventricle
Apex of the heart
Right ventricle
(b)

Heart and Coronary Vessels, Posterior
Figure 15.9 b

Time (seconds)
0
0.3
0.6
0.9
120
100
80
60
40
20
0
Pressure (mm Hg)
Aortic valve opens
Aortic valve closes
Aortic pressure
A-V valve closes
A-V valve opens
Atrial pressure
Ventricular pressure
160
120
80
Volume (ml)
Ventricular volume
R
P
T
Q
S
Electrocardiogram
Lub
Dup
Heart sounds
Ventricular systole
Ventricular diastole
Ventricular systole

Events in the Cardiac Cycle
Figure 15.11

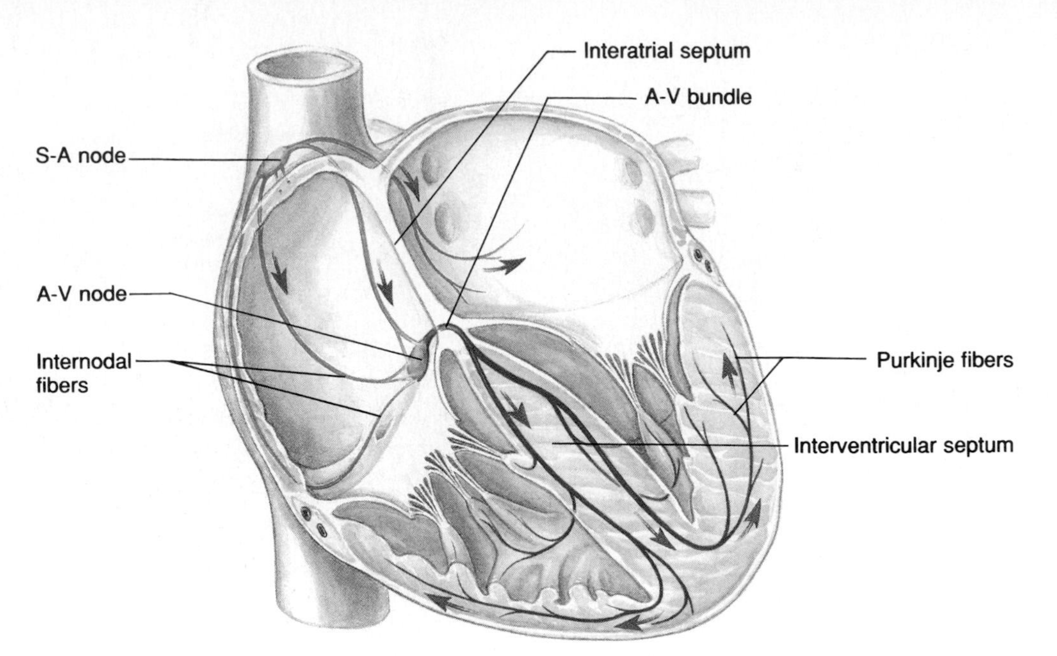
Interatrial septum
A-V bundle
S-A node
A-V node
Internodal fibers
Purkinje fibers
Interventricular septum

Cardiac Conduction System
Figure 15.12

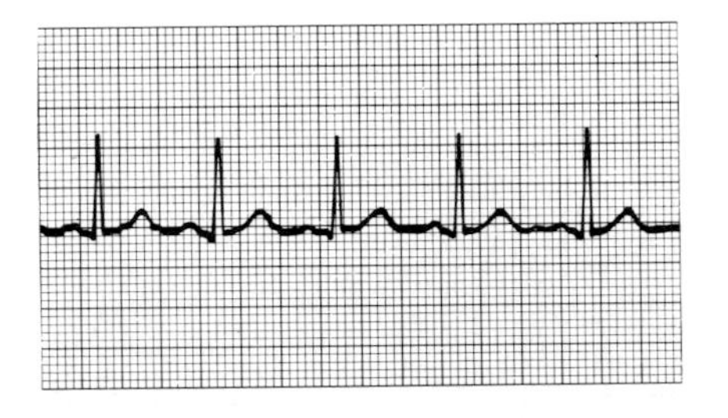

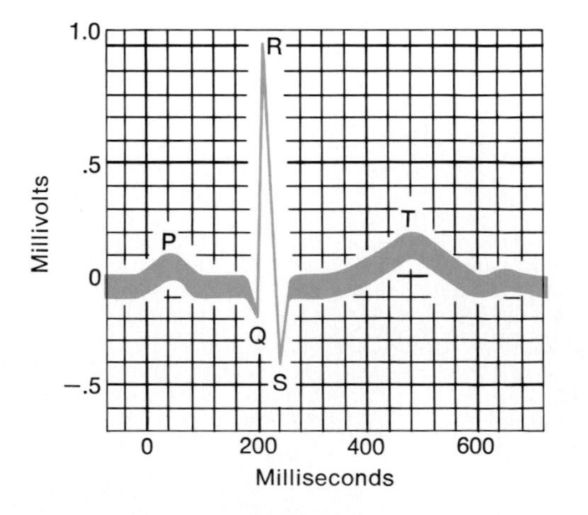
1.0
.5
0
−.5
Millivolts
R
P
T
Q
S
0
200
400
600
Milliseconds

ECG Pattern
Figure 15.15 (a–b)

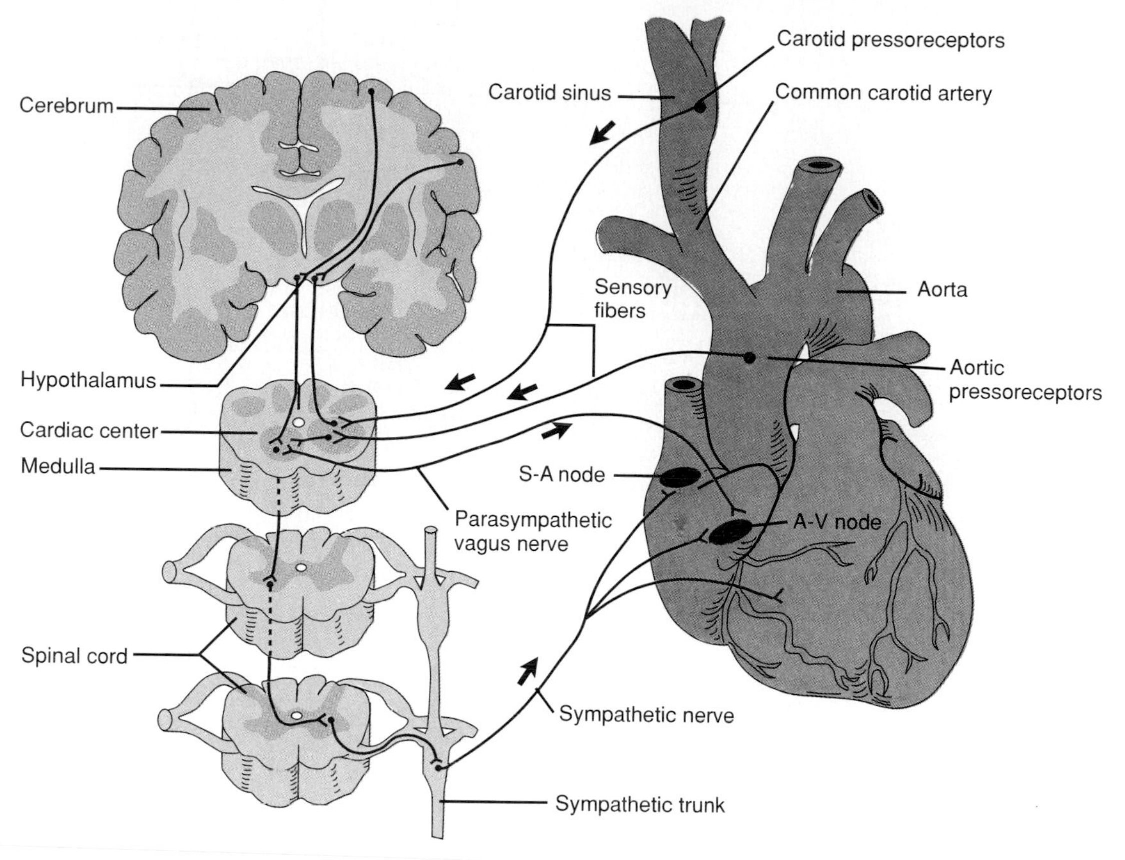

Carotid pressoreceptors
Carotid sinus
Common carotid artery
Cerebrum
Sensory fibers
Aorta
Hypothalamus
Aortic pressoreceptors
Cardiac center
Medulla
S-A node
Parasympathetic vagus nerve
A-V node
Spinal cord
Sympathetic nerve
Sympathetic trunk

Heart Rate Control
Figure 15.17

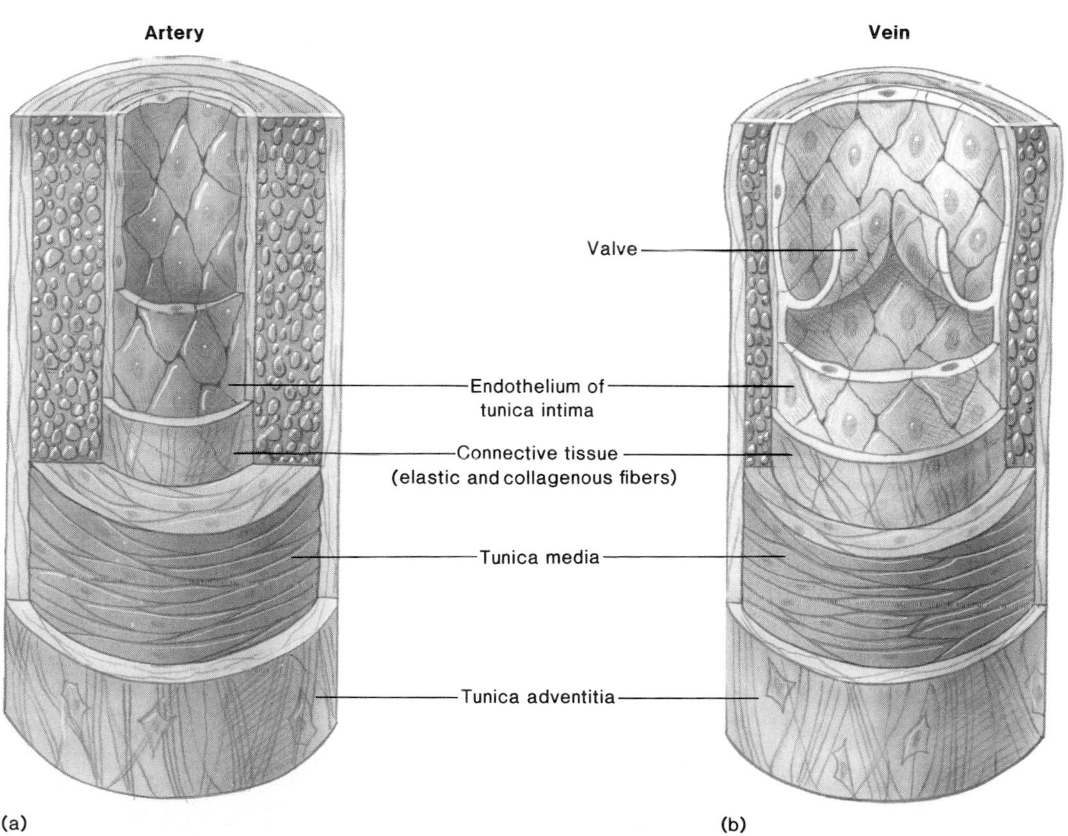

Artery
Vein
Valve
Endothelium of tunica intima
Connective tissue (elastic and collagenous fibers)
Tunica media
Tunica adventitia
(a)
(b)

Wall of an Artery and a Vein
Figure 15.18 (a–b)

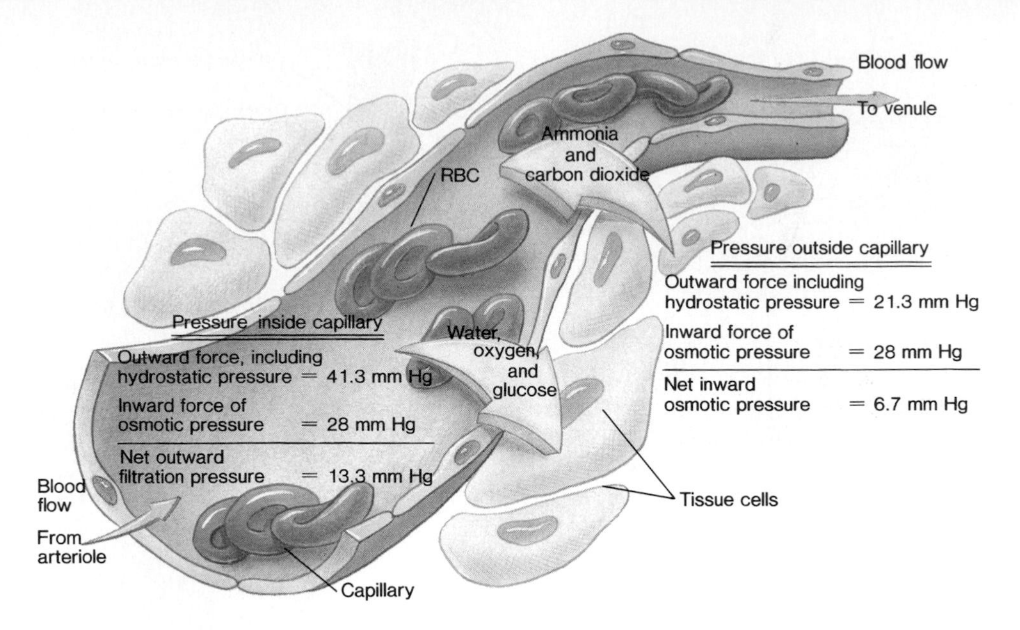
Blood flow
To venule
Ammonia
and
carbon dioxide
RBC
Pressure outside capillary
Outward force including
hydrostatic pressure = 21.3 mm Hg
Inward force of
osmotic pressure = 28 mm Hg
Net inward
osmotic pressure = 6.7 mm Hg
Pressure inside capillary
Outward force, including
hydrostatic pressure = 41.3 mm Hg
Inward force of
osmotic pressure = 28 mm Hg
Net outward
filtration pressure = 13.3 mm Hg
Water,
oxygen,
and
glucose
Blood
flow
From
arteriole
Tissue cells
Capillary

Capillary Exchanges
Figure 15.22

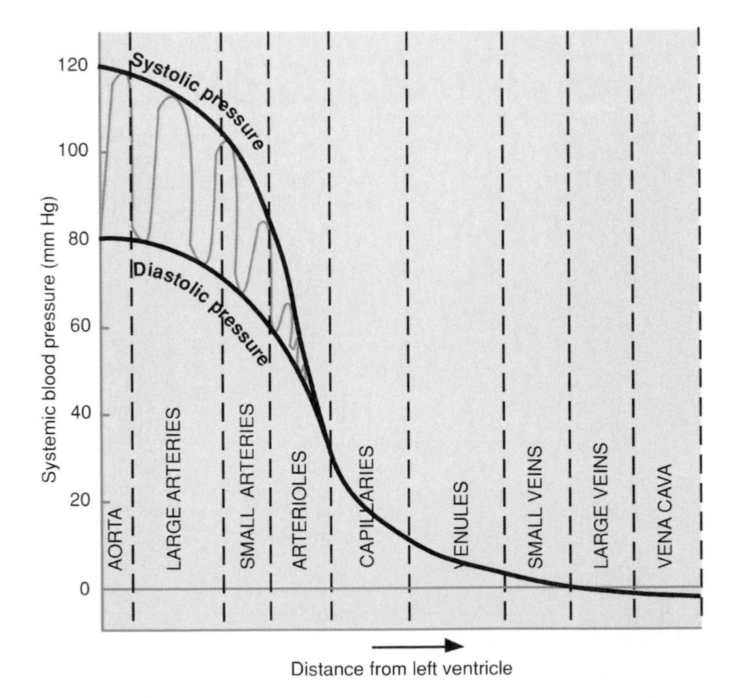
Systolic pressure
Diastolic pressure
Systemic blood pressure (mm Hg)
120
100
80
60
40
20
0
AORTA
LARGE ARTERIES
SMALL ARTERIES
ARTERIOLES
CAPILLARIES
VENULES
SMALL VEINS
LARGE VEINS
VENA CAVA
Distance from left ventricle

Blood Pressure Changes within Blood Vessels
Figure 15.25

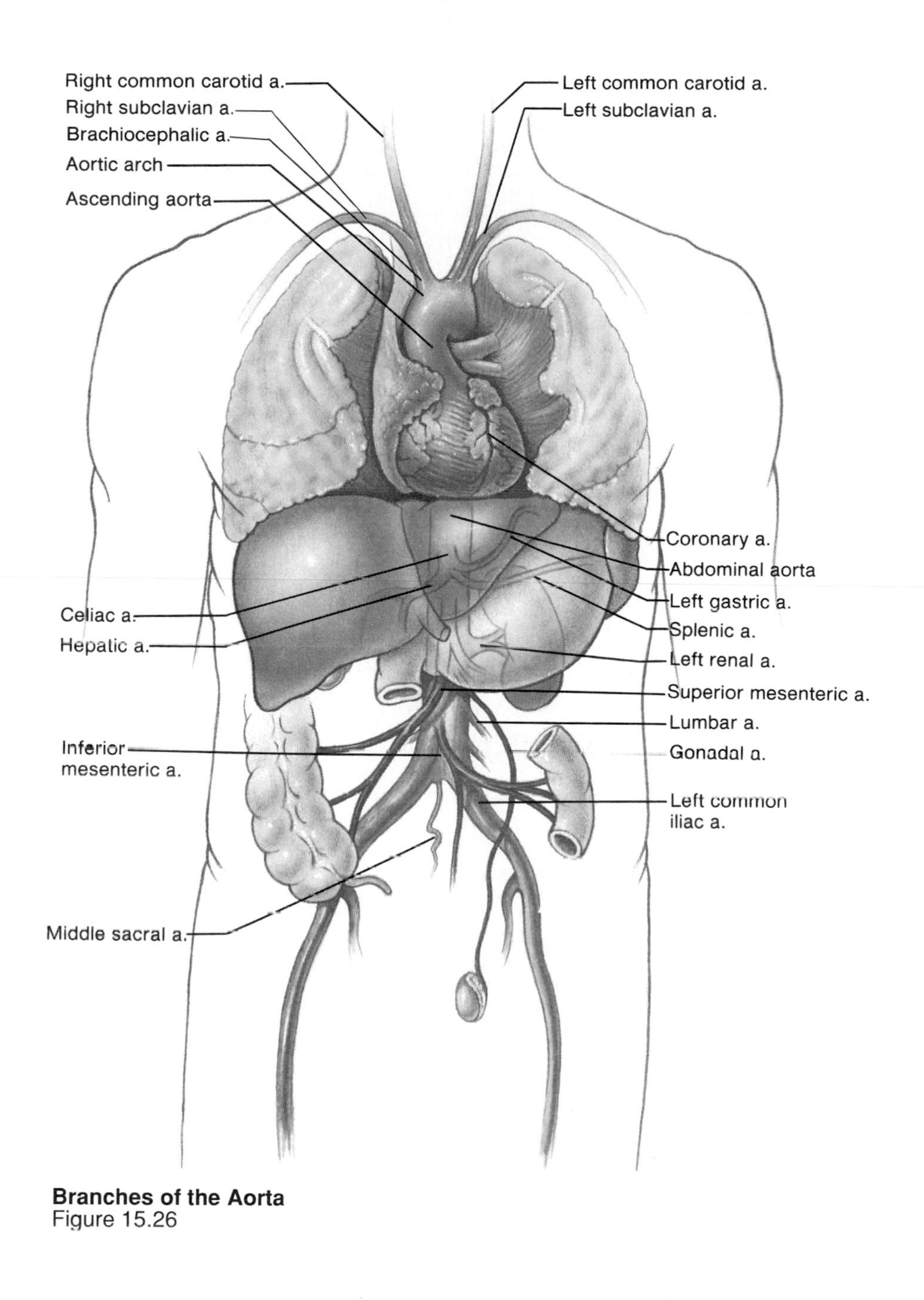

Right common carotid a.
Right subclavian a.
Brachiocephalic a.
Aortic arch
Ascending aorta
Left common carotid a.
Left subclavian a.
Coronary a.
Abdominal aorta
Left gastric a.
Splenic a.
Left renal a.
Superior mesenteric a.
Lumbar a.
Gonadal a.
Left common iliac a.
Celiac a.
Hepatic a.
Inferior mesenteric a.
Middle sacral a.

Branches of the Aorta
Figure 15.26

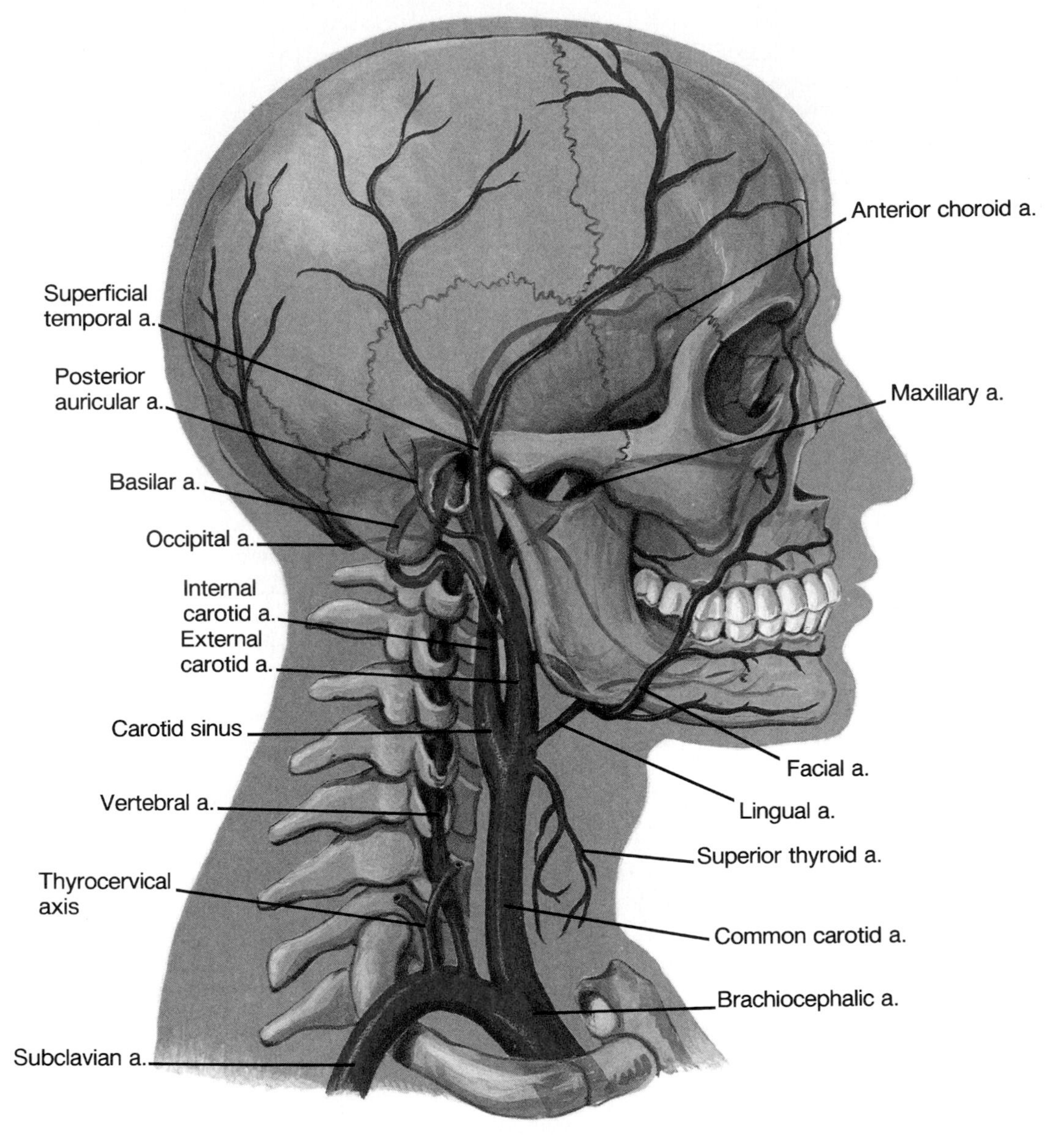
Anterior choroid a.
Superficial temporal a.
Posterior auricular a.
Maxillary a.
Basilar a.
Occipital a.
Internal carotid a.
External carotid a.
Carotid sinus
Facial a.
Lingual a.
Vertebral a.
Superior thyroid a.
Thyrocervical axis
Common carotid a.
Brachiocephalic a.
Subclavian a.

Arteries of the Brain, Head, and Neck
Figure 15.27

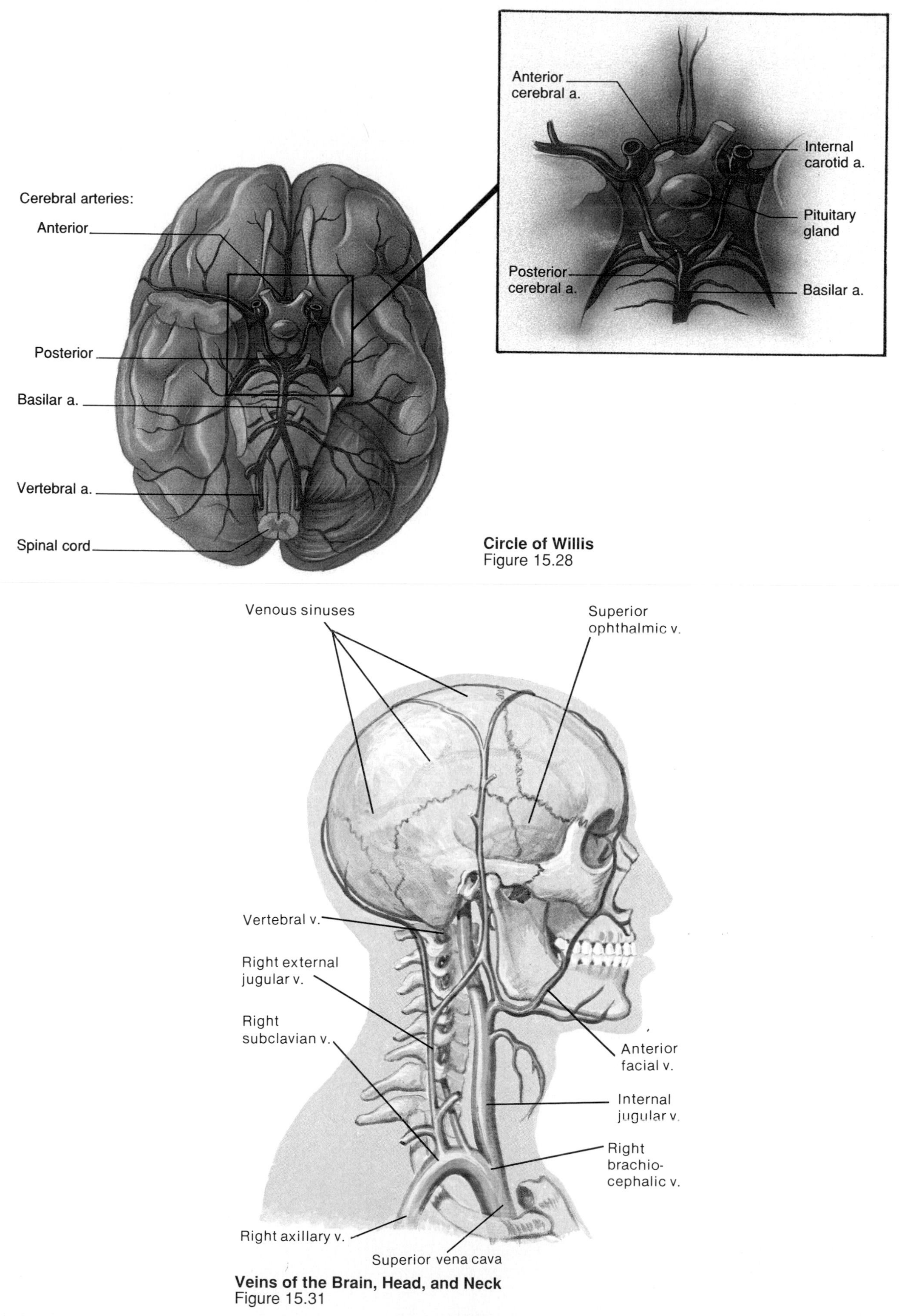
Anterior cerebral a.
Internal carotid a.
Pituitary gland
Posterior cerebral a.
Basilar a.
Cerebral arteries:
Anterior
Posterior
Basilar a.
Vertebral a.
Spinal cord

Circle of Willis
Figure 15.28

Venous sinuses
Superior ophthalmic v.
Vertebral v.
Right external jugular v.
Right subclavian v.
Anterior facial v.
Internal jugular v.
Right brachio-cephalic v.
Right axillary v.
Superior vena cava

Veins of the Brain, Head, and Neck
Figure 15.31

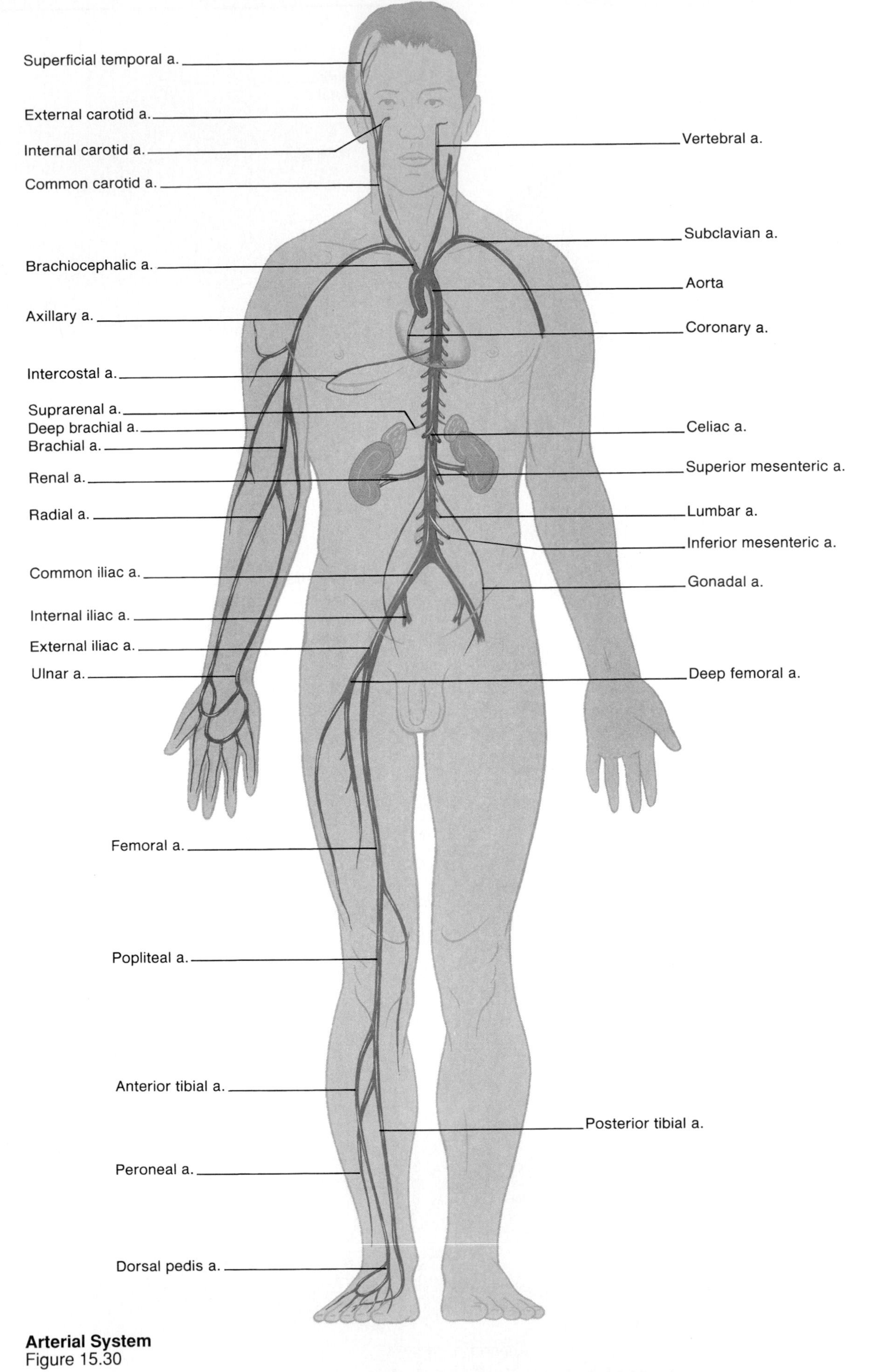
Superficial temporal a.
External carotid a.
Internal carotid a.
Common carotid a.
Brachiocephalic a.
Axillary a.
Intercostal a.
Suprarenal a.
Deep brachial a.
Brachial a.
Renal a.
Radial a.
Common iliac a.
Internal iliac a.
External iliac a.
Ulnar a.
Femoral a.
Popliteal a.
Anterior tibial a.
Peroneal a.
Dorsal pedis a.
Vertebral a.
Subclavian a.
Aorta
Coronary a.
Celiac a.
Superior mesenteric a.
Lumbar a.
Inferior mesenteric a.
Gonadal a.
Deep femoral a.
Posterior tibial a.

Arterial System
Figure 15.30

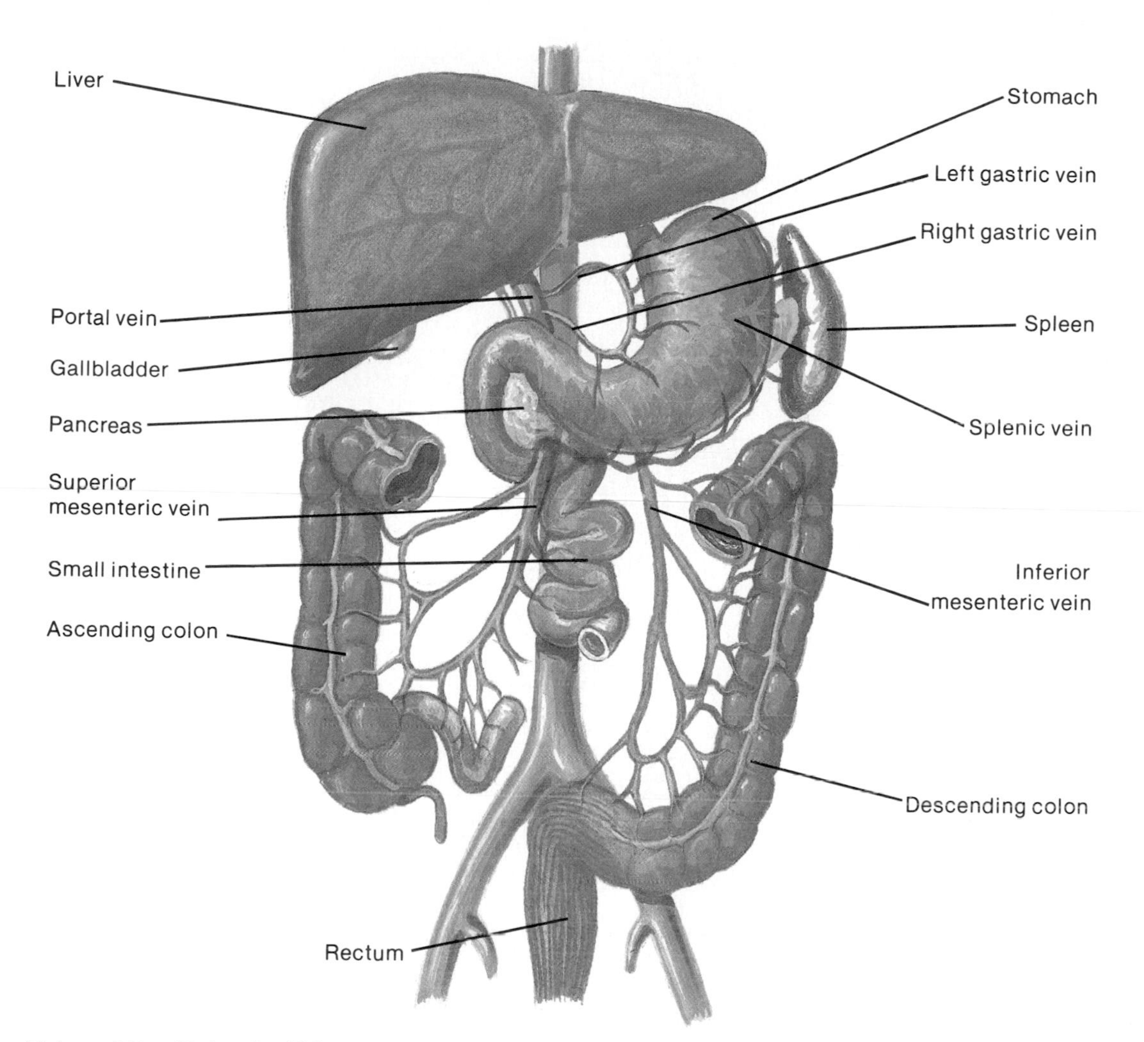
Liver
Stomach
Left gastric vein
Right gastric vein
Portal vein
Spleen
Gallbladder
Pancreas
Splenic vein
Superior
mesenteric vein
Small intestine
Inferior
mesenteric vein
Ascending colon
Descending colon
Rectum

Veins of the Abdominal Viscera
Figure 15.33

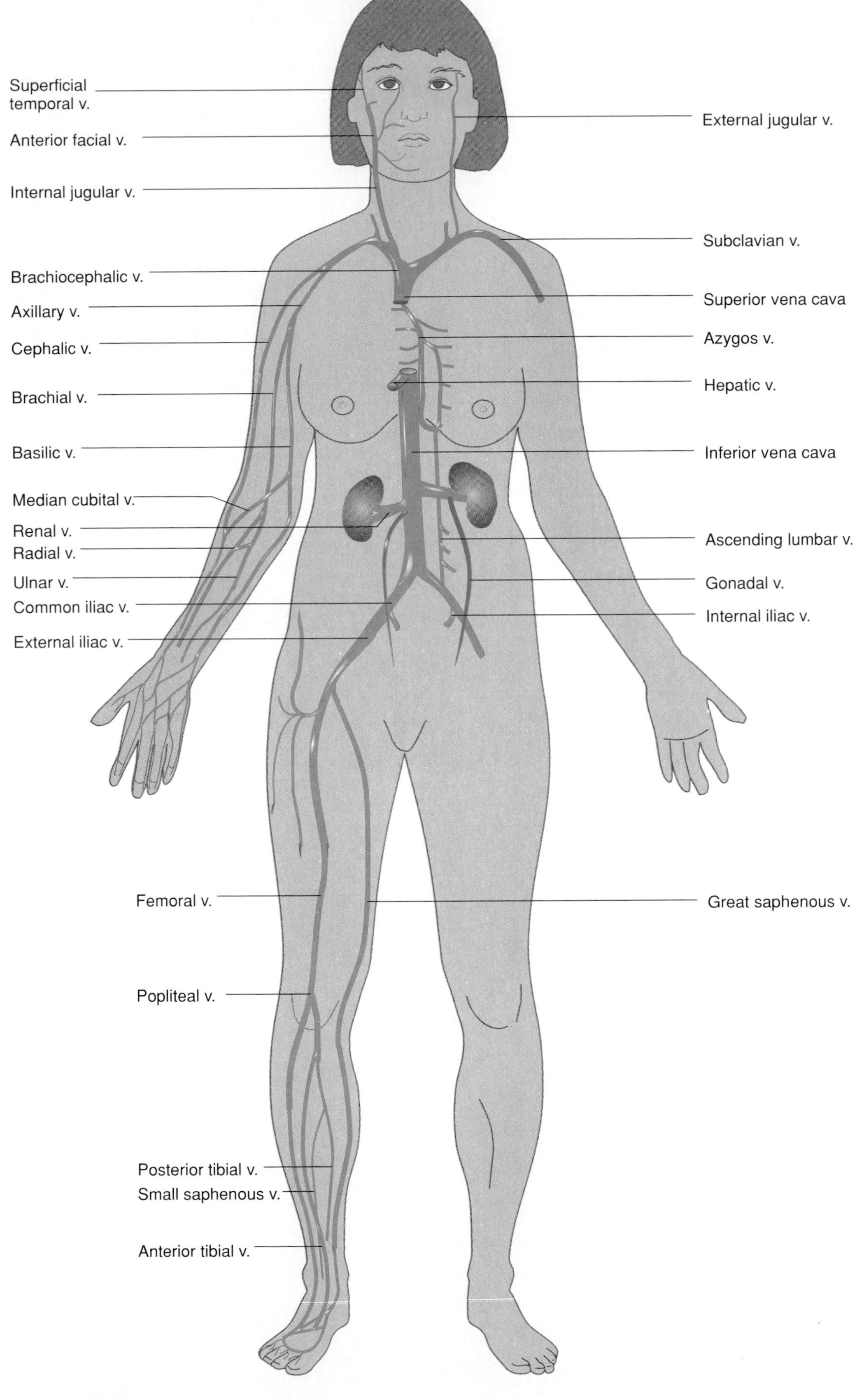
Superficial temporal v.
Anterior facial v.
Internal jugular v.
Brachiocephalic v.
Axillary v.
Cephalic v.
Brachial v.
Basilic v.
Median cubital v.
Renal v.
Radial v.
Ulnar v.
Common iliac v.
External iliac v.
Femoral v.
Popliteal v.
Posterior tibial v.
Small saphenous v.
Anterior tibial v.
External jugular v.
Subclavian v.
Superior vena cava
Azygos v.
Hepatic v.
Inferior vena cava
Ascending lumbar v.
Gonadal v.
Internal iliac v.
Great saphenous v.

Venous System
Figure 15.34

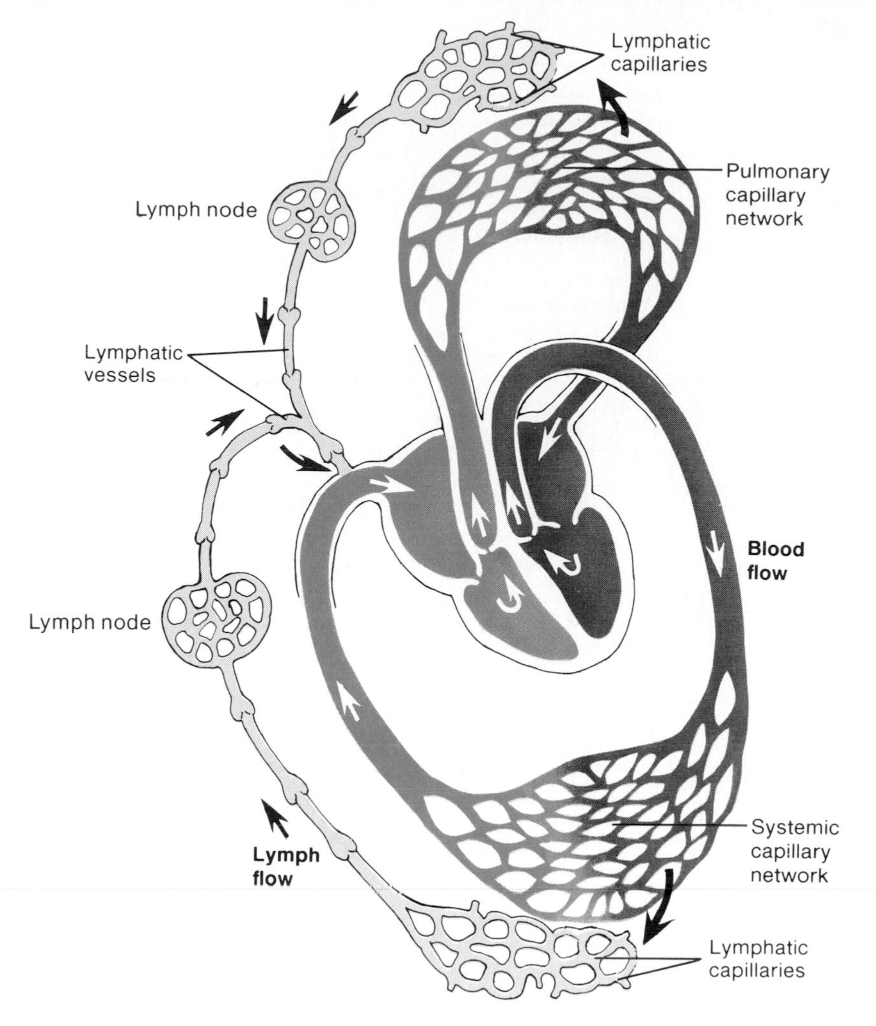
Lymphatic capillaries
Pulmonary capillary network
Lymph node
Lymphatic vessels
Blood flow
Lymph node
Systemic capillary network
Lymph flow
Lymphatic capillaries

Lymphatic System
Figure 16.1

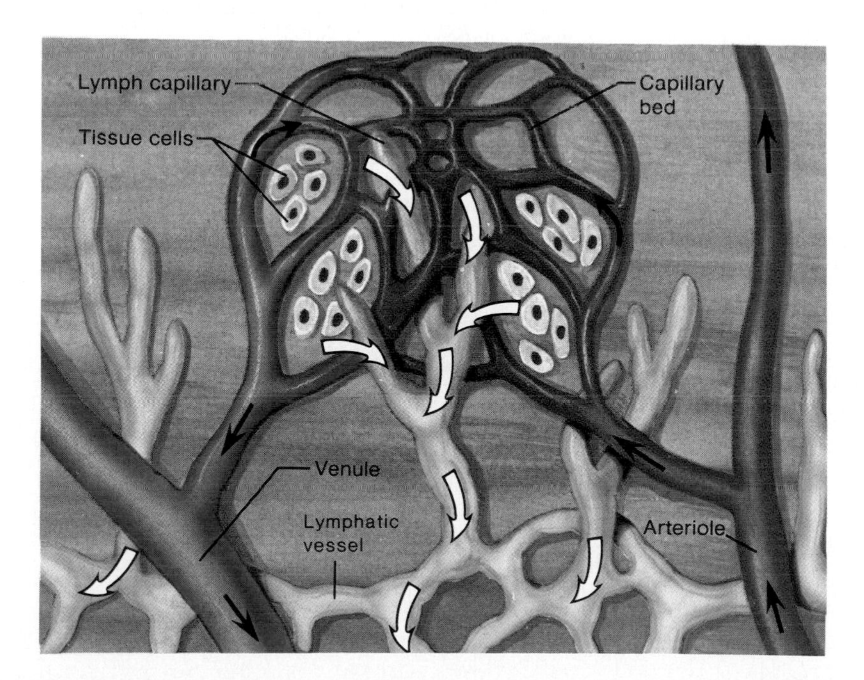
Lymph capillary
Capillary bed
Tissue cells
Venule
Lymphatic vessel
Arteriole

Lymph Formation
Figure 16.2

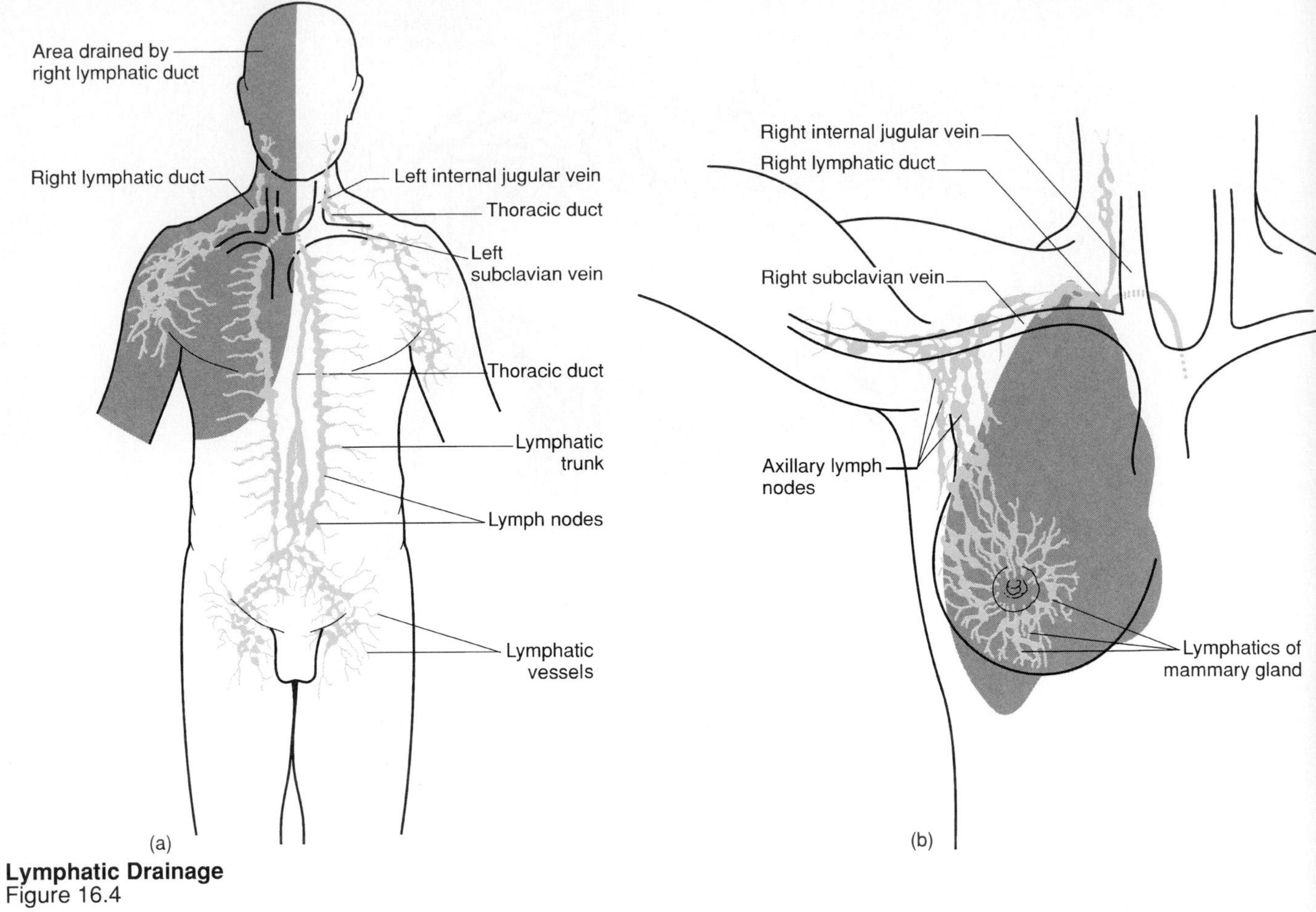

Area drained by
right lymphatic duct
Right lymphatic duct
Left internal jugular vein
Thoracic duct
Left
subclavian vein
Thoracic duct
Lymphatic
trunk
Lymph nodes
Lymphatic
vessels
(a)
Right internal jugular vein
Right lymphatic duct
Right subclavian vein
Axillary lymph
nodes
Lymphatics of
mammary gland
(b)

Lymphatic Drainage
Figure 16.4

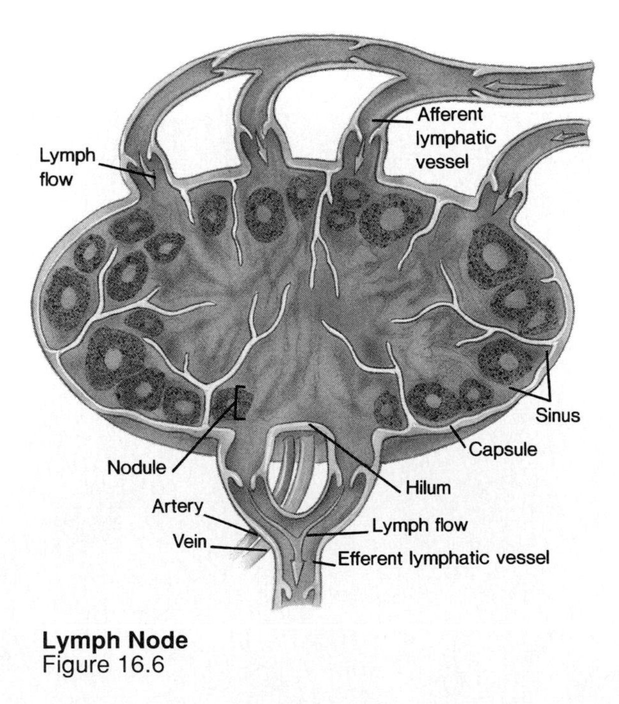

Afferent
lymphatic
vessel
Lymph
flow
Sinus
Capsule
Nodule
Hilum
Artery
Vein
Lymph flow
Efferent lymphatic vessel

Lymph Node
Figure 16.6

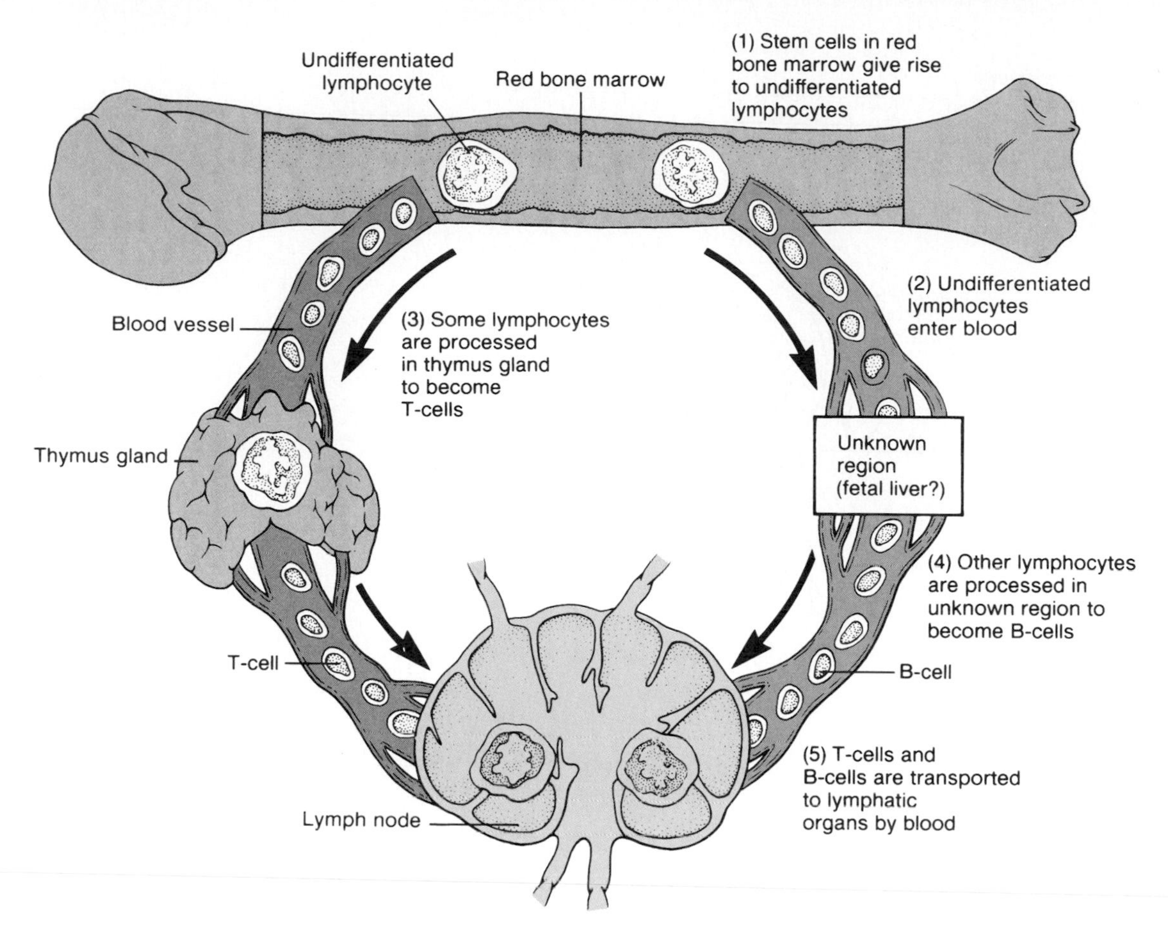

Undifferentiated lymphocyte
Red bone marrow
(1) Stem cells in red bone marrow give rise to undifferentiated lymphocytes
(2) Undifferentiated lymphocytes enter blood
Blood vessel
(3) Some lymphocytes are processed in thymus gland to become T-cells
Thymus gland
Unknown region (fetal liver?)
(4) Other lymphocytes are processed in unknown region to become B-cells
T-cell
B-cell
(5) T-cells and B-cells are transported to lymphatic organs by blood
Lymph node

B-cell and T-cell Origins
Figure 16.11

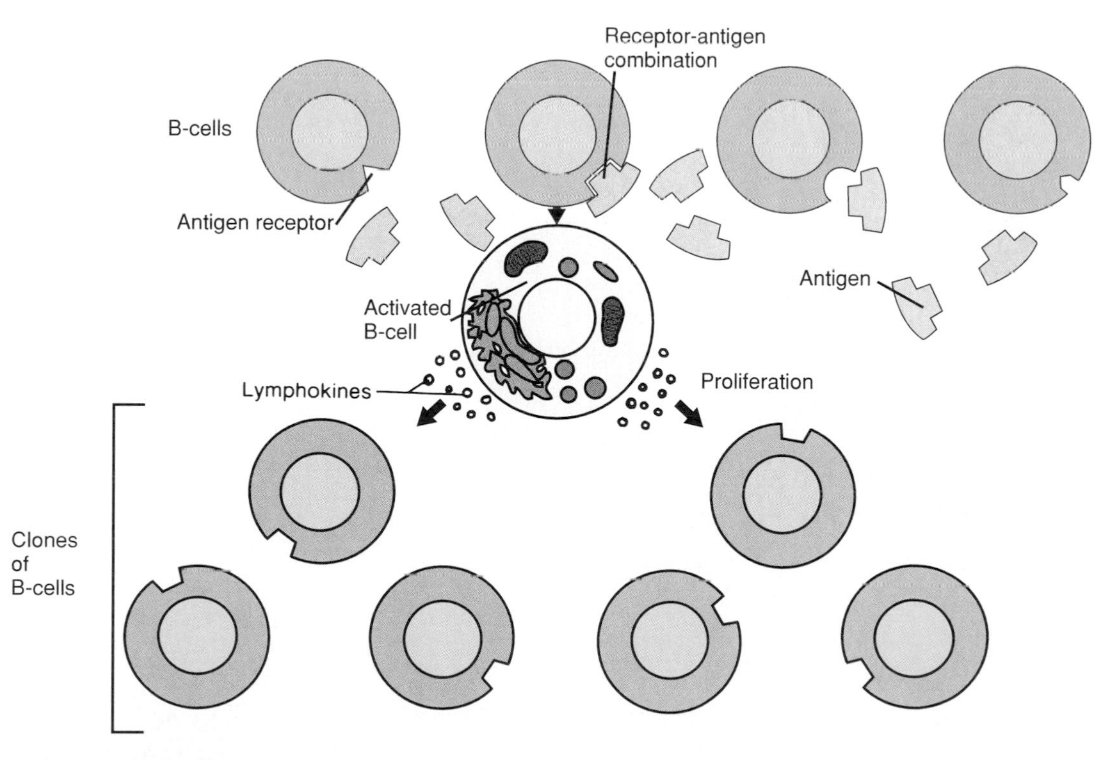

Receptor-antigen combination
B-cells
Antigen receptor
Antigen
Activated B-cell
Lymphokines
Proliferation
Clones of B-cells

Enlargement of B-cell Clone
Figure 16.13

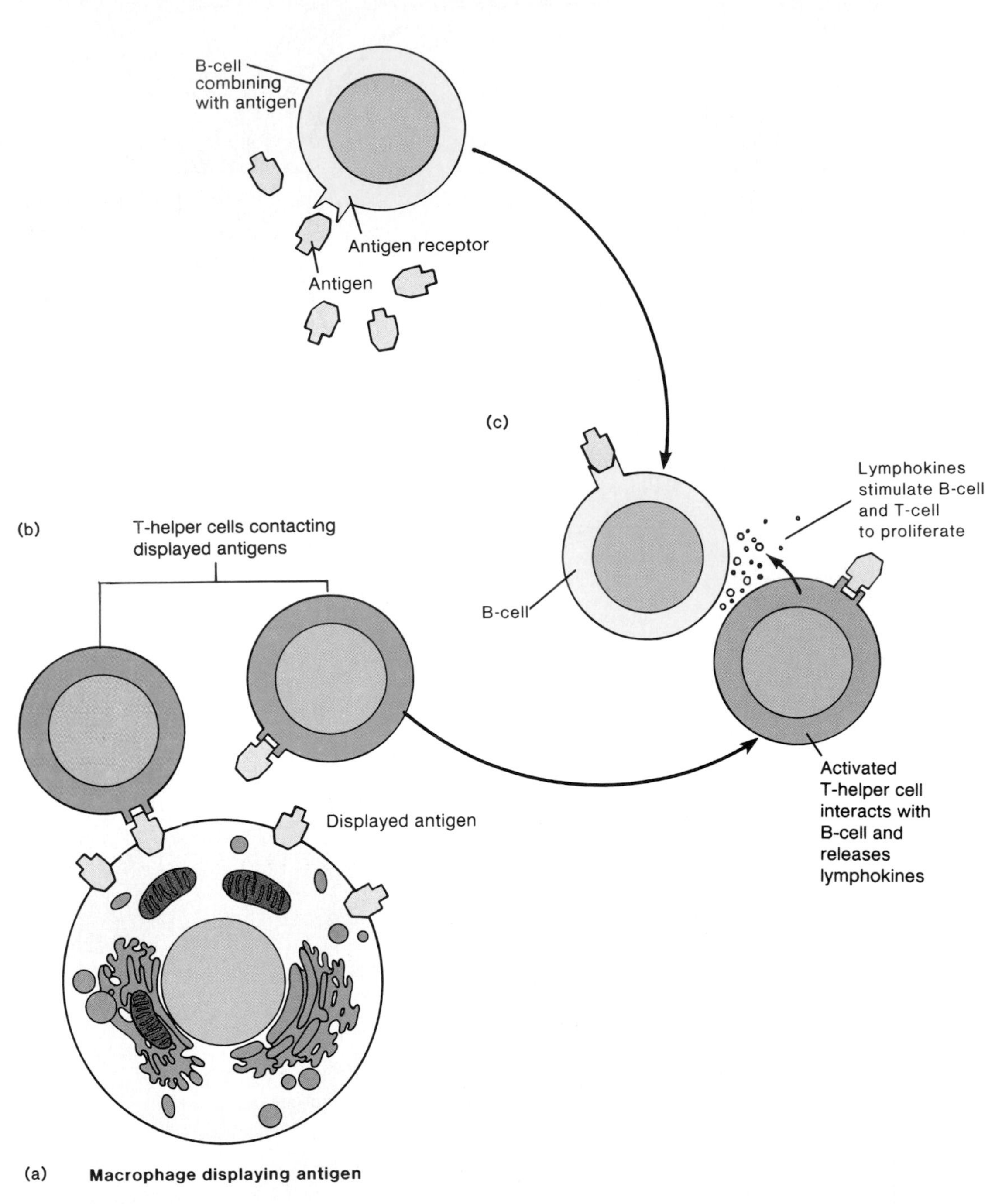
B-cell
combining
with antigen
Antigen receptor
Antigen
(c)
Lymphokines
stimulate B-cell
and T-cell
to proliferate
(b)
T-helper cells contacting
displayed antigens
B-cell
Activated
T-helper cell
interacts with
B-cell and
releases
lymphokines
Displayed antigen
(a)
Macrophage displaying antigen

B-cell and T-cell Interaction
Figure 16.15

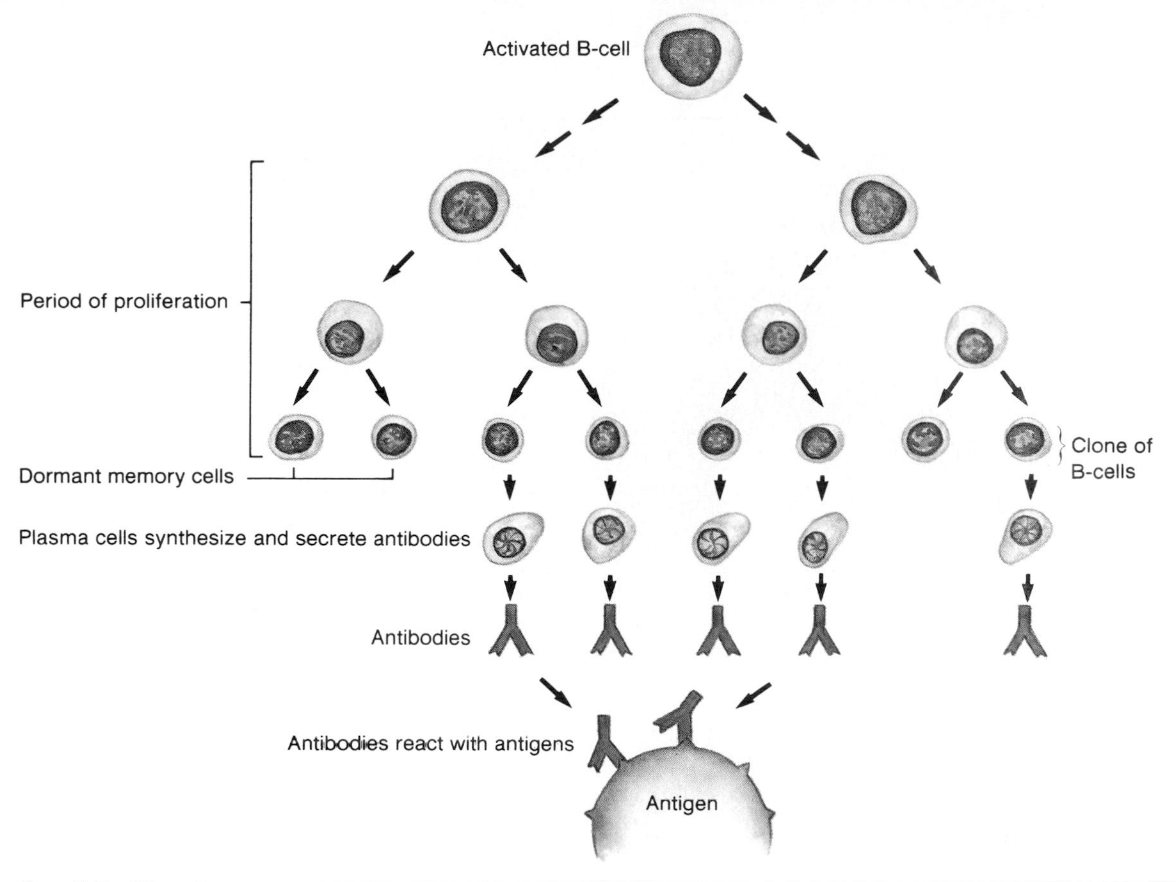

Activated B-cell
Period of proliferation
Dormant memory cells
Clone of B-cells
Plasma cells synthesize and secrete antibodies
Antibodies
Antibodies react with antigens
Antigen

B-cell Proliferation
Figure 16.16

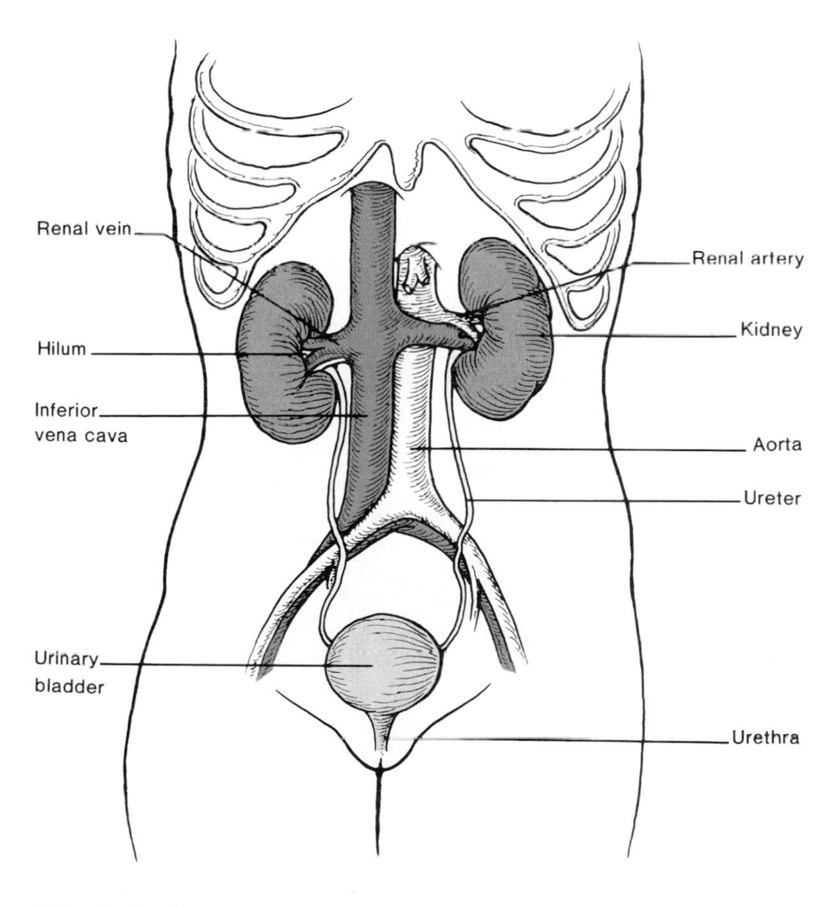

Renal vein
Renal artery
Kidney
Hilum
Inferior vena cava
Aorta
Ureter
Urinary bladder
Urethra

Urinary System
Figure 17.1

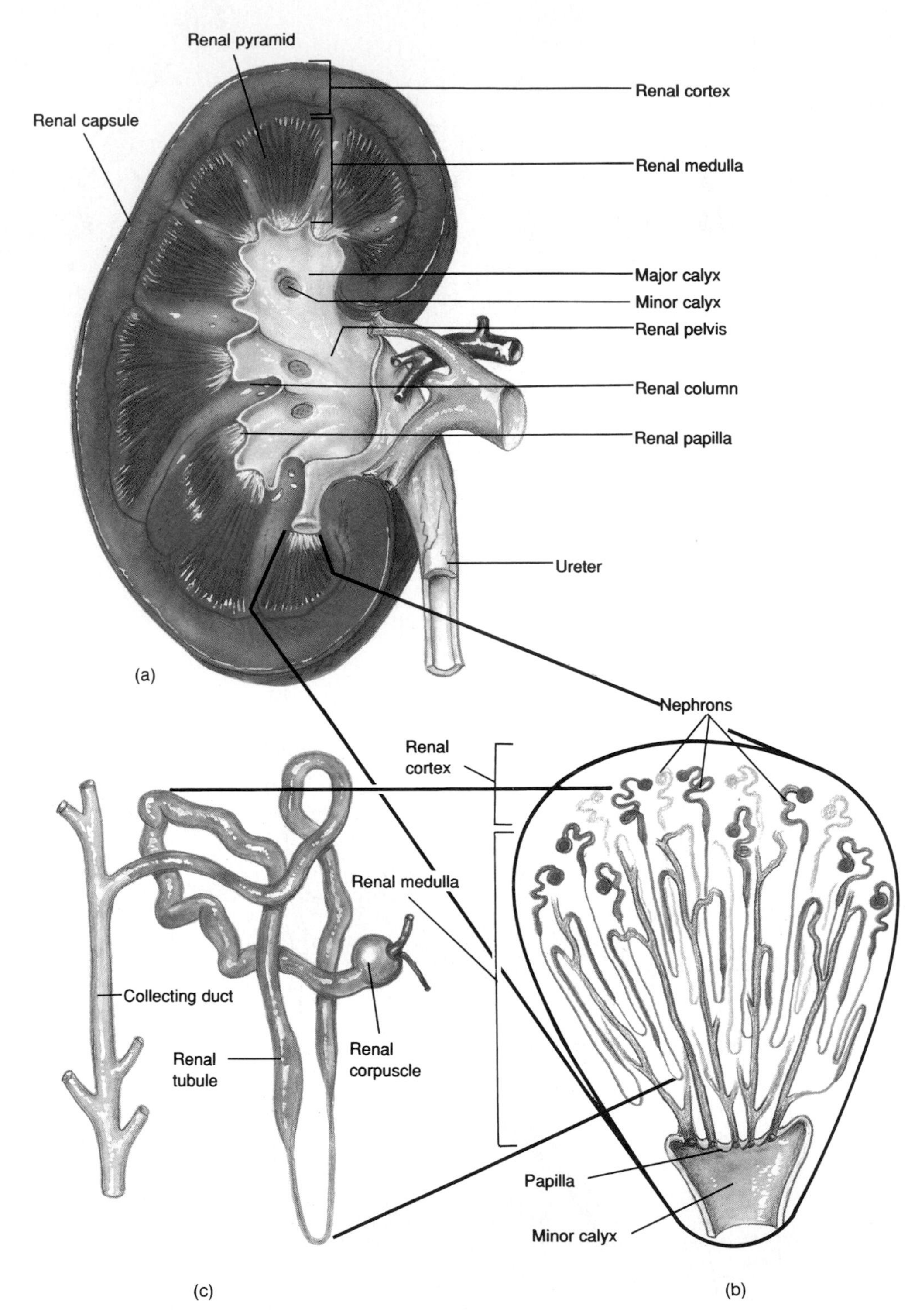
Renal pyramid
Renal cortex
Renal capsule
Renal medulla
Major calyx
Minor calyx
Renal pelvis
Renal column
Renal papilla
Ureter
(a)
Nephrons
Renal cortex
Renal medulla
Collecting duct
Renal tubule
Renal corpuscle
Papilla
Minor calyx
(c)
(b)

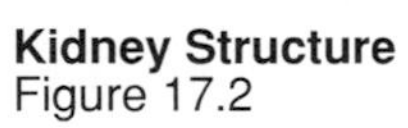

Kidney Structure
Figure 17.2

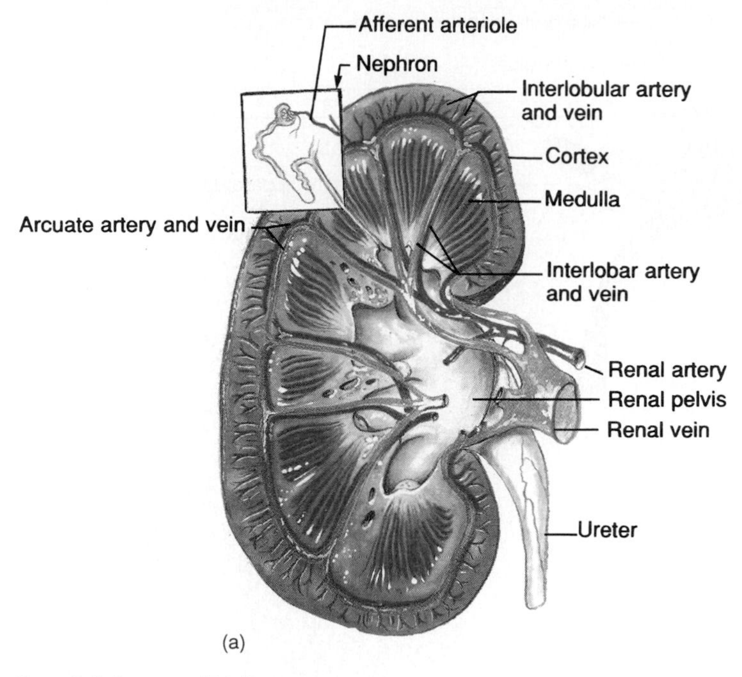
Afferent arteriole
Nephron
Interlobular artery and vein
Cortex
Medulla
Arcuate artery and vein
Interlobar artery and vein
Renal artery
Renal pelvis
Renal vein
Ureter
(a)

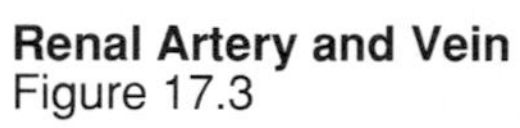

Renal Artery and Vein
Figure 17.3

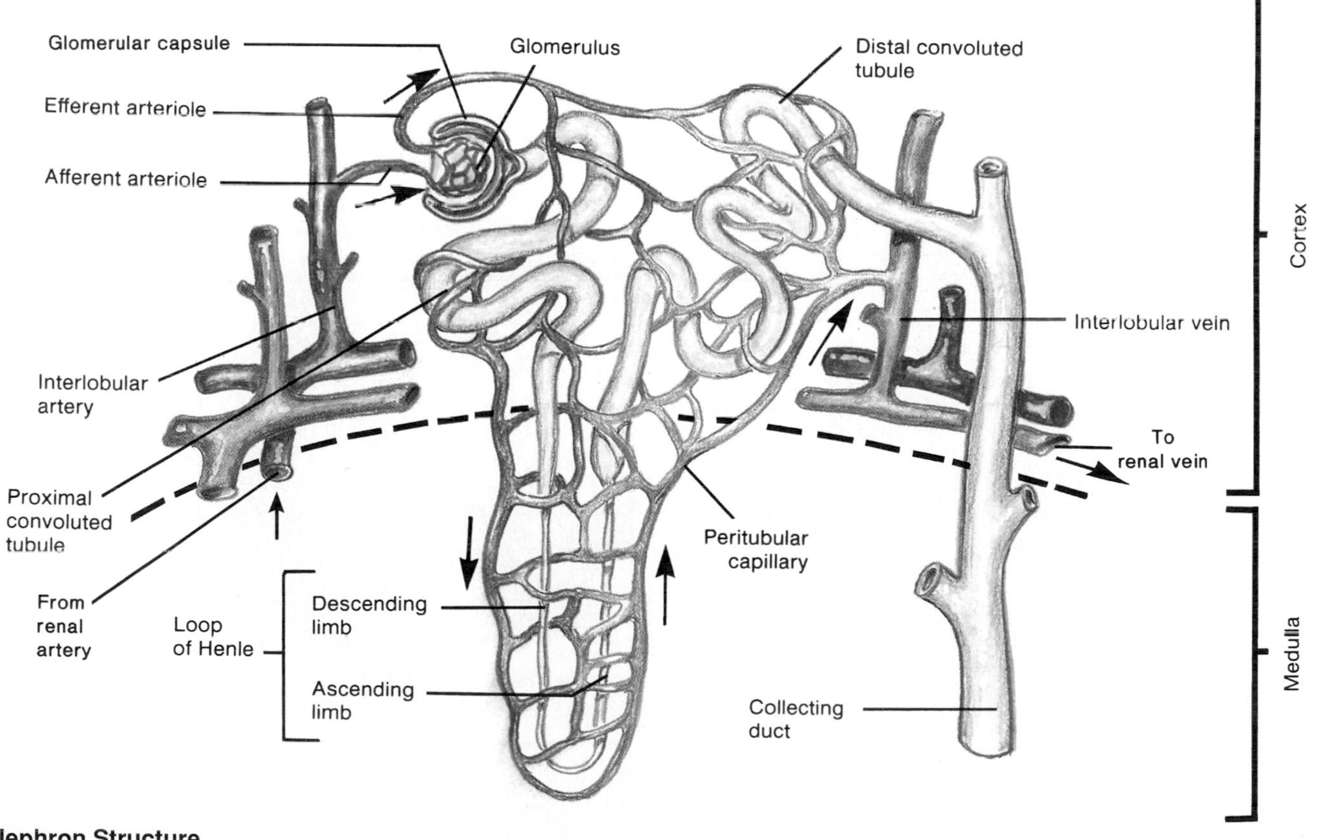
Glomerular capsule
Glomerulus
Distal convoluted tubule
Efferent arteriole
Afferent arteriole
Cortex
Interlobular vein
Interlobular artery
To renal vein
Proximal convoluted tubule
Peritubular capillary
From renal artery
Loop of Henle
Descending limb
Ascending limb
Collecting duct
Medulla

Nephron Structure
Figure 17.6

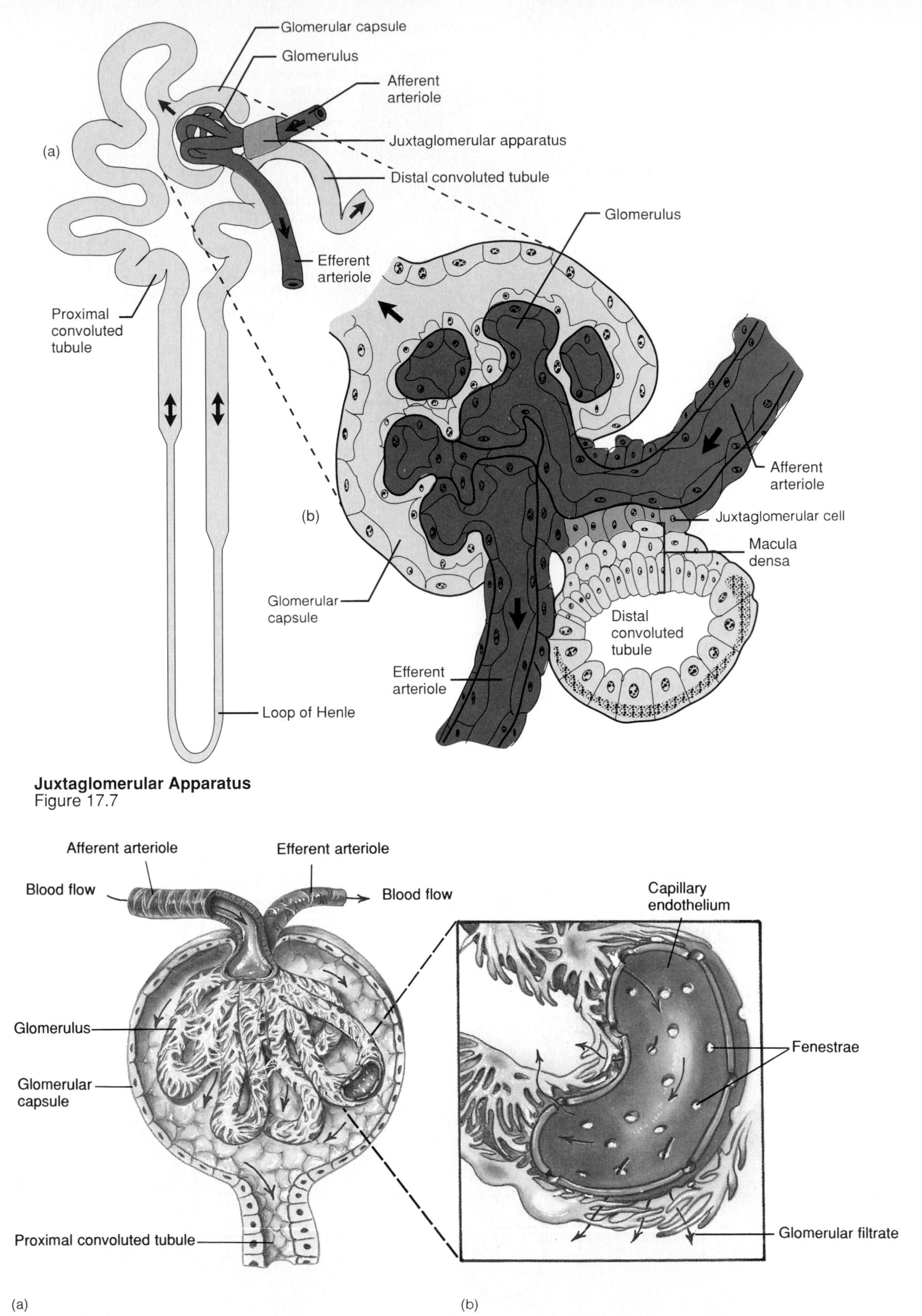
Glomerular capsule
Glomerulus
Afferent arteriole
Juxtaglomerular apparatus
Distal convoluted tubule
(a)
Efferent arteriole
Proximal convoluted tubule
Loop of Henle
Glomerulus
Afferent arteriole
Juxtaglomerular cell
Macula densa
(b)
Glomerular capsule
Distal convoluted tubule
Efferent arteriole

Juxtaglomerular Apparatus
Figure 17.7

Afferent arteriole
Efferent arteriole
Blood flow
Blood flow
Capillary endothelium
Glomerulus
Fenestrae
Glomerular capsule
Proximal convoluted tubule
Glomerular filtrate
(a)
(b)

Renal Filtration
Figure 17.8

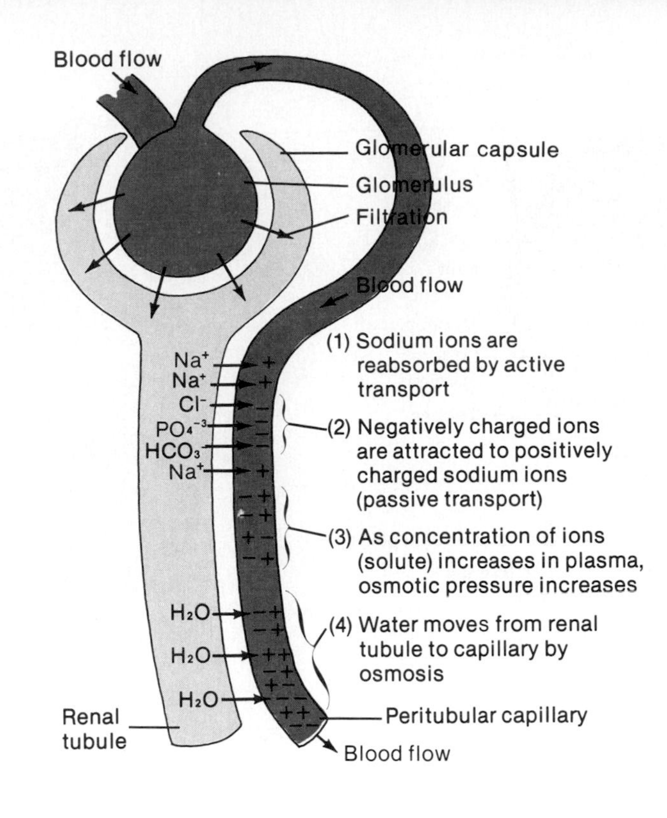
Blood flow
Glomerular capsule
Glomerulus
Filtration
Blood flow
(1) Sodium ions are reabsorbed by active transport
Na+
Na+
Cl-
PO4-3
HCO3-
Na+
(2) Negatively charged ions are attracted to positively charged sodium ions (passive transport)
(3) As concentration of ions (solute) increases in plasma, osmotic pressure increases
H2O
H2O
H2O
(4) Water moves from renal tubule to capillary by osmosis
Renal tubule
Peritubular capillary
Blood flow

Water Reabsorption and Passive Secretion
Figures 17.11 and 17.13

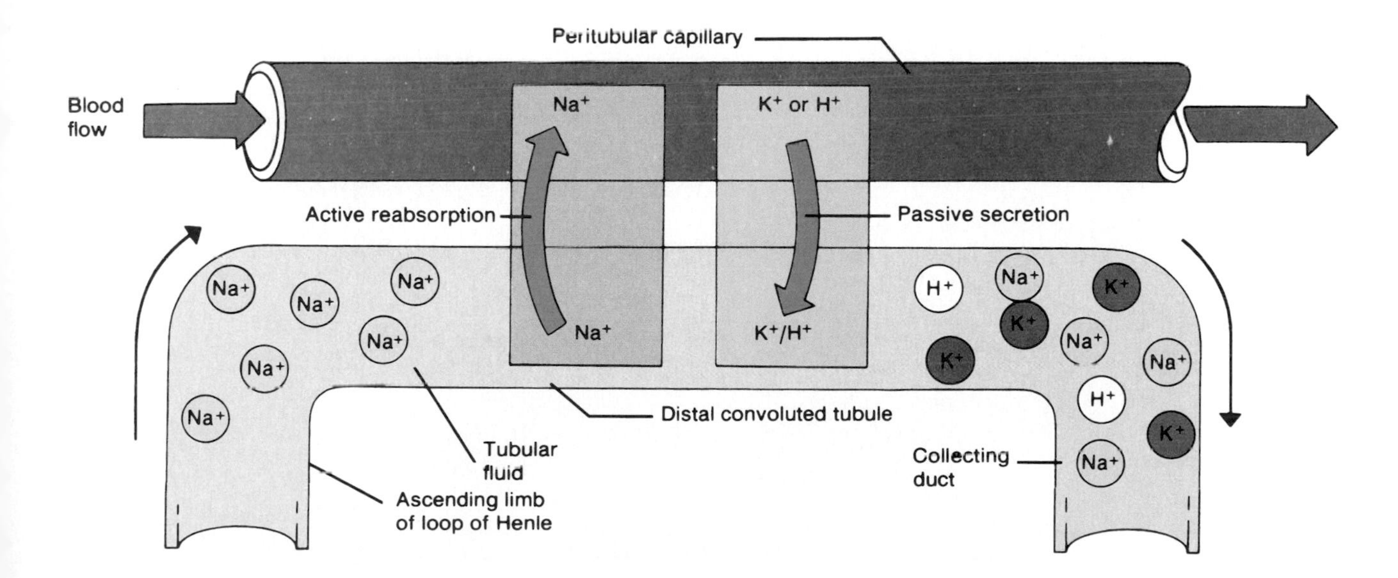
Peritubular capillary
Blood flow
Na+
K+ or H+
Active reabsorption
Passive secretion
Na+
K+/H+
Distal convoluted tubule
Tubular fluid
Ascending limb of loop of Henle
Collecting duct

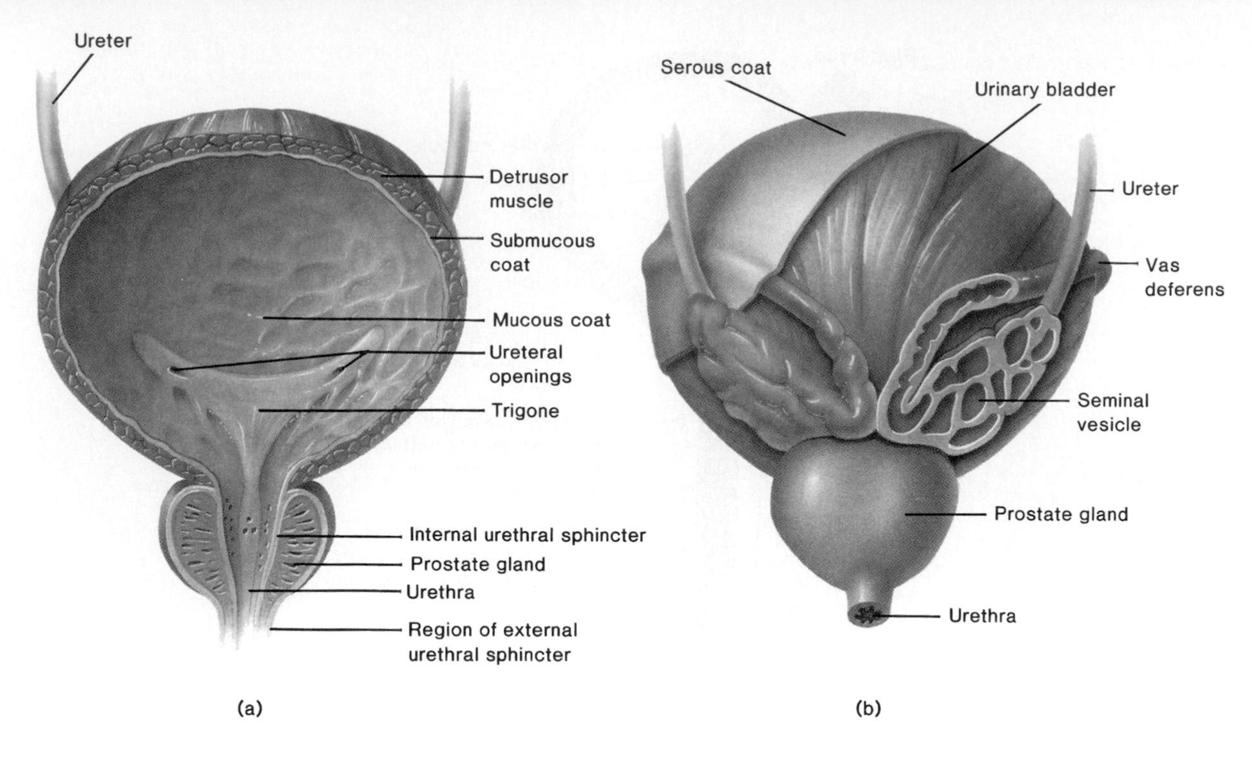
Ureter
Detrusor muscle
Submucous coat
Mucous coat
Ureteral openings
Trigone
Internal urethral sphincter
Prostate gland
Urethra
Region of external urethral sphincter
(a)
Serous coat
Urinary bladder
Ureter
Vas deferens
Seminal vesicle
Prostate gland
Urethra
(b)

Urinary Bladder
Figure 17.15

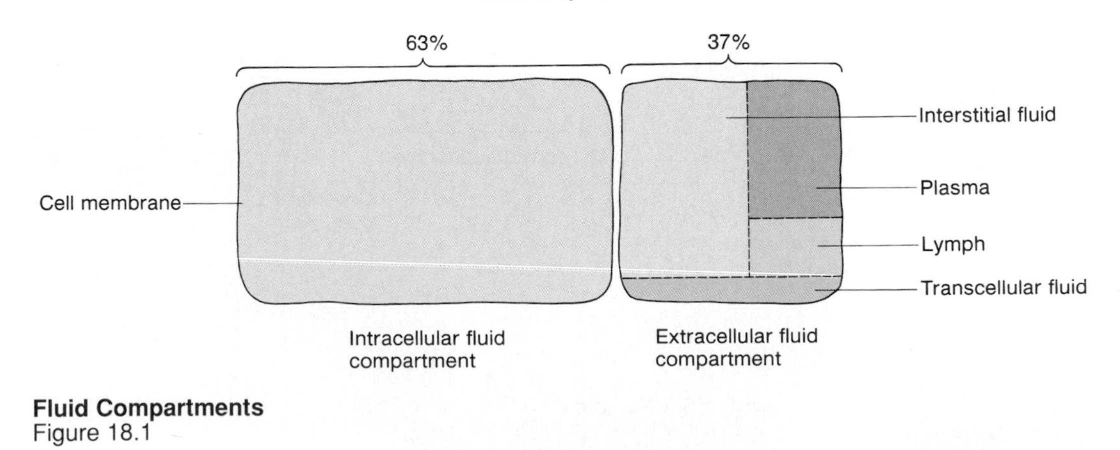
Total body water
63%
37%
Interstitial fluid
Plasma
Cell membrane
Lymph
Transcellular fluid
Intracellular fluid compartment
Extracellular fluid compartment

Fluid Compartments
Figure 18.1

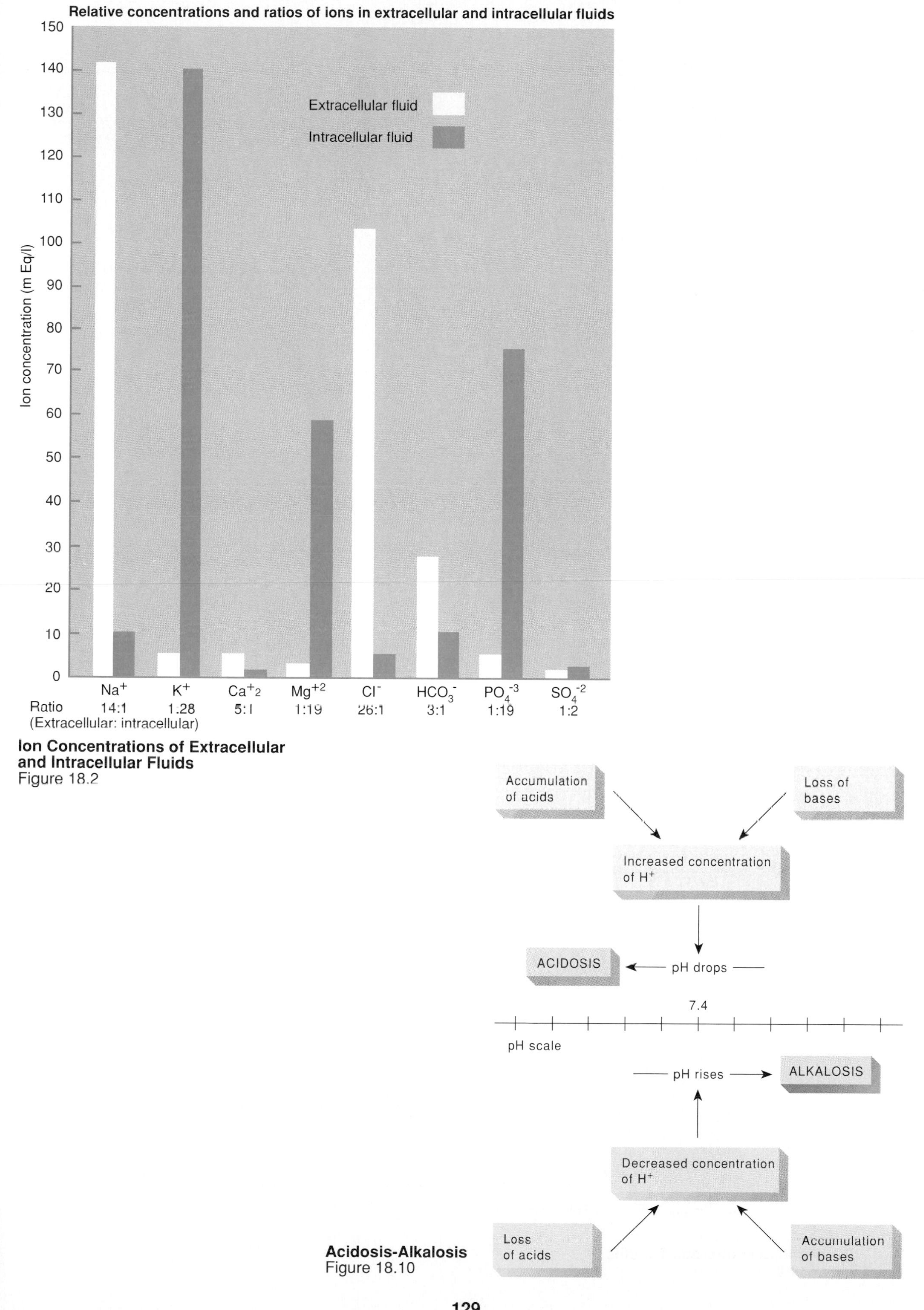

Relative concentrations and ratios of ions in extracellular and intracellular fluids
Extracellular fluid
Intracellular fluid
Ion concentration (m Eq/l)
150
140
130
120
110
100
90
80
70
60
50
40
30
20
10
0
Na+
K+
Ca+2
Mg+2
Cl-
HCO3-
PO4-3
SO4-2
Ratio
14:1
1.28
5:1
1:19
26:1
3:1
1:19
1:2
(Extracellular: intracellular)

Ion Concentrations of Extracellular and Intracellular Fluids
Figure 18.2

Accumulation of acids
Loss of bases
Increased concentration of H+
ACIDOSIS
pH drops
7.4
pH scale
pH rises
ALKALOSIS
Decreased concentration of H+
Loss of acids
Accumulation of bases

Acidosis-Alkalosis
Figure 18.10

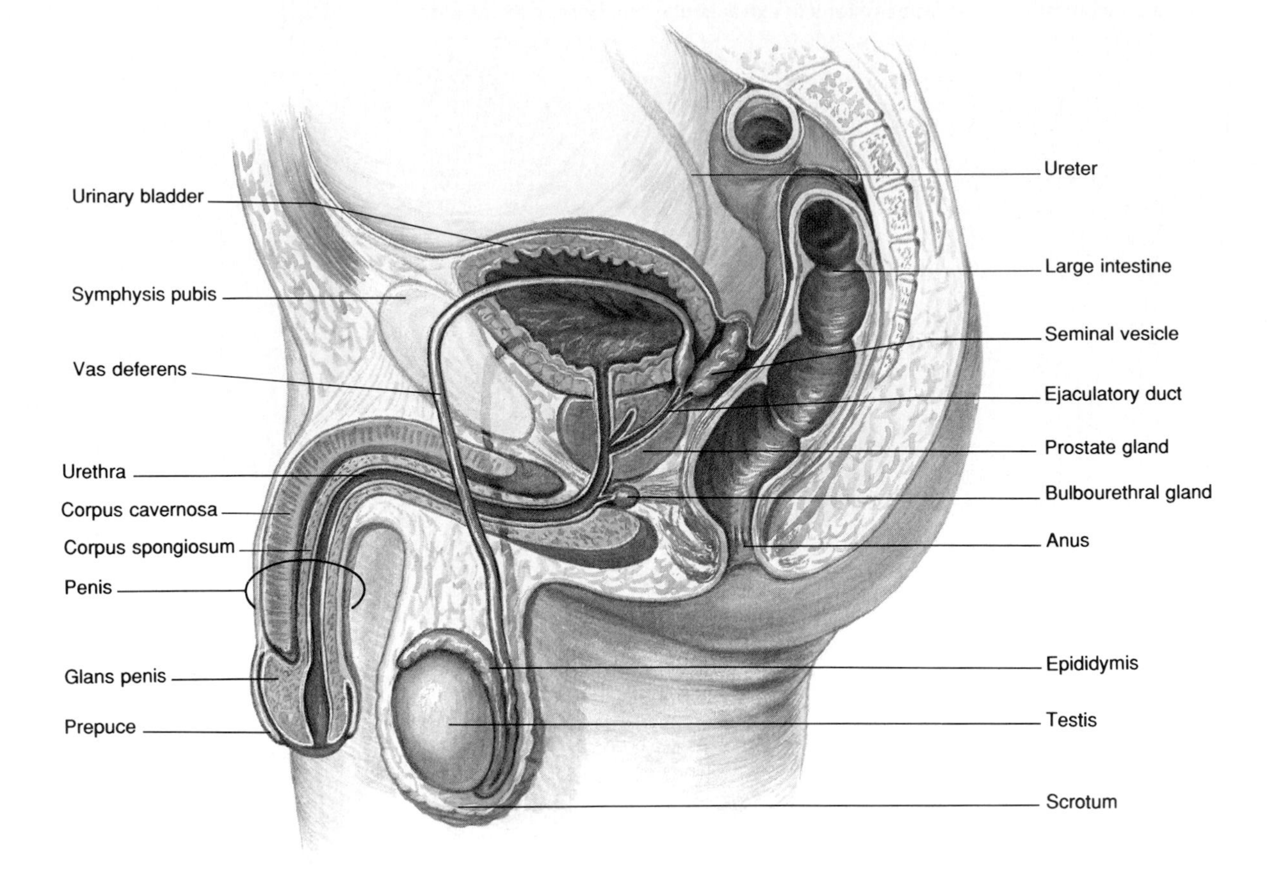
Urinary bladder
Symphysis pubis
Vas deferens
Urethra
Corpus cavernosa
Corpus spongiosum
Penis
Glans penis
Prepuce
Ureter
Large intestine
Seminal vesicle
Ejaculatory duct
Prostate gland
Bulbourethral gland
Anus
Epididymis
Testis
Scrotum

Male Reproductive System
Figure 19.1

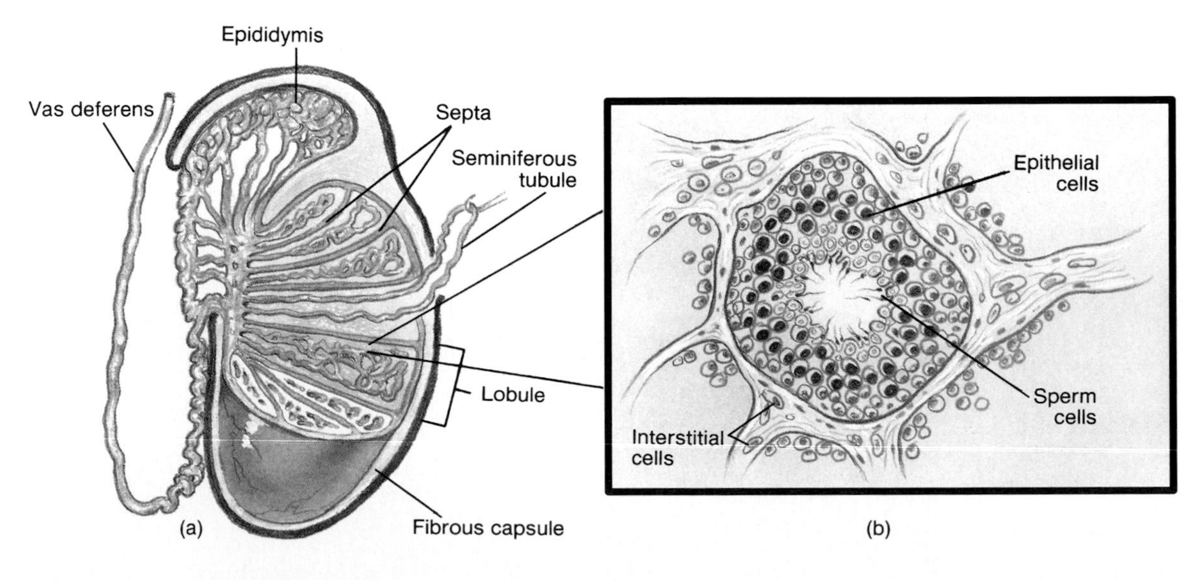
Epididymis
Vas deferens
Septa
Seminiferous tubule
Lobule
Fibrous capsule
(a)
Epithelial cells
Sperm cells
Interstitial cells
(b)

Testis and Seminiferous Tubules
Figure 19.2

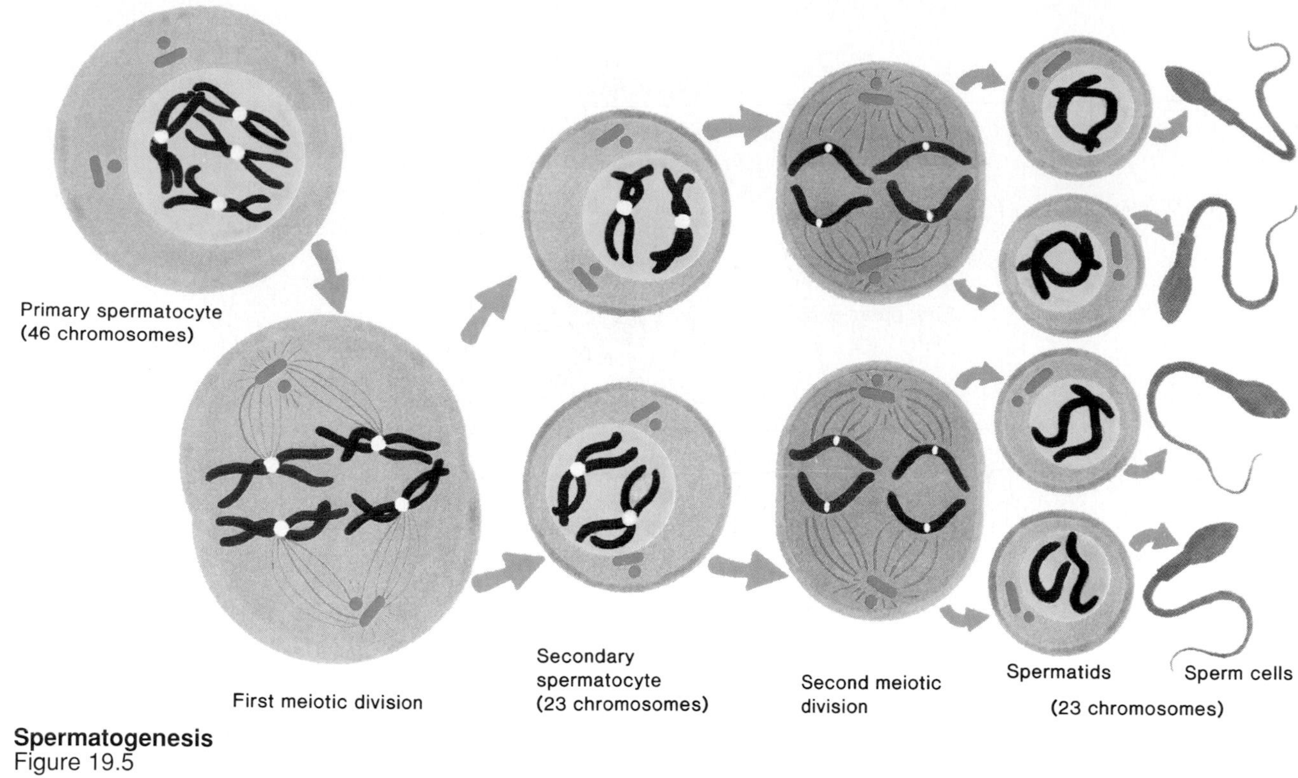
Primary spermatocyte
(46 chromosomes)
First meiotic division
Secondary
spermatocyte
(23 chromosomes)
Second meiotic
division
Spermatids
Sperm cells
(23 chromosomes)

Spermatogenesis
Figure 19.5

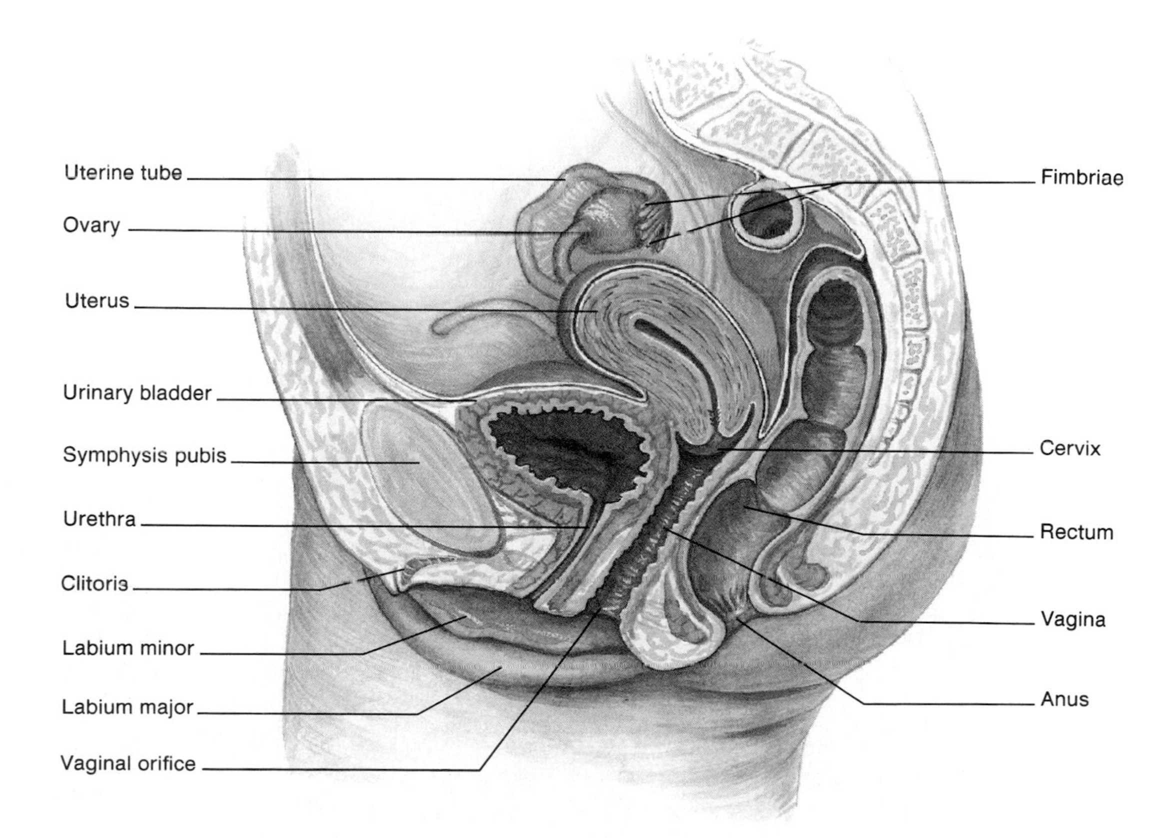
Uterine tube
Ovary
Uterus
Urinary bladder
Symphysis pubis
Urethra
Clitoris
Labium minor
Labium major
Vaginal orifice
Fimbriae
Cervix
Rectum
Vagina
Anus

Female Reproductive System
Figure 19.7

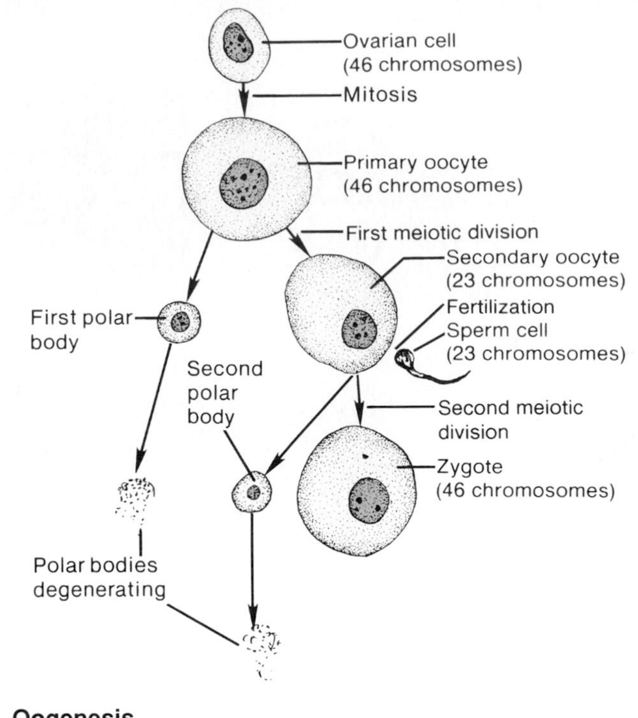
Ovarian cell
(46 chromosomes)
Mitosis
Primary oocyte
(46 chromosomes)
First meiotic division
Secondary oocyte
(23 chromosomes)
First polar body
Fertilization
Sperm cell
(23 chromosomes)
Second polar body
Second meiotic division
Zygote
(46 chromosomes)
Polar bodies degenerating

Oogenesis
Figure 19.8

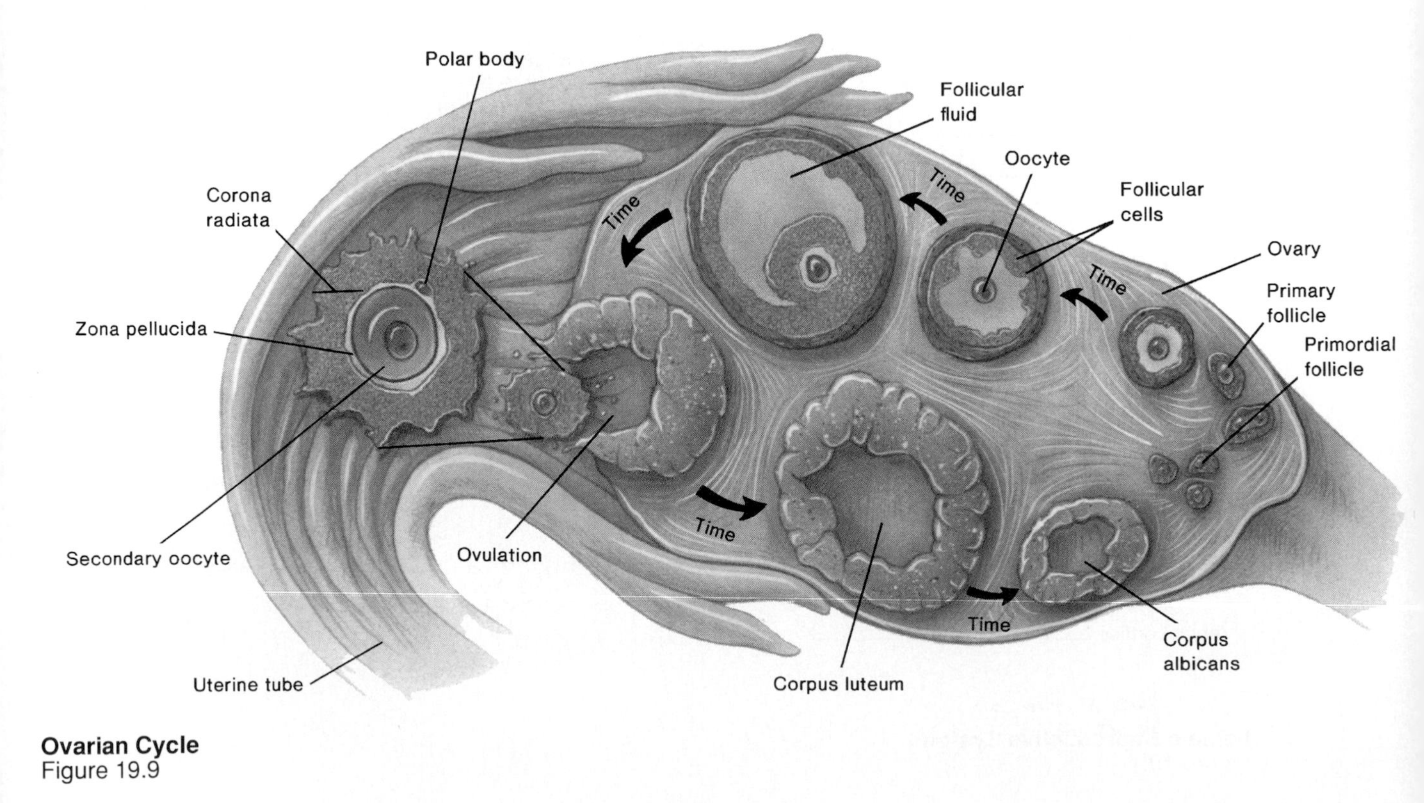
Polar body
Follicular fluid
Oocyte
Follicular cells
Corona radiata
Time
Time
Time
Ovary
Primary follicle
Primordial follicle
Zona pellucida
Time
Time
Secondary oocyte
Ovulation
Uterine tube
Corpus luteum
Corpus albicans

Ovarian Cycle
Figure 19.9

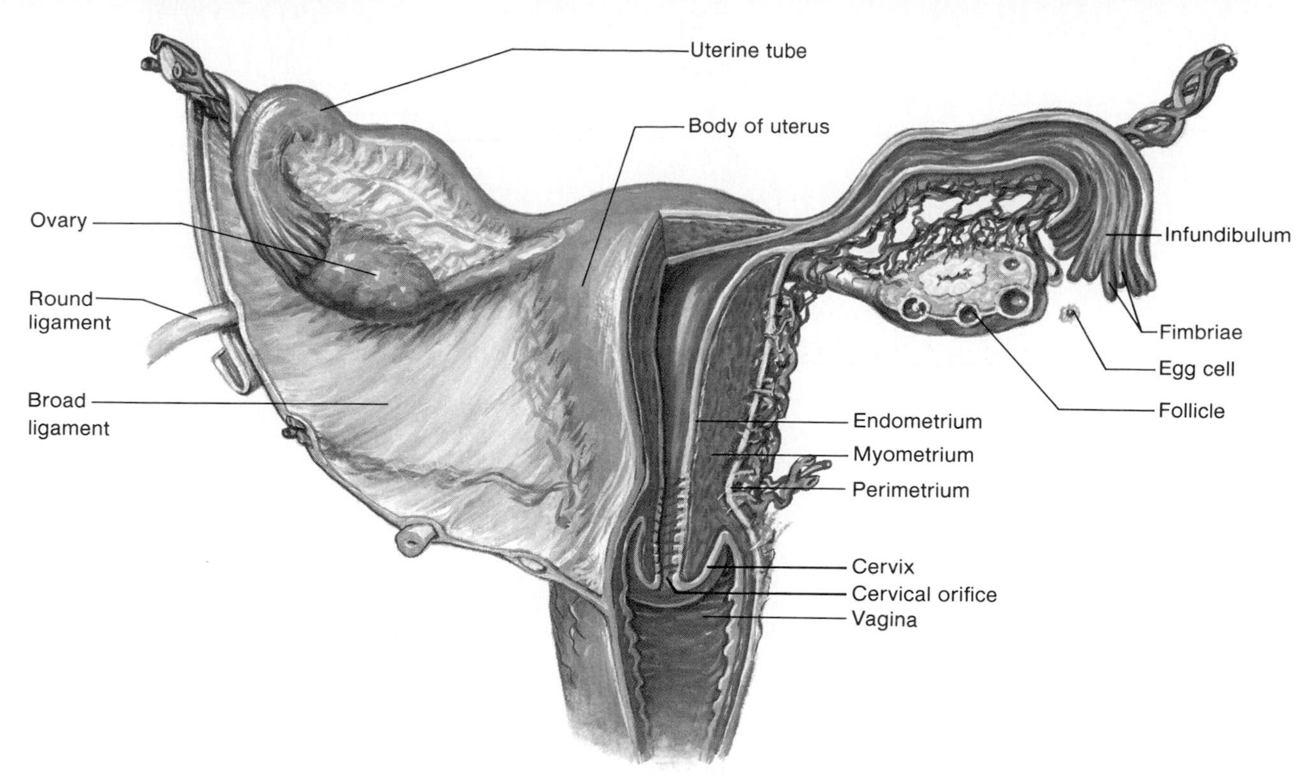
Uterine tube
Body of uterus
Ovary
Round ligament
Broad ligament
Infundibulum
Fimbriae
Egg cell
Follicle
Endometrium
Myometrium
Perimetrium
Cervix
Cervical orifice
Vagina

Female Internal Reproductive Organs
Figure 19.11

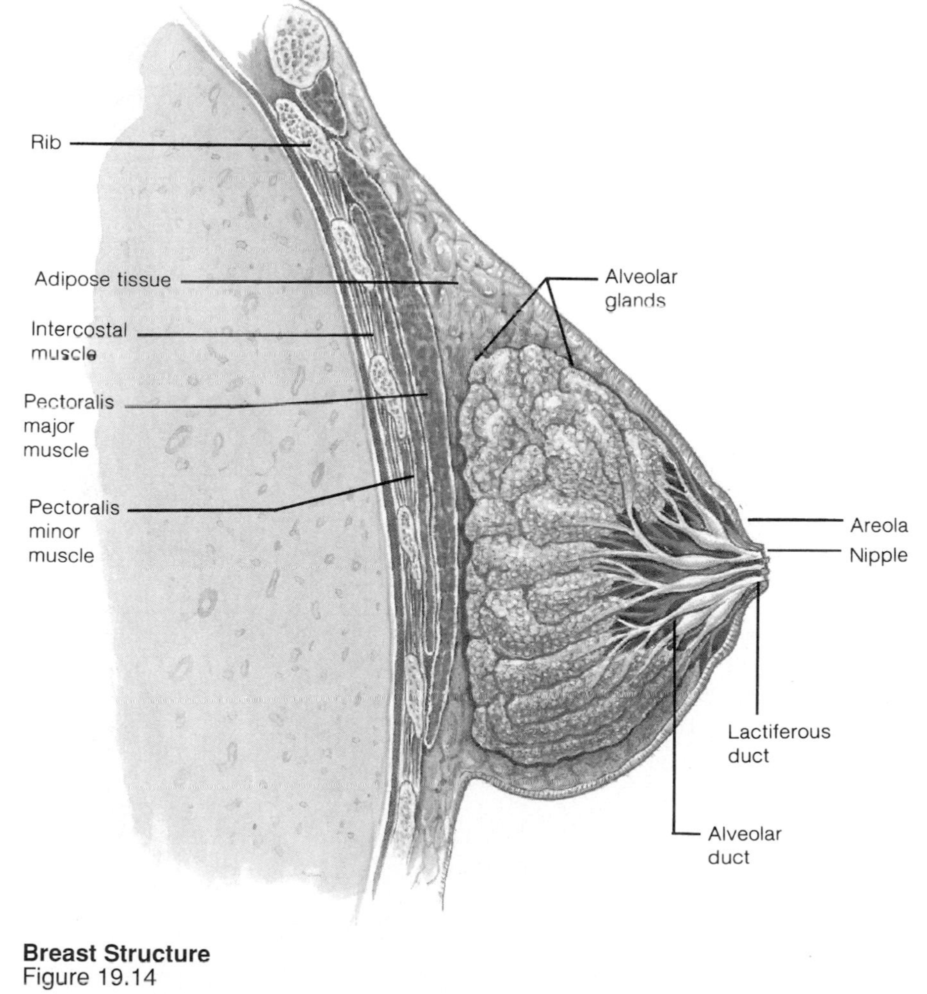
Rib
Adipose tissue
Intercostal muscle
Pectoralis major muscle
Pectoralis minor muscle
Alveolar glands
Areola
Nipple
Lactiferous duct
Alveolar duct

Breast Structure
Figure 19.14

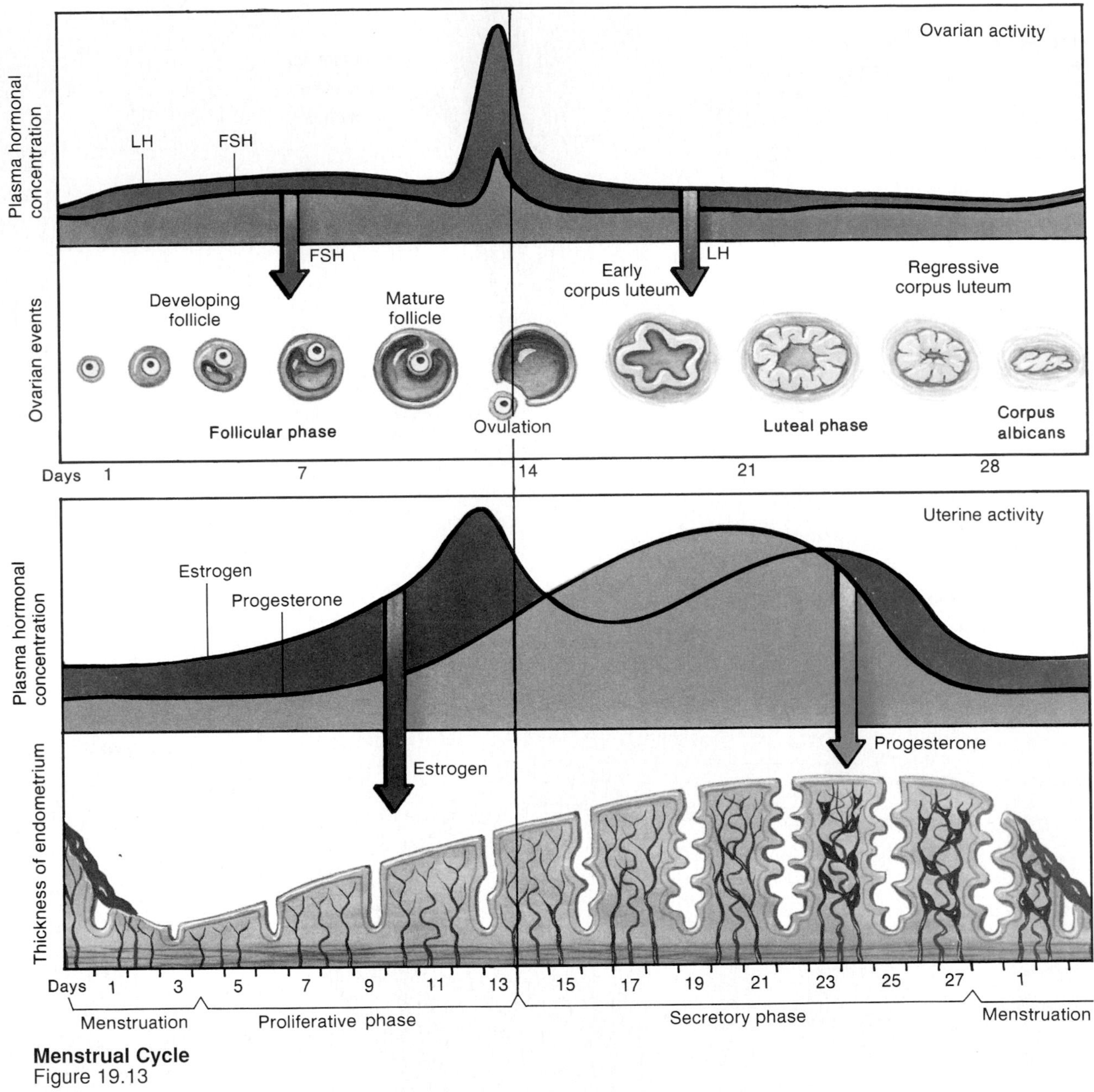

Ovarian activity
Plasma hormonal concentration
LH
FSH
FSH
LH
Ovarian events
Developing follicle
Mature follicle
Early corpus luteum
Regressive corpus luteum
Follicular phase
Ovulation
Luteal phase
Corpus albicans
Days 1
7
14
21
28
Uterine activity
Plasma hormonal concentration
Estrogen
Progesterone
Progesterone
Estrogen
Thickness of endometrium
Days 1 3 5 7 9 11 13 15 17 19 21 23 25 27 1
Menstruation
Proliferative phase
Secretory phase
Menstruation

Menstrual Cycle
Figure 19.13

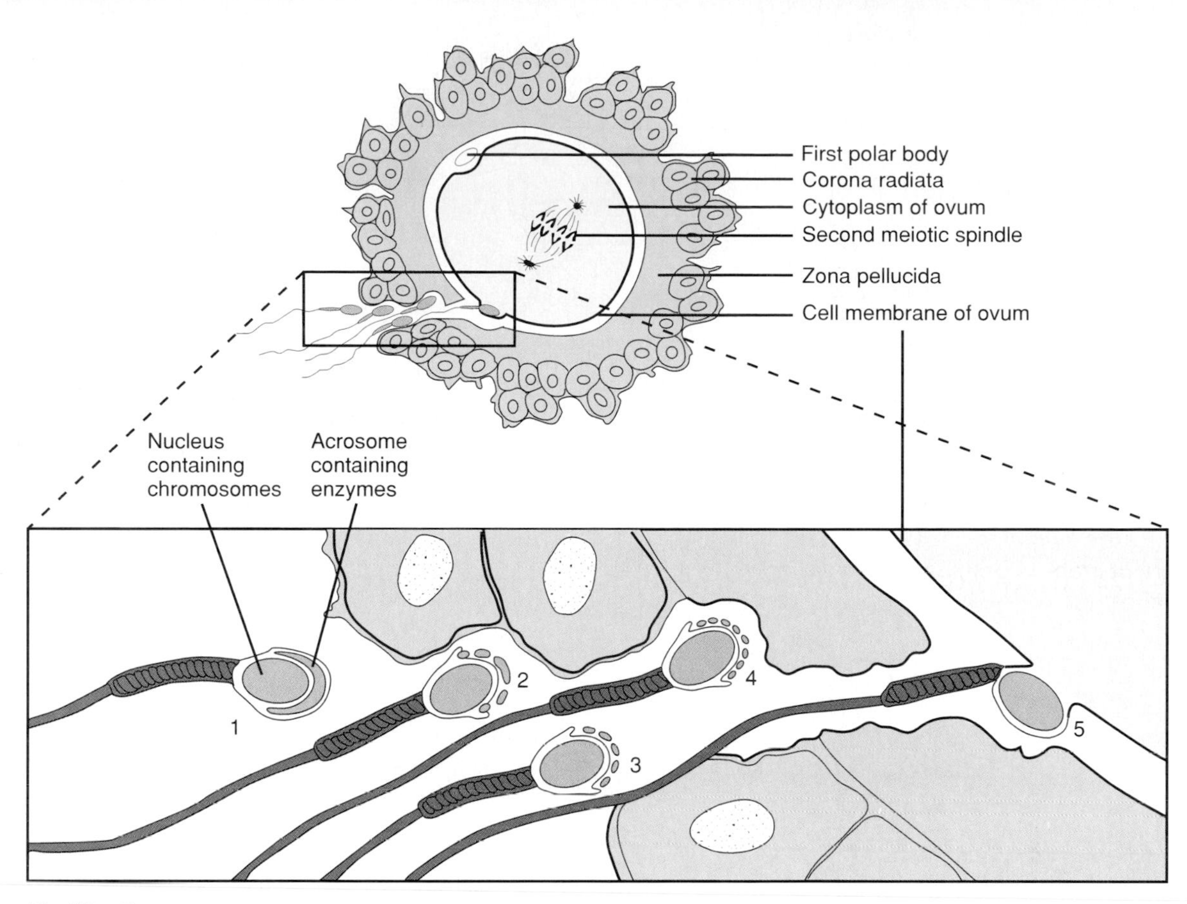

First polar body
Corona radiata
Cytoplasm of ovum
Second meiotic spindle
Zona pellucida
Cell membrane of ovum
Nucleus containing chromosomes
Acrosome containing enzymes
1
2
3
4
5

Fertilization
Figure 20.2

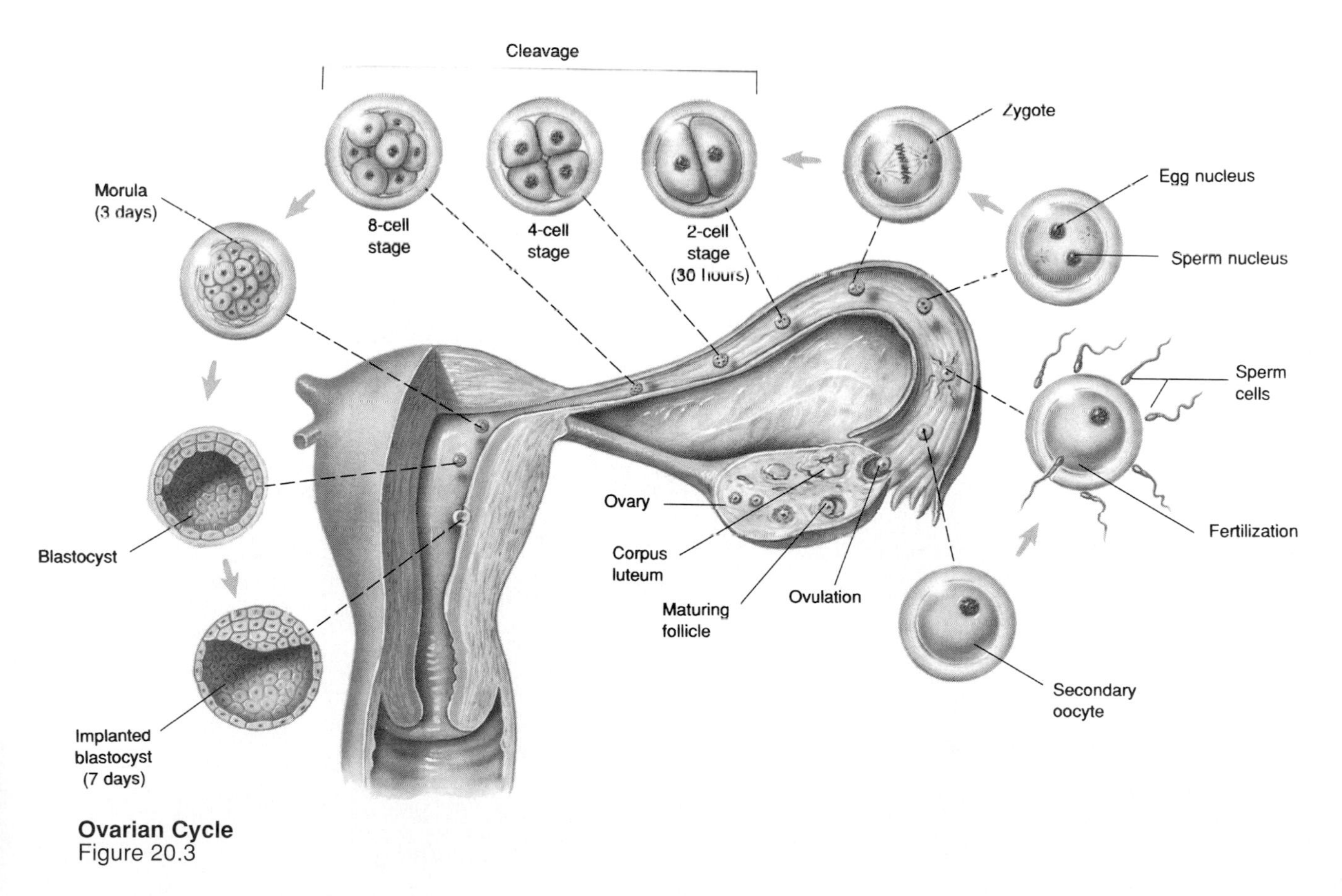

Cleavage
Zygote
Egg nucleus
Sperm nucleus
Sperm cells
Fertilization
Secondary oocyte
Morula (3 days)
8-cell stage
4-cell stage
2-cell stage (30 hours)
Blastocyst
Implanted blastocyst (7 days)
Ovary
Corpus luteum
Maturing follicle
Ovulation

Ovarian Cycle
Figure 20.3

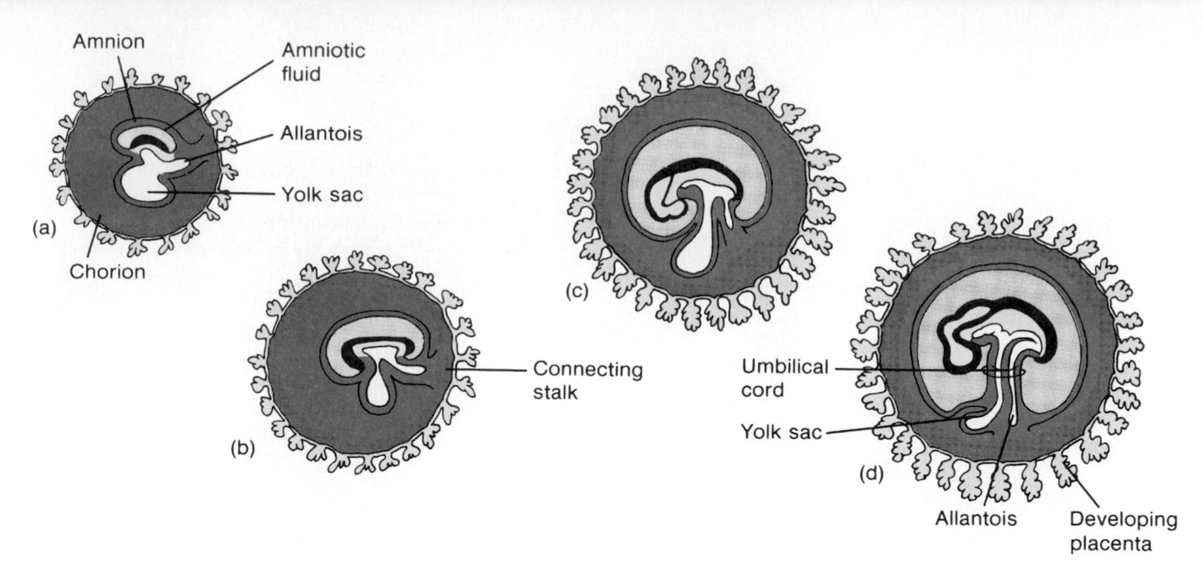
Amnion
Amniotic fluid
Allantois
Yolk sac
(a)
Chorion
(b)
Connecting stalk
(c)
Umbilical cord
Yolk sac
(d)
Allantois
Developing placenta

Development of Umbilical Cord
Figure 20.11

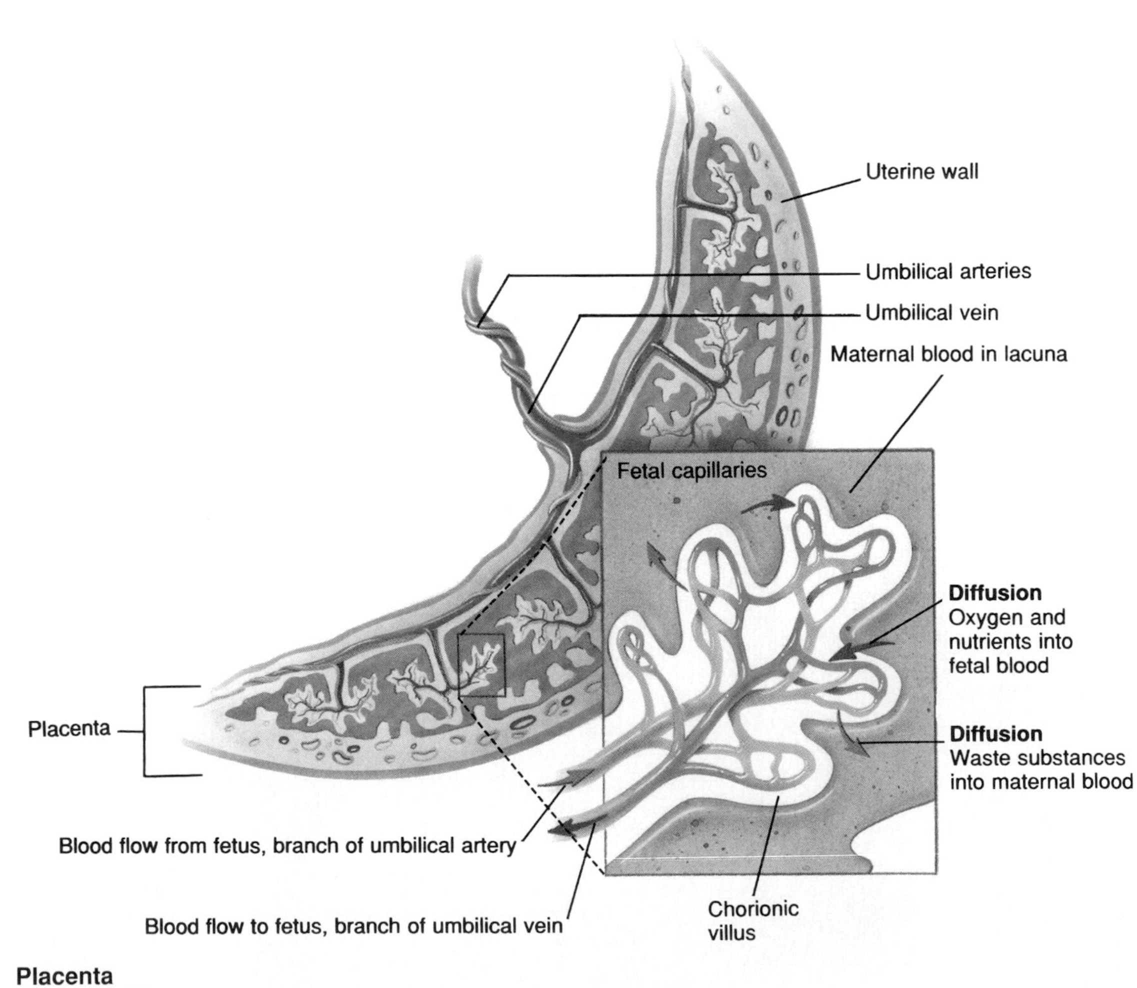
Uterine wall
Umbilical arteries
Umbilical vein
Maternal blood in lacuna
Fetal capillaries
Diffusion
Oxygen and nutrients into fetal blood
Diffusion
Waste substances into maternal blood
Placenta
Blood flow from fetus, branch of umbilical artery
Blood flow to fetus, branch of umbilical vein
Chorionic villus

Placenta
Figure 20.13

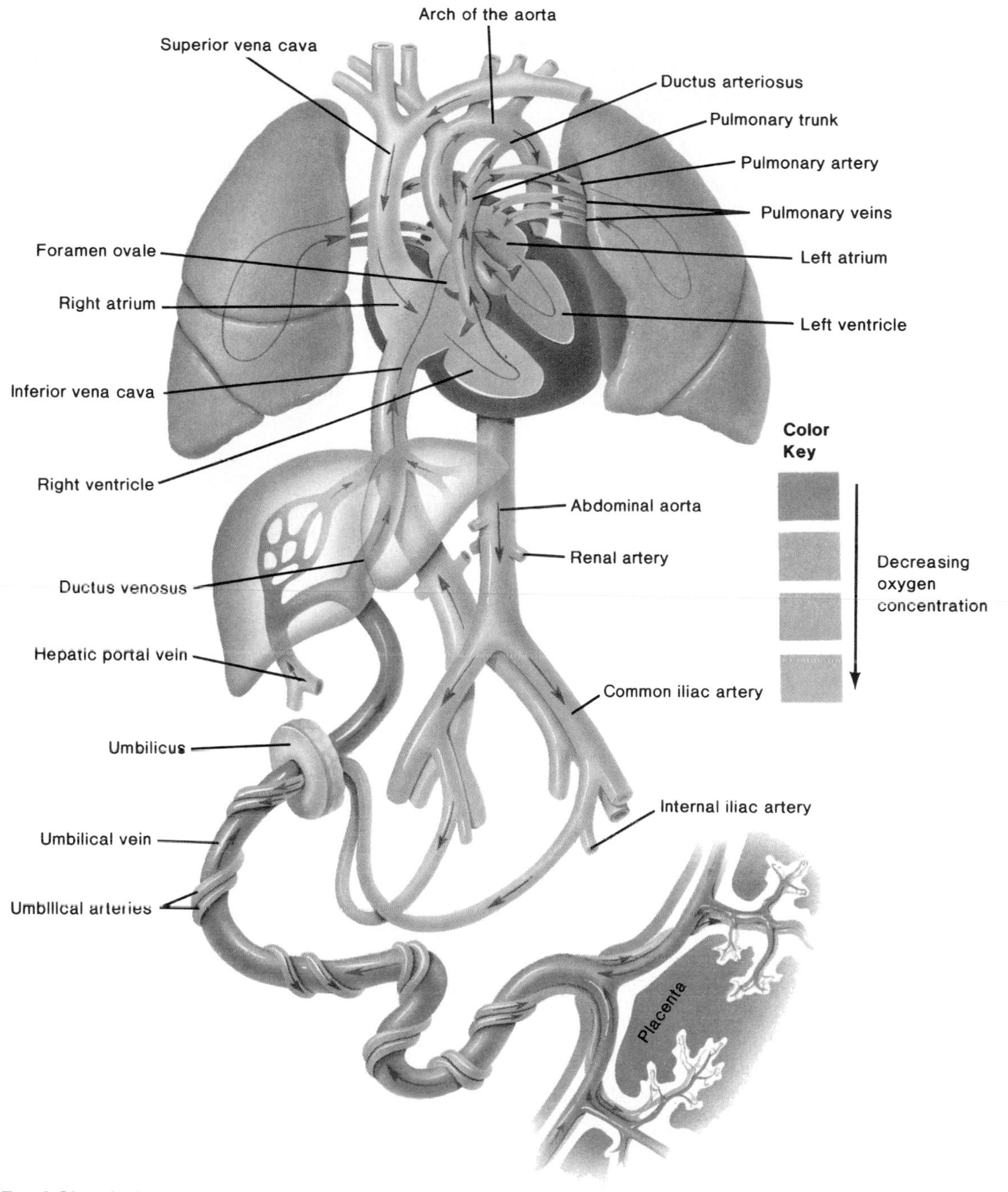

Arch of the aorta
Superior vena cava
Ductus arteriosus
Pulmonary trunk
Pulmonary artery
Pulmonary veins
Foramen ovale
Left atrium
Right atrium
Left ventricle
Inferior vena cava
Color Key
Right ventricle
Abdominal aorta
Renal artery
Decreasing oxygen concentration
Ductus venosus
Hepatic portal vein
Common iliac artery
Umbilicus
Internal iliac artery
Umbilical vein
Umbilical arteries
Placenta

Fetal Circulation
Figure 20.16

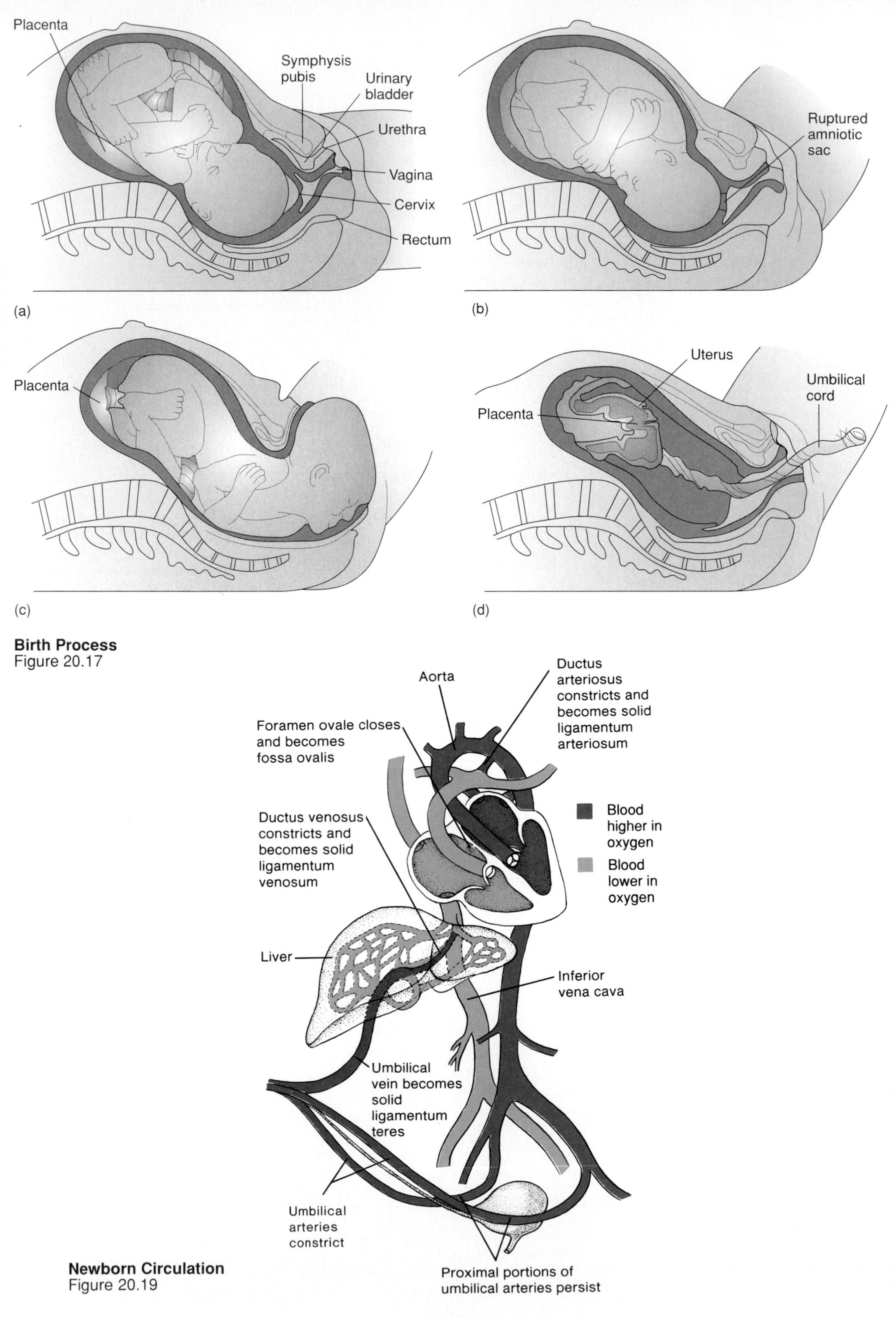
Placenta
Symphysis pubis
Urinary bladder
Urethra
Vagina
Cervix
Rectum
(a)
Ruptured amniotic sac
(b)
Placenta
(c)
Uterus
Umbilical cord
Placenta
(d)

Birth Process
Figure 20.17

Aorta
Ductus arteriosus constricts and becomes solid ligamentum arteriosum
Foramen ovale closes and becomes fossa ovalis
Ductus venosus constricts and becomes solid ligamentum venosum
Blood higher in oxygen
Blood lower in oxygen
Liver
Inferior vena cava
Umbilical vein becomes solid ligamentum teres
Umbilical arteries constrict
Proximal portions of umbilical arteries persist

Newborn Circulation
Figure 20.19

CREDITS

Photo

Fig. 5.1b: © Ed Reschke
Fig. 5.2b: © Ed Reschke
Fig. 5.3b: © Manfred Kage/Peter Arnold, Inc.
Fig. 5.4b: © Ed Reschke
Fig. 5.5b: © Fred Hossler/Visuals Unlimited
Fig. 5.6a: © Ed Reschke
Fig. 5.11b: © Ed Reschke
Fig. 5.12b: © Ed Reschke
Fig. 5.13b: © Ed Reschke
Fig. 5.14b: © Ed Reschke
Fig. 5.15b: © Ed Reschke
Fig. 5.16b: © John Cunningham/Visuals Unlimited
Fig. 5.17b: © Victor B. Eichler, Ph.D.
Fig. 5.18b: © Ed Reschke
Fig. 5.19b: © Ed Reschke
Fig. 5.20b: © Ed Reschke
Fig. 5.21b: © Manfred Kage/Peter Arnold, Inc.
Fig. 5.22b: © Manfred Kage/Peter Arnold, Inc.

Line Art

Fig. 1.6: From Kent M. Van De Graaff, *Human Anatomy,* 3d ed. Copyright © 1992 Wm. C. Brown Communications, Inc., Dubuque, Iowa. All Rights Reserved. Reprinted by permission.

Fig. 1.7: From Kent M. Van De Graaff and Stuart Ira Fox, *Concepts of Human Anatomy and Physiology.* Copyright © 1986 Wm. C. Brown Communications, Inc., Dubuque, Iowa. All Rights Reserved. Reprinted by permission.

Fig. 3.16: From Stuart Ira Fox, *Human Physiology,* 3d ed. Copyright © 1990 Wm. C. Brown Communications, Inc., Dubuque, Iowa. All Rights Reserved. Reprinted by permission.

Fig. 4.18: From Stuart Ira Fox, *Human Physiology,* 3d ed. Copyright © 1990 Wm. C. Brown Communications, Inc., Dubuque, Iowa. All Rights Reserved. Reprinted by permission.

Fig. 5.7: From Kent M. Van De Graaff, *Human Anatomy,* 3d ed. Copyright © 1992 Wm. C. Brown Communications, Inc., Dubuque, Iowa. All Rights Reserved. Reprinted by permission.

Fig. 7.8a: From Kent M. Van De Graaff, *Human Anatomy,* 3d ed. Copyright © 1992 Wm. C. Brown Communications, Inc., Dubuque, Iowa. All Rights Reserved. Reprinted by permission.

Fig. 7.8b: From Kent M. Van De Graaff, *Human Anatomy,* 3d ed. Copyright © 1992 Wm. C. Brown Communications, Inc., Dubuque, Iowa. All Rights Reserved. Reprinted by permission.

Fig. 7.9: From Kent M. Van De Graaff, *Human Anatomy,* 3d ed. Copyright © 1992 Wm. C. Brown Communications, Inc., Dubuque, Iowa. All Rights Reserved. Reprinted by permission.

Fig. 7.11: From Kent M. Van De Graaff, *Human Anatomy,* 3d ed. Copyright © 1992 Wm. C. Brown Communications, Inc., Dubuque, Iowa. All Rights Reserved. Reprinted by permission.

Fig. 7.12: From Kent M. Van De Graaff, *Human Anatomy,* 3d ed. Copyright © 1992 Wm. C. Brown Communications, Inc., Dubuque, Iowa. All Rights Reserved. Reprinted by permission.

Fig. 7.13: From Kent M. Van De Graaff, *Human Anatomy,* 3d ed. Copyright © 1992 Wm. C. Brown Communications, Inc., Dubuque, Iowa. All Rights Reserved. Reprinted by permission.

Fig. 7.14: From Kent M. Van De Graaff, *Human Anatomy,* 3d ed. Copyright © 1992 Wm. C. Brown Communications, Inc., Dubuque, Iowa. All Rights Reserved. Reprinted by permission.

Fig. 8.14: From Kent M. Van De Graaff and Stuart Ira Fox, *Concepts of Human Anatomy and Physiology,* 3d ed. Copyright © 1992 Wm. C. Brown Communications, Inc., Dubuque, Iowa. All Rights Reserved. Reprinted by permission.

Fig. 8.15: From Kent M. Van De Graaff and Stuart Ira Fox, *Concepts of Human Anatomy and Physiology,* 3d ed. Copyright © 1992 Wm. C. Brown Communications, Inc., Dubuque, Iowa. All Rights Reserved. Reprinted by permission.

Fig. 9.15: From Kent M. Van De Graaff, *Human Anatomy,* 3d ed. Copyright © 1992 Wm. C. Brown Communications, Inc., Dubuque, Iowa. All Rights Reserved. Reprinted by permission.

Fig. 9.22a: From Kent M. Van De Graaff, *Human Anatomy,* 3d ed. Copyright © 1992 Wm. C. Brown Communications, Inc., Dubuque, Iowa. All Rights Reserved. Reprinted by permission.

Fig. 9.26: From Kent M. Van De Graaff, *Human Anatomy,* 3d ed. Copyright © 1992 Wm. C. Brown Communications, Inc., Dubuque, Iowa. All Rights Reserved. Reprinted by permission.

Fig. 9.27: From Kent M. Van De Graaff and Stuart Ira Fox, *Concepts of Human Anatomy and Physiology,* 2d ed. Copyright © 1989 Wm. C. Brown Communications, Inc., Dubuque, Iowa. All Rights Reserved. Reprinted by permission.

Fig. 10.2: From Kent M. Van De Graaff and Stuart Ira Fox, *Concepts of Human Anatomy and Physiology,* 3d ed. Copyright © 1992 Wm. C. Brown Communications, Inc., Dubuque, Iowa. All Rights Reserved. Reprinted by permission.

Fig. 10.3: From Kent M. Van De Graaff and Stuart Ira Fox, *Concepts of Human Anatomy and Physiology,* 2d ed. Copyright © 1989 Wm. C. Brown Communications, Inc., Dubuque, Iowa. All Rights Reserved. Reprinted by permission.

Fig. 10.10a: From Kent M. Van De Graaff and Stuart Ira Fox, *Concepts of Human Anatomy and Physiology,* 3d ed. Copyright © 1992 Wm. C. Brown Communications, Inc., Dubuque, Iowa. All Rights Reserved. Reprinted by permission.

Fig. 10.10b: From Stuart Ira Fox, *Human Physiology,* 4th ed. Copyright © 1993 Wm. C. Brown Communications, Inc., Dubuque, Iowa. All Rights Reserved. Reprinted by permission.

Fig. 10.20: From Kent M. Van De Graaff and Stuart Ira Fox, *Concepts of Human Anatomy and Physiology,* 3d ed. Copyright © 1992 Wm. C. Brown Communications, Inc., Dubuque, Iowa. All Rights Reserved. Reprinted by permission.

Fig. 12.1: From Kent M. Van De Graaff, *Human Anatomy,* 3d ed. Copyright © 1992 Wm. C. Brown Communications, Inc., Dubuque, Iowa. All Rights Reserved. Reprinted by permission.

Fig. 12.3: From Kent M. Van De Graaff and Stuart Ira Fox, *Concepts of Human Anatomy and Physiology,* 3d ed. Copyright © 1992 Wm. C. Brown Communications, Inc., Dubuque, Iowa. All Rights Reserved. Reprinted by permission.

Fig. 12.9: From Kent M. Van De Graaff, *Human Anatomy,* 3d ed. Copyright © 1992 Wm. C. Brown Communications, Inc., Dubuque, Iowa. All Rights Reserved. Reprinted by permission.

Fig. 12.15: From Kent M. Van De Graaff, *Human Anatomy,* 3d ed. Copyright © 1992 Wm. C. Brown Communications, Inc., Dubuque, Iowa. All Rights Reserved. Reprinted by permission.

Fig. 12.28: From Kent M. Van De Graaff and Stuart Ira Fox, *Concepts of Human Anatomy and Physiology,* 3d ed. Copyright © 1992 Wm. C. Brown Communications, Inc., Dubuque, Iowa. All Rights Reserved. Reprinted by permission.

Fig. 13.1: From Kent M. Van De Graaff, *Human Anatomy,* 3d ed. Copyright © 1992 Wm. C. Brown Communications, Inc., Dubuque, Iowa. All Rights Reserved. Reprinted by permission.

Fig. 13.7: From Kent M. Van De Graaff, *Human Anatomy,* 3d ed. Copyright © 1992 Wm. C. Brown Communications, Inc., Dubuque, Iowa. All Rights Reserved. Reprinted by permission.

Fig. 15.17: From Kent M. Van De Graaff and Stuart Ira Fox, *Concepts of Human Anatomy and Physiology,* 3d ed. Copyright © 1992 Wm. C. Brown Communications, Inc., Dubuque, Iowa. All Rights Reserved. Reprinted by permission.

Fig. 15.28: From Kent M. Van De Graaff and Stuart Ira Fox, *Concepts of Human Anatomy and Physiology,* 3d ed. Copyright © 1992 Wm. C. Brown Communications, Inc., Dubuque, Iowa. All Rights Reserved. Reprinted by permission.

Fig. 16.2: From Kent M. Van De Graaff, *Human Anatomy,* 3d ed. Copyright © 1992 Wm. C. Brown Communications, Inc., Dubuque, Iowa. All Rights Reserved. Reprinted by permission.

Fig. 17.7: From Kent M. Van De Graaff and Stuart Ira Fox, *Concepts of Human Anatomy and Physiology,* 3d ed. Copyright © 1992 Wm. C. Brown Communications, Inc., Dubuque, Iowa. All Rights Reserved. Reprinted by permission.

Fig. 17.13: From Stuart Ira Fox, *Human Physiology,* 4th ed. Copyright © 1993 Wm. C. Brown Communications, Inc., Dubuque, Iowa. All Rights Reserved. Reprinted by permission.

Fig. 19.11: From Kent M. Van De Graaff and Stuart Ira Fox, *Concepts of Human Anatomy and Physiology,* 3d ed. Copyright © 1992 Wm. C. Brown Communications, Inc., Dubuque, Iowa. All Rights Reserved. Reprinted by permission.

Fig. 19.14: From Kent M. Van De Graaff and Stuart Ira Fox, *Concepts of Human Anatomy and Physiology,* 3d ed. Copyright © 1992 Wm. C. Brown Communications, Inc., Dubuque, Iowa. All Rights Reserved. Reprinted by permission.

Fig. 20.2: From Kent M. Van De Graaff, *Human Anatomy,* 3d ed. Copyright © 1992 Wm. C. Brown Communications, Inc., Dubuque, Iowa. All Rights Reserved. Reprinted by permission.

Fig. 20.3: From Kent M. Van De Graaff and Stuart Ira Fox, *Concepts of Human Anatomy and Physiology,* 3d ed. Copyright © 1992 Wm. C. Brown Communications, Inc., Dubuque, Iowa. All Rights Reserved. Reprinted by permission.

Fig. 20.17: From Kent M. Van De Graaff and Stuart Ira Fox, *Concepts of Human Anatomy and Physiology,* 3d ed. Copyright © 1992 Wm. C. Brown Communications, Inc., Dubuque, Iowa. All Rights Reserved. Reprinted by permission.